Memorial Volume for
Stanley Mandelstam

Memorial Volume for
STANLEY MANDELSTAM

Editors

Nathan Berkovits (São Paulo State University, Brazil)
Lars Brink (Chalmers University of Technology, Sweden)
Ling-Lie Chau (University of California, Davis, USA)
Kok Khoo Phua (Nanyang Technological University, Singapore)
Charles B Thorn (University of Florida, Gainesville, USA)

NEW JERSEY · LONDON · SINGAPORE · BEIJING · SHANGHAI · HONG KONG · TAIPEI · CHENNAI · TOKYO

Published by

World Scientific Publishing Co. Pte. Ltd.

5 Toh Tuck Link, Singapore 596224

USA office: 27 Warren Street, Suite 401-402, Hackensack, NJ 07601

UK office: 57 Shelton Street, Covent Garden, London WC2H 9HE

Library of Congress Cataloging-in-Publication Data
Names: Berkovits, N., editor. | Brink, Lars, 1945–   editor. | Chau, Ling-Lie, editor. | Phua, K. K., editor. |
    Thorn, Charles, 1946–   editor. | Mandelstam, Stanley, honouree.
Title: Memorial volume for Stanley Mandelstam / edited by
    Nathan Berkovits (São Paulo State University, Brazil),
    Lars Brink (Chalmers University of Technology, Sweden),
    Ling-Lie Chau (University of California, Davis, USA),
    Kok Khoo Phua (Nanyang Technological University, Singapore),
    Charles B. Thorn (University of Florida, Gainesville, USA).
Description: Singapore ; Hackensack, NJ : World Scientific, [2017]
Identifiers: LCCN 2017019093| ISBN 9789813207844 (hardcover) | ISBN 9813207841 (hardcover) |
    ISBN 9789813227903 (paperback) | ISBN 9813227907 (pbk.)
Subjects: LCSH: Mandelstam, Stanley. | Physicists--Biography. | Particles (Nuclear physics) | String models.
Classification: LCC QC16.M344 M46 2017 | DDC 530.092--dc23
LC record available at https://lccn.loc.gov/2017019093

British Library Cataloguing-in-Publication Data
A catalogue record for this book is available from the British Library.

The editors and publisher would like to thank the following publishers of the various journals for their assistance and permissions to include the selected reprints found in this volume:
The American Physical Society (*Physical Review*);
IOP Publishing (*Reports on Progress in Physics*);
Elsevier (*Annals of Physics*).

While every effort has been made to contact the publishers of reprinted papers prior to publication, we have not been successful in some cases. Where we could not contact the publishers, we have acknowledged the source of the material. Proper credit will be accorded to these publications in future editions of this work after permission is granted.

Copyright © 2017 by World Scientific Publishing Co. Pte. Ltd.

All rights reserved. This book, or parts thereof, may not be reproduced in any form or by any means, electronic or mechanical, including photocopying, recording or any information storage and retrieval system now known or to be invented, without written permission from the publisher.

For photocopying of material in this volume, please pay a copying fee through the Copyright Clearance Center, Inc., 222 Rosewood Drive, Danvers, MA 01923, USA. In this case permission to photocopy is not required from the publisher.

Contents

Preface ix

Photos of Stanley Mandelstam xi

1. Recollections of Stanley Mandelstam 1
   Geoffrey Chew

2. Scientific Biography of Stanley Mandelstam: 1955–1980 5
   Charles B. Thorn

3. Scientific Biography of Stanley Mandelstam: 1981–2016 35
   Nathan Berkovits

4. Stanley Mandelstam: Brief Biography and Selected Publications with Commentary 53
   Ling-Lie Chau

5. Stanley Mandelstam: The Early Years at a 'Most Stimulating Theoretical Group' 73
   Sabine Lee

6. The Guiding Influence of Stanley Mandelstam, from S-Matrix Theory to String Theory 79
   Peter Goddard

7. Remembering Stanley: From a Source of Inspiration to a Fair Strong Competitor 91
   G. Veneziano

8. Stanley Mandelstam and Me and Life on the Light-cone   97
   Lars Brink

9. Reminiscences of Stanley Mandelstam   113
   John H. Schwarz

10. Stanley Mandelstam and My Postdoctoral Years at Berkeley   119
    Steven Frautschi

11. Reminiscences on Stanley Mandelstam   123
    Korkut Bardakci

12. Remembering a Gentle Giant of Physics   125
    Charles Sommerfield

13. Grad School with Stanley Mandelstam   127
    Joseph Polchinski

14. Remembering a Gentle Giant of Physics   129
    Mary K. Gaillard

15. Mandelstam & NAL   131
    Pierre Ramond

16. The Influence of Stanley Mandelstam   133
    Michael B. Green

17. My Interaction with Stanley Mandelstam   137
    Paolo Di Vecchia

18. My Advisor Stanley   141
    Sang-Jin Sin

19. Stanley Mandelstam My Graduate Supervisor   145
    Arjun Berera

Reprints and Abstracts of Selected Publications*

The Mandelstam representations in the Mandelstam variables for S-matrices

20. Determination of the Pion–Nucleon Scattering Amplitude from Dispersion Relations and Unitarity. General Theory [*Phys. Rev.* **112** (1958) 1344–1360][Reprint]   151

21. Analytic Properties of Transition Amplitudes in Perturbation Theory [*Phys. Rev.* **115** (1959) 1741–1751][Abstract]   168

22. Two-Dimensional Representations of Scattering Amplitudes and Their Applications [in *Quantum Theory of Fields. Proceedings of the Twelfth (1961) Solvay Conference on Physics*, pp. 209–233][Abstract]   169

The S-matrix approach

23. Theory of Low-Energy Pion–Pion Interactions [*Phys. Rev.* **119** (1960) 467–477][Reprint]   170

24. Dispersion Relations in Strong-Coupling Physics [*Reports on Progress in Physics* **XXV** (1962) 99–162][Reprint]   181

The Mandelstam path-field formulation for quantum gauge theories and Feynman rules

25. Quantum Electrodynamics Without Potentials [*Annals Phys.* **19** (1962) 1–24][Reprint]   245

26. Feynman Rules for Electromagnetic and Yang–Mills Fields from the Gauge-Independent Field-Theoretic Formalism [*Phys. Rev.* **175** (1968) 1580–1623][Abstract]   269

The Mandelstam path-field formulation for quantum general relativity and the Feynman rules

27. Quantization of the Gravitational Field [*Annals Phys.* **19** (1962) 25–66][Reprint]   270

---

*These publications are organized into nine subjects of Mandelstam's major contributions as given by L. L. Chau in her paper (paper 4 of this volume) and selected from the selected publications in her paper. S. Mandelstam was the single author of all the selected publications here, except for paper 23 which was co-authored by G. F. Chew and S. Mandelstam.

28. Feynman Rules for the Gravitational Field from the Coordinate-Independent Field-Theoretic Formalism [*Phys. Rev.* **175** (1968) 1604–1623][Abstract]   312

The precursor for the discovery of string theory

29. Dynamics Based on Rising Regge Trajectories [*Phys. Rev.* **166** (1968) 1539–1552][Reprint]   313

The elucidation of mechanisms for quark confinement in QCD

30. Vortices and Quark Confinement in Non-Abelian Gauge Theories [*Phys. Lett. B* **53** (1975) 476–478][Abstract]   327

31. Charge-Monopole Duality and the Phases of Non-Abelian Gauge Theories [*Phys. Rev. D* **19** (1979) 2391–2401][Abstract]   328

32. General Introduction to Confinement [*Phys. Rept.* **67** (1980) 109–121][Abstract]   329

Non-perturbative constructions of the bosonization (or fermionization) in $(1+1)$-dimension QFTs

33. Soliton Operators for the Quantized sine-Gordon Equation [*Phys. Rev. D* **11** (1975) 3026–3030][Abstract]   330

The proof of perturbative ultraviolet finiteness and $\beta = 0$, in any gauge and to all orders, of $N = 4$ SYM

34. Light-cone Superspace and the Ultraviolet Finiteness of the $N=4$ Model [*Nucl. Phys. B* **213** (1983) 149–168][Abstract]   331

Many important publications on string theory and eventually the first long-awaited proof of pertubative ultraviolet finiteness of superstring theory so it can be considered as a contender for being the theory of quantum gravity

35. Dual-Resonance Models [*Phys. Rept.* **13** (1974) 259–353][Abstract]   332

36. The n-loop String Amplitude: Explicit Formulas, Finiteness and Absence of Ambiguities [*Phys. Lett. B* **277** (1992) 82–88][Abstract]   333

Preface

Stanley Mandelstam (1928–2016) was one of the leaders in the dramatic development of the theory of elementary particles, a leader who was unique in his quiet, gentle and humane manner. Growing up in South Africa he earned a degree in chemical engineering, but his mind was set on theoretical physics, and he left for Cambridge to continue his studies there. He took a BA from Trinity College in 1954, sharing the university prize for the best performance in applied mathematics and theoretical physics in the final examination with Jeffrey Goldstone, another legendary physicist also at Trinity. He then quickly earned a Ph.D. in 1956 from Birmingham with Rudolf Peierls. After postgraudate work at Birmingham, Columbia University and the University of California at Berkeley, in 1960 he returned to Birmingham as a professor, then in 1963 back to Berkeley, where he remained for the rest of his life.

Together with Geoffrey Chew, he established the Berkeley group as leaders in the quickly developing theory of particle physics especially in the analytic S-matrix approach to strong interactions. Stanley Mandelstam's seminal contributions to S-matrix theory were strongly informed by his deep understanding of quantum field theory, to which he also made fundamental contributions. His contributions to string theory were also seminal and profound. As briefly outlined on the back cover of this volume, he was a supreme master of QFT, the S-matrix approach, and string theory. His influences on theoretical and mathematical physics are deep and diverse, in almost all the current major research efforts trying to deepen our understanding of the physical universe. His influences live on.

Stanley Mandelstam was a gentleman in physics. He had a razor sharp mind which became obvious in any physics discussion, but as a person he was soft-spoken and friendly; he was not a man who imposed his ideas on others.

We, the editors, and World Scientific Publishing take great pleasure in presenting this volume, in which we have collected scientific biographies as well as personal reminiscences and contributions by friends, colleagues and former associates, and students, celebrating the life and science of Stanley Mandelstam. We have also included a section reprinting some of his early seminal papers and abstracts of selected papers representing the full spectrum of his contributions. We hope and believe that readers will not only find this volume useful for reviewing a glorious period of modern physics but will also cherish the memories it evokes of a universally liked and admired leader in theoretical physics.

The Editors
N. Berkovits, L. Brink, L. L. Chau, K. K. Phua and C. Thorn

Photos of Stanley Mandelstam*

Photo courtesy of Mandelstam's family.

*See more photos in the articles by L. L. Chau and A. Berera.

Photo courtesy of Department of Physics, UC Berkeley.

Recollections of Stanley Mandelstam

Geoffrey Chew
*Department of Physics, University of California,
Berkeley, CA 94720-7300, USA
gfchew@lbl.edu*

This article is based upon the audio recording taken on September 9th 2016 by my children. I thank Chee Hok Lim for writing the first transcript in the World Scientific format and Ling-Lie Chau for comments and footnotes.

1. First Encounter with Stanley Mandelstam in 1958 and the Few Years Following that Encounter

I had come back to the UC Berkeley in 1957 after six years at the University of Illinois. In the spring of 1958 I attended a meeting of the APS in Washington DC. It was in a hotel, maybe Sheraton. At this point in my own work, I've been focusing on the notion of scattering amplitudes, elements of the scattering matrix or S-matrix. It was very interesting to explore the consequences of the scattering amplitudes being analytic functions. At that point, working with other people, notably, Francis Low and Murray Gell-Mann, I had arrived at some notions which seemed relevant to what experiments were showing.

Presumably, at this meeting in Washington I delivered a paper talking about my own ideas. But I noticed, in the program of the meeting, someone whom I had never heard of, named Stanley Mandelstam, was giving a paper in the same area. I can't recall what the title of his paper was, but it was close enough to my area. I attended the talk and my impression was that Stanley Mandelstam was unaware of some of the difficulties in the subject he was dealing with. He was making statements which I couldn't believe. Since I had never heard of Stanley Mandelstam, I assumed that this was because of his lack of experience. So after his talk, I went up to him directly in the auditorium where he gave the talk, and asked him questions. He gave his answers, but I didn't understand his answers. So I suggested that we go to another location in the hotel and sit together. At that point, I was persuaded that I was going to be teaching Stanley about some issues that he was unaware of. So we went to one of the lobbies in the hotel and sat for a couple of hours talking about what he was attempting to do. At the end of the two hours, I did not understand what he was saying, but I was persuaded that he knew what he was talking about even though I didn't understand it. I asked Stanley where he had done his work, and learned from him that he had recently come to the US from England where he did his graduate study, first in Cambridge then in Birmingham. After getting his PhD in England, he got a postdoctoral fellowship at Columbia University and had been there about three months. He told me that there was no one at Columbia University with whom he was having conversations. I had become persuaded during our two-hour conversation that, although I did not understand what Stanley was doing, he understood it and that he had an extraordinary talent. Although I didn't understand what he had accomplished, I felt sure he was doing something important. I wanted to continue these conversations, and asked him if he would consider relocating from Columbia University to Berkeley, California, where I had resumed my position and worked there for about a year. He said Yes. So I went to a telephone in the lobby and called up the individual who was in charge of the government contract that was supporting theoretical particle physics at Berkeley at that time. I asked if we had enough fund to support Stanley and I was told, Yes. So I came back and asked Stanley if he would like to shift from Columbia to Berkeley. He said Yes without hesitation. That was the beginning of my life with Stanley, and after that I had a great deal of contact with him.

What Stanley had claimed to show was that scattering amplitudes were analytic functions of more than one complex variable. Up till then, it had been taken for granted that the scattering amplitudes are analytic functions of one complex variable, which was usually considered to be the energy of the scattering. But Stanley was proposing that not only the energy but also the angle involved in the scattering was part of the functions of two complex variables. I became more and more persuaded that he was right, and our association continued. At the same time there was much skepticism among colleagues, questioning whether what he was proposing was correct. I can remember that another member of the faculty at Berkeley whose name was Eyvind Wichmann, who was especially skeptical. He was a very formal theoretical physicist and he simply could not believe what Stanley was proposing could be correct. He made a bet with me that within a year it would be shown that Stanley's idea was foundationally incorrect. As the year went along, more and more people became interested in what Stanley was proposing. I remember at the end of the year, Eyvind conceded (although he still did not believe that Stanley was correct) that I had won the bet, because Stanley was taken seriously by other very respectable theoretical physicists. Eyvind wrote on a sheet of paper that I had won the bet, and I believe that the bet was for 50 cents. So he attached the 50 cents to the sheet of paper and posted it on my office door where it stayed for about a year. I remember it with much amusement.

During that year Stanley and I worked together and wrote a paper.[a] By then there were many other people doing similar research. Stanley rapidly became famous, which culminated in his being invited to the 1961 Solvay conference. I think it was organized by Oppenheimer. There was the photo of attendees at the conference, which is still sitting in my office here in Berkeley. There were three rows. The first two rows are completely filled with extremely famous individuals, most of whom had won the Nobel Prize. In the back row Stanley and I appeared side by side, together with a few other junior theoretical physicists. It was astonishing how rapidly he had become famous.

2. Later Years

We continued in close contact over the years. After couple of years at Berkeley, I believe, Stanley was given a faculty appointment in Britain. But only after about two years there, he was invited to come back to Berkeley and accepted a regular faculty appointment here. He has continued to be on the faculty of Berkeley right until his death in the middle of June 2016.

I can also recall that in my contacts with Stanley, he repeatedly explained to me things which I had not previously appreciated. We did not work closely together after that two-year initial interval when he himself was just a postdoc. But we continued to have a relatively close relationship over the years. I had lots of contact with Stanley in Europe as well as in Berkeley. I spent a good time in France. In particular he lived in an apartment in Paris which my wife Denyse and I were purchasing. So we had a good deal of contact all over the places with Stanley. These contacts for me were a very important part of my life.

At some point Ling-Lie Chau, who had been one of my students in the early 1960s, had worked on aspects related to Stanley's thinking. As we were growing older together, she held get-togethers, including those on New Year's Eve every year since 1999, which were extremely enjoyable. I think Ling-Lie will have more accurate recollections. She will also have more stories to tell about Stanley that I forgot or only learned from her after Stanley's death.[b] For example, she reminded

---

[a]Note by L. L. Chau: There were two papers of the same title co-authored with Mandelstam alone, and two more with another author, so totally four papers with Mandelstam.
G. F. Chew and S. Mandelstam, Theory of low-energy pion–pion interactions, *Phys. Rev.* **119**, 467 (1960); Theory of low-energy pion–pion interactions II, *Nuovo Cim.* **19**, 752 (1961).
G. F. Chew, S. Mandelstam and H. P. Noyes, S-wave dominant solutions of the pion–pion integral equations, *Phys. Rev.* **119**, 478 (1960).
G. F. Chew, S. C. Frautschi and S. Mandelstam, Regge poles in pi–pi scattering, *Phys. Rev.* **126**, 1202 (1962).
[b]See L. L. Chau, Endearing memories of Stanley Mandelstam, to appear in *the Memorial Volume for Stanley Mandelstam*, to be published by World Scientific.

me about the embarrassing story that I forgot to attend the retirement party for him given by the Physics Department for which I was supposed to be the main speaker![c]

---

[c]Note by L. L. Chau: To make up for that mishap, Geoff generously held another retirement party for Stanley, inviting the faculty members and their families of the whole Physics Department to an upscale restaurant for dinner!

Scientific Biography of Stanley Mandelstam: 1955–1980

Charles B. Thorn

*Institute for Fundamental Theory, Department of Physics,
University of Florida, Gainesville, FL 32611, USA*
thorn@phys.ufl.edu

I review Stanley Mandelstam's many contributions to particle physics, quantum field theory and string theory covering the years 1955 through 1980. His more recent work will be reviewed by Nathan Berkovits.

1. Introduction

Stanley Mandelstam's career in theoretical physics spans nearly 60 years of pathbreaking research in the theory of elementary particles. He was an undisputed master of quantum field theory (QFT), which, to this day, is accepted as the indispensable framework for describing relativistic quantum mechanics,

By the time Stanley began his research in the mid-1950s,[1] quantum electrodynamics (QED), the QFT of the electromagnetic field and charged particles, was well-understood in the context of perturbation theory. This success was based on the good fortune that the expansion parameter $\alpha = e^2/(4\pi\hbar c) \approx 1/137$ is quite small. From the beginning, the problems Stanley set out to illuminate involved aspects of QFT that were not amenable to perturbation theory. In the early 1950s it was evident that the strong interactions (the physics of the nuclear force), if described by QFT, would involve an expansion parameter (analogous to α) which was not small. Thus the strong interactions became an early focus of Stanley's considerable talents.

His deep understanding of the analytic structure of the individual terms of perturbation theory (Feynman diagrams) led him to a proposal for a representation of the scattering amplitude for pions and nucleons, now known as the Mandelstam representation.[2,3] This proposal became the mainstay of the S-matrix approach to the theory of strong interactions espoused by Geoffrey Chew. Together with Geoff, Stanley played a pivotal role in laying the foundations for this program.[5–7]

Although initially S-matrix theory focused on seeking self-consistent solutions based on Lorentz invariance, unitarity, crossing symmetry and the analyticity of the S matrix in the Lorentz scalars $p_k \cdot p_l$,[a] the ideas of Tullio Regge on the analyticity of partial wave amplitudes in complex angular momentum (J) quickly became a source of deeper insights, especially with respect to the high energy behavior of scattering amplitudes.[8,16–21]

Stanley's unmatched understanding of the analytic structure of Feynman diagrams soon elucidated just how intricate the singularity structure of the complex J-plane was.[16,17] There were cuts in J which shielded the physics from the dire consequences of the essential singularity found by Gribov and Pomeranchuk. Furthermore, using his eponymous representation, he showed that all of these complications required the presence of the "third double spectral function" ρ_{tu}. He then suggested a new approximation scheme in which ρ_{tu} is neglected in first instance[24] and then brought back in perturbatively.[b] This was a crucial insight. The second and third double spectral functions are absent from the sun of all planar Feynman diagrams. So a systematic approximation scheme could be formulated: First sum all the planar diagrams to a given physical process, and then bring in nonplanar diagrams in a perturbative expansion. It was at least consistent with known facts to hypothesize that the sum of planar diagrams has only simple Regge poles in the J-plane. Stanley proposed[23,27] that one seek an approximation in which Regge trajectories are linear and, simultaneously, resonances are narrow. The latter means that the only singularities of scattering amplitudes are poles in the momentum variables $p_k \cdot p_l$, located on the real axis.

A "bootstrap" solution based on these principles was then the goal of many young theorists, including Dolen, Horn, and Schmid from a phenomenological point of view.[71] Ademollo, Rubinstein, Veneziano, and Virasoro[72] were soon in hot pursuit of such a bootstrap. This activity culminated in Veneziano's discovery[73] of his four point amplitude and the rest, as they say, is history. String theory was born and in the ensuing decades has developed into a promising approach to quantum gravity.

Stanley embraced these exciting developments whole-heartedly.[30–32] He sketched how generalizations of Veneziano's formula could be used to build a relativistic quark model of mesons and baryons.[34–37] He built on the light-cone quantization of the Nambu–Goto string[86] to show that the dual resonance models were indeed the scattering amplitudes of strings.[42] He used his light-cone interacting string formalism to accomplish a theoretical tour-de-force: the calculation of scattering amplitudes involving four or more fermions,[43,65] and then to complete the calculation of all multi-loop diagrams in string theory.[66–68]

---

[a]Minkowski scalar products formed from the 4-momenta p_k of the scattered particles in the initial and final states.
[b]Years later (1973) Gerard 't Hooft[70] identified the expansion parameter for non-planar contributions as $1/N$ for an SU(N) non-Abelian gauge theory.

These were all valiant efforts in aiding string theory *per se*. But Stanley never dropped his fascination for the strong interactions, which had indirectly led to all these theoretical developments. By the early 1970s, with the discovery of asymptotic freedom and charm, it became abundantly clear that the strong interactions should be described by a non-Abelian gauge theory of quarks and gluons (QCD). The catch was that quarks and gluons should be permanently trapped inside of baryons and mesons — the hypothesis of quark confinement. So it was natural for Stanley to attack the problem of quark confinement. Motivated by an analogy with a superconductor's confinement of magnetic monopoles due to the condensation of a charged field, he developed a "dual" scheme for the condensation of color-magnetic monopoles in QCD.[46-53]

But still more achievements were in store. As a spinoff from his work on confinement, Stanley illuminated the equivalence of the two-dimensional massive Thirring model to the two-dimensional sine–Gordon QFT by constructing an explicit operator mapping between the two QFT's.[54,55] This nice piece of mathematical physics provided a toy prototype of the duality between magnetic and electric confinement in non-Abelian gauge theories. Then supersymmetry became a hot topic in the late 1970s and early 1980s. There were exciting conjectures, such as that $\mathcal{N}=4$ supersymmetric non-Abelian gauge theory should be finite in the ultraviolet. Naturally, Stanley was one of the first to construct a proof of this deep conjecture.

In Secs. 2–7 below I endeavor to explain Stanley's important achievements through the 1970s. The final Sec. 8 is my personal tribute to him emphasizing his role as one of the pioneers of string theory. My colleague Nathan Berkovits will write Part II of his scientific biography, describing Mandelstam's work since 1980.

2. Double Dispersion Relations

A general scattering process is a reaction

$$P_1 + P_2 \longrightarrow P_3 + P_4 + \cdots P_N \tag{1}$$

where each letter P_k stands for a different particle participating in the process either in the initial or final state. It is described by an amplitude $\mathcal{A}$ which is a function of the Lorentz scalars $p_i \cdot p_j$ formed from the energies and momenta of all the particles. Here $-p_i^2 = m_i^2$ where m_i is the mass of the ith particle which is not variable, so we can, without loss of information, take $i < j$. In addition, energy and momentum is conserved so that there are only $N-1$ independent p_i. If we eliminate $p_N = -p_1 - \cdots - p_{N-1}$ then the mass shell constraint $p_N^2 = -m_N^2$ can be rewritten

$$-\sum_{i<j<N}(p_i+p_j)^2 = \sum_{i=1}^{N} m_i^2. \tag{2}$$

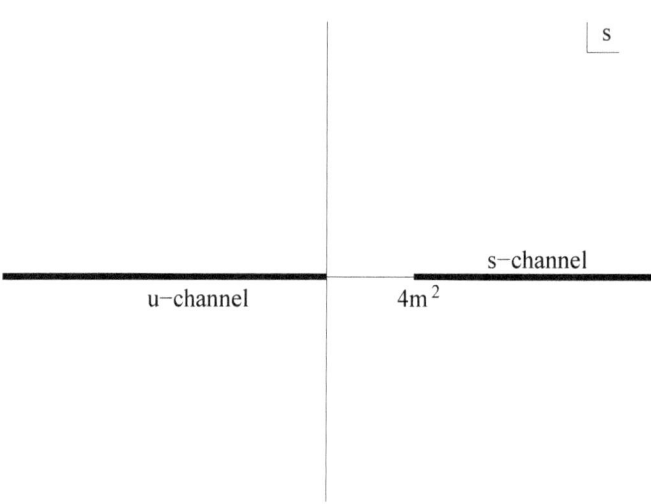

Fig. 1. The complex s plane at $t = 0$. The thick lines indicate branch cuts. The right branch point is due to the threshold at $s = 4m^2$, the left branch point is due to threshold at $u = 4m^2 = 4m^2 - s - t$, and is shown for $t = 0$.

Mandelstam focused on the case $N = 4$ and defined the three invariant variables

$$s = -(p_1 + p_2)^2, \quad t = -(p_2 + p_3)^2, \quad u = -(p_1 + p_3)^2, \tag{3}$$

$$s + t + u = m_1^2 + m_2^2 + m_3^2 + m_4^2 \tag{4}$$

any two of which can be chosen as independent variables.

For simplicity let us assume that all masses are the same, so that $s+t+u = 4m^2$. Then crossing symmetry is the statement that $\mathcal{A}$ describes three possible scattering processes, depending on the range of s, t, u. The s-channel is $P_1 + P_2 \to P_3 + P_4$, which requires $s > 4m^2$ and $t, u < 0$; the t-channel is $P_1 + \bar{P}_4 \to \bar{P}_2 + P_3$, which requires $t > 4m^2$ and $s, u < 0$; and the u channel is $P_1 + \bar{P}_3 \to \bar{P}_2 + P_4$ which requires $u > 4m^2$ and $s, t < 0$. In this reaction notation the bar denotes the antiparticle. Since the three channels are in nonoverlapping regions of s, t, u, applying crossing symmetry requires continuing somehow between these regions. In S-matrix theory one postulates that $\mathcal{A}$ is an analytic function of s, t, u in a domain of the complex planes which contains all three regions. For example at $t = 0$ the postulated domain of analyticity in the s plane is shown in Fig. 1. A single dispersion relation is obtained from the analyticity hypothesis by first writing the Cauchy formula

$$A(s) = \frac{1}{2\pi i} \oint_C ds' \frac{A(s')}{s' - s} \tag{5}$$

where s is in the domain of analyticity, and the contour C is an infinitesimal circle centered on s. Then one deforms C to large semicircles at infinity plus straight lines just above and just below the cuts. The result is a formula like

$$A(s) = \frac{1}{2\pi i} \int_{4m^2}^\infty ds' \frac{\mathcal{D}_R}{s' - s} + \frac{1}{2\pi i} \int_{-\infty}^0 ds' \frac{\mathcal{D}_L}{s' - s} + \text{semicircle terms} \tag{6}$$

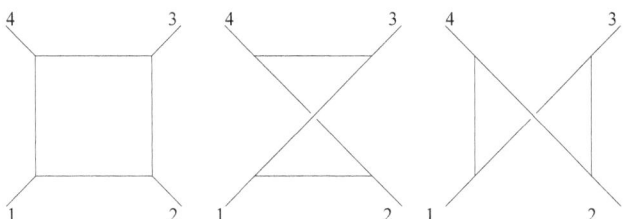

Fig. 2. The three single loop box Feynman diagrams, which are the simplest possessing the nonzero double spectral functions, ρ_{st}, ρ_{su}, or ρ_{tu} respectively. Relative to the s channel, the last one is the "third" double spectral function. Note that since the cyclic order of the particle labels is the same in all three diagrams, the second and third diagrams have crossed lines.

where $\mathcal{D}_{R,L}$ are the discontinuities across the right and left cuts. Such relations have experimental implications. For example for $t = 0$ the discontinuities can be related by the optical theorem to total cross-sections in the respective scattering channels and the semicircle terms will go to zero if $\mathcal{A}$ vanishes sufficiently rapidly at infinity. The dispersion relation then expresses $\mathcal{A}(s,0)$ in terms of an integral over the total cross-section.

Mandelstam's seminal contribution[2,3] was to exploit the postulate of simultaneous analyticity in more than one variable — two in the case of $N = 4$ to obtain double dispersion relations. The right sides are now double integrals, and the Mandelstam representation for equal masses takes the form

$$\mathcal{A} = \frac{1}{\pi^2} \int_{4m^2}^{\infty} ds'\, dt'\, \frac{\rho_{st}(s',t')}{(s'-s)(t'-t)} + \frac{1}{\pi^2} \int_{4m^2}^{\infty} ds'\, dt'\, \frac{\rho_{su}(s',u')}{(s'-s)(u'-u)}$$
$$+ \frac{1}{\pi^2} \int_{4m^2}^{\infty} ds'\, dt'\, \frac{\rho_{tu}(t',u')}{(t'-t)(u'-u)} \,. \qquad (7)$$

Here we have assumed sufficient fall off at infinity and have not included pole terms. The analyticity assumed here is not universally true: there are situations where singularities occur at complex values of s, t, u for which the integrals would not be over real variables. But it does reflect the analyticity of low orders in perturbation theory, for example that of the box diagrams in Fig. 2. The Mandelstam representation explicitly exhibits the singularities in s, t, u that are expected on the basis of particle thresholds and experience with Feynman diagrams. It therefore proved an indispensable tool in the development of Chew's program of S-matrix theory.[5–7]

3. Complex Angular Momentum

The striking way the Mandelstam representation separates the singularities of the four point amplitude into three distinct terms, characterized by the double spectral functions ρ_{st}, ρ_{su}, and ρ_{tu}, has also played a central role in understanding the singularity structure of partial wave amplitudes continued into the complex angular momentum plane, as envisioned by Tullio Regge.

Partial wave amplitudes trade dependence on angle for dependence on angular momentum. In scattering by a potential the scattering amplitude $f(E,\theta)$ can be expanded in Legendre polynomials thus

$$f(E,\theta) = \sum_{J=0}^{\infty} a_J(E) P_J(\cos\theta) \qquad (8)$$

where a_J are the partial wave amplitudes, which completely characterize the scattering. When J is an integer, $a_J(E)$ has a pole at the energy $E_{J,n}$ of each bound state with angular momentum J:

$$a_J \sim \frac{R_{Jn}}{E - E_{Jn}}, \qquad E \sim E_{Jn}. \qquad (9)$$

When a_J is continued into the complex J-plane, E_{Jn} becomes an analytic function of J, and the pole location moves. Viewed as a pole in the J plane, it becomes a Regge pole $\alpha(E)$ whose location depends on energy. The contribution of a Regge pole to the scattering amplitude is proportional to $P_{\alpha(E)}(\cos\theta)$ which has the large $\cos\theta$ behavior $(\cos\theta)^{\alpha(E)}$, known as Regge asymptotic behavior.

Chew, Frautschi, and Mandelstam[8] realized the importance of these ideas in relativistic S-matrix theory. The momentum transfer invariant $t = (s-4m^2)(\cos\theta-1)/2$, so large $\cos\theta$ is identified with large t. The invariant s is related to the scattering energy, so in a relativistic context, Regge behavior is interpreted as

$$A(s,t) \sim \beta(s)(t)^{\alpha(s)}, \quad \text{as} \quad t \to \infty. \qquad (10)$$

In t channel scattering this links high energy (large t) to the exchange of the Regge trajectory $J = \alpha(s)$ in the momentum transfer variable ($s < 0$ in the t-channel). This was not only a breakthrough in theory, but it also had direct experimental implications. One could extract the trajectories $\alpha(s)$ for $s < 0$ by careful measurement of high energy (large t) cross-sections at varying momentum transfer s. The trajectory $\alpha(s)$ is an extrapolation of the relation between angular momentum and mass squared of the particles at discrete positive values of s. These can be dramatically presented on a Chew–Frautschi plot as shown in the Fig. 3 for which we have switched the roles of s and t. While physical angular momentum is necessarily nonnegative, the extrapolation $\alpha(t)$ to negative t could well become negative, with welcome implications for the convergence of dispersion relations.

A subtlety of the continuation to complex angular momentum in the relativistic case is that even and odd angular momentum must, in principle, be continued separately. This is due to the presence of both st and su double spectral functions. As a consequence Regge trajectories in the even signature (odd signature) continuation only imply particles at even (odd) values of J. There is no *a priori* relation between the even and odd trajectories. However, the Chew–Frautschi plots for various trajectories (see Fig. 4) showed near coincidence of even and odd trajectories. This exchange degeneracy suggested that some processes were dominated by only one double spectral function.

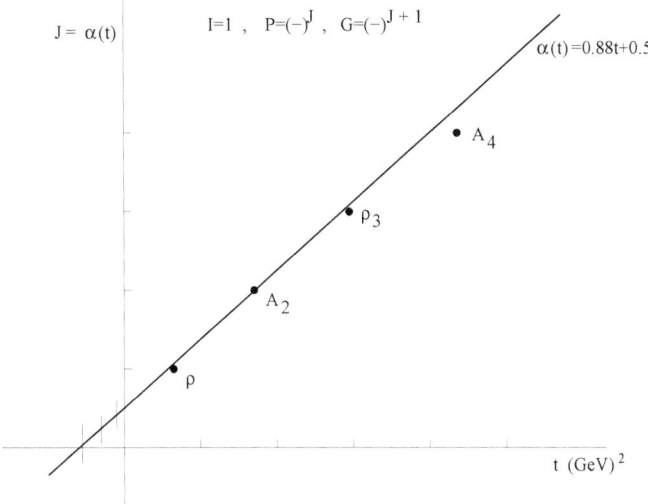

Fig. 3. Chew–Frautschi plot for isospin 1 exchange in the t-channel. The only data points for $t > 0$ are the angular momentum of particles of mass $\sqrt{t}$ with the quantum numbers ($I = 1$) of the t-channel. In the early sixties the only particle on the plot was the ρ meson. For $t < 0$, t is in the physical region of s channel scattering, and $\alpha(t)$ can be inferred from experiments at large positive s. The error bars are only to roughly indicate the measurement uncertainties. The experimental near linearity of Regge trajectories, especially for those with non-vacuum quantum numbers was very suggestive.

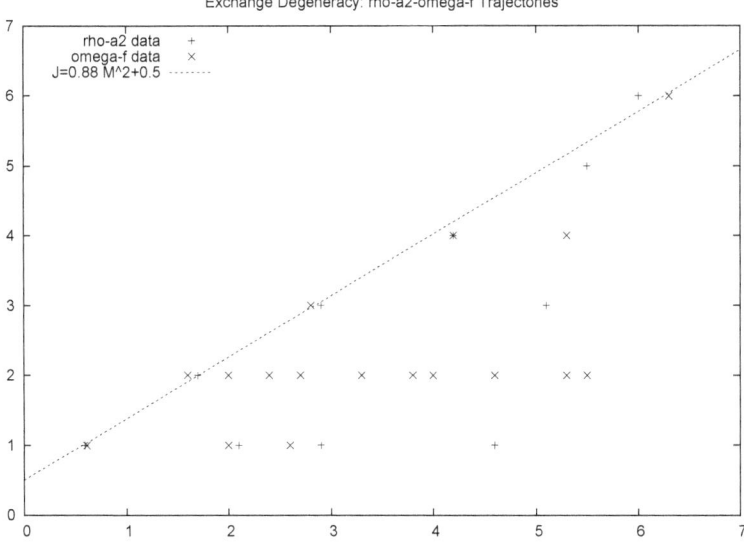

Fig. 4. Evidence for exchange degeneracy. We plot angular momentum versus mass squared for known resonances with isospin 1 and 0 and both even and odd angular momentum. The ones with highest J fall very close to the same linear Regge trajectory. Exchange degeneracy would be a consequence of the dominance of a single double spectral function, or, alternatively, of the dynamics of planar Feynman diagrams.

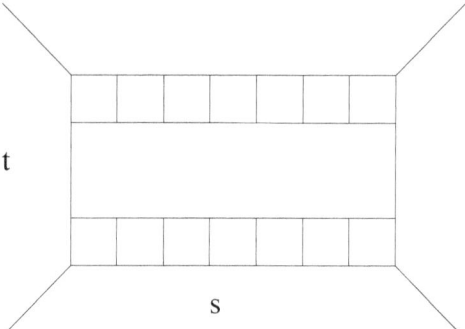

Fig. 5. Feynman diagrams reflecting the AFS argument for Regge cuts. The ladder structures are meant to be summed over all numbers of rungs so that they represent a composite structure describing the exchange of a Regge pole. AFS argued that the exchange of two Regge poles would produce a Regge cut. Mandelstam showed that the apparent cut in these diagrams is not present on the physical sheet due to the planarity of the diagrams.

Although the promising phenomenological success of Regge pole phenomenology generated much excitement, it soon became clear that the complex J plane has a much more complicated singularity structure than simply Regge poles. In spring 1962 Amati, Fubini, and Stranghellini (AFS) argued for the presence of Regge cuts in processes involving the exchange of two Regge poles. Later in the same year Gribov and Pomeranchuk (GP) uncovered the apparent need for an essential singularity in the complex J plane, in the form of an accumulation point of Regge poles, near $J = -1$ for scattering of scalar (spin 0) particles. In Ref. 16, Mandelstam showed that this phenomenon occurs at nonnegative $J = 2s - 1$ for processes involving particles with spin s. Besides the difficulties these complications caused for the experimental interpretation of Regge poles, they also raised serious theoretical questions.

Mandelstam, in a tour-de-force, clarified these issues in a way that indicated how one might be able to gain control over the ensuing complications.[17] He first showed that the Feynman diagrams considered by AFS did not in fact possess the angular momentum cuts they had argued for. Those diagrams involved the exchange of two ladder sub-diagrams as shown in Fig. 5. And because they were planar the potential cuts necessarily cancelled out on the physical sheet, and, moreover were not afflicted with the GP phenomenon. He identified the key reason, namely that planar diagrams lacked third double spectral functions. He then showed that with sufficient nonplanarity as in Fig. 6 the cuts remained on the physical sheet. Finally he showed that the cuts of this type screened the physical sheet from the essential singularities discovered by GP. These two papers definitively resolved the theoretical issues posed by AFS and GP.

Moreover, he pinned down how these complications of complex angular momentum arise: they appear only in amplitudes possessing third double spectral functions. If such contributions to a given process were relatively small, Regge pole

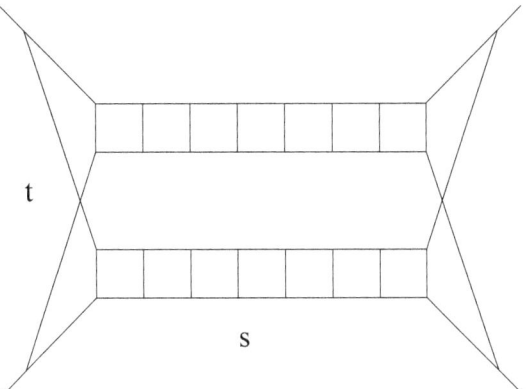

Fig. 6. Diagrams for which a Regge cut is present on the physical sheet. The nonplanar structures on either side of the diagram reflect the contribution of the third double spectral function ρ_{su} in the t-channel. If one perturbs in nonplanarity, e.g. in powers of the ρ_{su}, Regge cuts first appear at second order $O(\rho_{su}^2)$.

phenomenology might be viable. In a later paper with Chau (formerly Wang)[24] they clarified further the Gribov–Pomeranchuk phenomenon making use of precisely this strategy–perturbing in the effects of the third double spectral function. Feynman diagrams which involve such effects have some degree of nonplanarity, as is evident from Figs. 5 and 6, suggesting that one way to perturb in them is to take as first approximation the sum of only planar Feynman diagrams.

The most promising experimental processes for neglect of nonplanar effects involve nonzero quantum number exchange, a famous example of which is the charge exchange reaction $\pi^- p \to \pi^0 n$. An insightful analysis of this process was reported in the summer of 1967 by Dolen, Horn, and Schmid (DHS)[71] on the assumption that Regge poles control the high energy behavior. The success of this analysis gave impetus to the work that led to dual resonance models (DRM).

4. Narrow-Resonance/Regge Bootstrap

In a seminal paper,[27] announced earlier at a conference,[23] Mandelstam laid the theoretical foundation which paved the way toward DRM. He proposed that the bootstrap program in S-matrix theory be implemented in stages. Motivated by the near linearity of the Regge trajectories from analyses such as DHS, he noted that exactly linear trajectories would imply that resonances lying on the trajectory would be narrow (i.e. long-lived). This was because the trajectory functions should satisfy a dispersion relation of the form

$$\alpha(s) = as + b + \frac{1}{\pi} \int ds' \frac{\operatorname{Im} \alpha}{s' - s}. \tag{11}$$

Exact linearity would then require $\operatorname{Im} \alpha = 0$, and since the resonance widths are proportional to $\operatorname{Im} \alpha$, they would have to be zero as well. Of course such an approximation means that unitarity is only approximate. Assuming that third double spectral effects like Regge cuts and the GP phenomenon were absent and also that all particles lie on Regge trajectories, the analyticity of the S-matrix could be implemented in two ways. Focusing on an amplitude with singularities in the $s > 0$ and $t < 0$ channels, one could either write the amplitude as a sum of zero width resonance poles *or* write the same amplitude as a sum of t-channel Regge poles. These two expansions should agree: which gives the approximate bootstrap equation

$$\sum_k \frac{R_k(t)}{s - m_k^2} = \sum_l \beta_l(t) s^{\alpha_l(t)} . \qquad (12)$$

This equality (called *duality* at the time) should then lead to restrictions on the Regge parameters and the particle masses. He went further by assuming only a finite number of terms (typically 1 or 2) on each side leading to some reasonable relations. The DHS analysis had confronted the data of pi nucleon charge exchange with either one (or two) resonances or one (or two) Regge trajectories, and showed that either described the data well. But the path-breaking proposal of Mandelstam's paper was that it should be possible to devise a narrow resonance approximation to Regge-behaved scattering amplitudes provided that the Regge trajectories were exactly linear. Relaxing exact unitarity made this bootstrap problem much more tractable than the original one, which insisted on exact unitarity. Indeed roughly a year later Veneziano discovered an exact solution of Mandelstam's narrow resonance bootstrap: the celebrated Veneziano formula:

$$\mathcal{A} = g^2 \frac{\Gamma(-\alpha(s))\Gamma(-\alpha(t))}{\Gamma(-\alpha(s) - \alpha(t))}, \quad \alpha(x) = \alpha' x + \alpha_0 . \qquad (13)$$

This formula is manifestly crossing symmetric under $s \leftrightarrow t$, has only poles in s or t at $\alpha(s) = -n$, $n = 0, 1, \ldots$, and, using Stirling's formula can be shown to have Regge behavior $\mathcal{A} \sim \Gamma(-\alpha(t))(-s)^{\alpha(t)}$ as $s \to -\infty$ at fixed t.

Over the following year Mandelstam employed the Veneziano formula in his narrow resonance Regge bootstrap program sketched above. One of the salient features of the formula is the absence of u channel poles: in effect it possesses only the double spectral function ρ_{st}. This implies a planar structure in whatever underlying dynamics might produce it. Meson spectroscopy is well explained by thinking of mesons as quark–antiquark composites. To incorporate this quark structure in his bootstrap, Mandelstam imagined the planar structure to be bounded by a quark line, which should inject spin and flavor quantum numbers. In this language the Veneziano model itself would correspond to spinless flavor singlet quarks. It should then be supplemented with spin and flavor factors corresponding to spin 1/2 quarks carrying flavor SU(3) quantum numbers. In Ref. 34, Mandelstam determined these factors by imposing crossing symmetry and factorization in both t and s channels. As explained in the paper, an inevitable defect of such a construction, due to the requirement of Lorentz invariance, is a doubling of trajectories half of which are ghosts, meaning that the particle pole residues due to them have the wrong sign. Not yet realizing that ghost-free narrow resonance models would eventually be found, Mandelstam speculated that the narrow resonance approximation was responsible for such defects. In any case the relativistic quark bootstrap was compared to experiment under the assumption that these ghost trajectories would not play a significant role, with qualitative success.

According to the quark model, baryons are composites of three quarks, so that baryonic processes cannot be described by purely planar models. In Ref. 35, Mandelstam constructed what he called minimal nonplanar dual resonance models which could describe the three quark structure of baryonic amplitudes. He then used them to extend his relativistic quark model bootstrap to include such processes.[36,37] The minimal nonplanar dual models constructed here have received little attention in the literature, and their further study might well lead to interesting insights.

5. Dual Resonance Models

Veneziano's discovery precipitated an explosion of research activity directed toward discovering and developing the underlying theory of the Veneziano amplitude. Many researchers participated in the development of generalizations. Mandelstam embraced, guided, and participated in these rapid developments. Early on, he explored possible modifications of the original 4 particle Veneziano amplitude.[30] He also clarified Virasoro's new four particle amplitude by giving it a two-dimensional integral representation analogous to a one-dimensional integral representation of Veneziano's amplitude.[31] This development presaged the eventual closed string interpretation of the Virasoro model.

Of course, with his early appreciation of the importance of analyticity in multiple invariants describing a process, Mandelstam immediately realized that there should be generalizations to narrow resonance amplitudes involving any number of particles. The physical requirement of unitarity played an important guiding role in these generalizations. The narrow resonance approximation certainly abandons exact unitarity, but scattering amplitudes are still subject to unitarity constraints. The approximation amounts to the assertion that scattering amplitudes are meromorphic in the invariants $p_k \cdot p_l$, that is: the only singularities are simple poles. In this limit the content of unitarity is that the residues of each such pole in a $N+M$ particle amplitude must factorize into a product of scattering amplitudes for $N+1$ and $M+1$ particles. Unitarity further requires these residues to obey positivity constraints Fig. 7.

This program took the young particle theorists of the time by storm, whose intense activity led to a rapid development from 5 particle amplitudes to N particle amplitudes in a matter of several months. The amplitudes were discovered by others, but Mandelstam and his colleague Bardakci,[32] and, independently, Fubini and Veneziano[74] were the first to prove the necessary factorization properties of the N particle amplitude, including a determination of the degeneracies of the mass

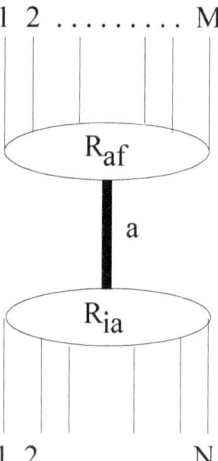

Fig. 7. Unitarity in the narrow resonance limit dictates that a resonance pole in a process of N particles to M particles factor: $\mathcal{A} \sim \sum_a R_{ia} R_{af}/(s - m_R^2)$ where m_R is the mass of the resonance, R_{ia} is the scattering amplitude for the process $i \to a$ and R_{af} is the scattering amplitude for the process $a \to f$. In the case of forward elastic scattering ($f = i$) $R_{ai} = R_{ia}^*$ and each factored term must be positive (No ghosts).

levels. More precisely they determined an upper bound on the degeneracies, because in their initial factorization some of the factored terms were not positive in the elastic forward case. If there were to be no ghosts in the factorization there should be an alternate factorization with a smaller degeneracy. If no such factorization were possible the model would have ghosts and therefore violate unitarity. In their work, Bardakci and Mandelstam found the first of an infinite string of Ward-like identities which could enable this alternate factorization.

Miguel Virasoro was the one who identified the necessary Ward-like identities.[76] They were described in the operator formalism developed by Fubini, Gordon, and Veneziano, in terms of a set of operators L_n which obeyed the Virasoro algebra

$$[L_n, L_m] = (n-m)L_{n+m} + \frac{D}{12}(n^3 - n)\delta_{n,-m} \qquad (14)$$

where D is the space time dimension. Provided that the intercept of the leading Regge trajectory is unity, the operators $A_n = L_n - L_0 + n - 1$ for $n > 0$ annihilated physical states. The identity found by Bardakci and Mandelstam was the $n = 1$ case A_1.[77] The requirement of unit intercept was an early disappointing sign that the known dual resonance amplitudes could not give a precise description of strong interactions. Indeed, it predicted a massless vector meson, which was a far cry from the mass of the ρ meson, the lowest mass spin 1 hadron. On the other hand the corresponding intercept for the Virasoro dual resonance models was 2, corresponding to a massless spin 2 particle. For hadrons that was even worse, but it was a tantalizing early hint that dual models might have something to say about quantum gravity!

Although Virasoro had found enough Ward-like identities to remove ghosts from the DRM, it took a while before it was finally proved that's what they did.[80,81] Those proofs made clear that the ghost issue was highly nontrivial, because their absence required $D \leq 26$ for the generalized Veneziano and Virasoro models. DRM were inconsistent if the space–time dimension was too high. At first glance this would not seem to be a serious problem since $D = 4$ in our universe. But difficulties with perturbative unitarity only disappeared if $D = 26$. Some of these dimensions could be compact, but they would still make the particle spectrum much too rich for our world.

The model just discussed is the bosonic DRM. Despite its freedom from ghosts for $\alpha_0 = 1$ and $D = 26$, it still possessed a flaw: a particle of negative mass squared, a tachyon, when $\alpha_o(m_o^2) = \alpha' m_o^2 + 1 = 0$ in the open (Veneziano) sector and $\alpha_c(m_c^2) = \alpha' m_c^2/2 + 2 = 0$ in the closed (Virasoro) sector. Sometimes a tachyon in an approximate spectrum signals that one is perturbing about the wrong ground state, so that the flaw is not necessarily fatal. But the problem of finding an alternative ground state has so far evaded solution.

One can, alternatively, attribute the problem to the absence of spin degrees of freedom in the model. All of the angular momentum of the massless spin 1 particle is from orbital motion so there is a "lighter" state, the tachyon. In addition there was no place for spin 1/2 fermions in the bosonic DRM. The introduction of spin degrees of freedom into DRM thus become a major focus of activity. Since we were no longer tolerating ghosts of any kind, Mandelstam's scheme of appending spin factors to the bosonic amplitudes, as in Refs. 34–37, would not do. The way the Dirac particle avoids such ghost states is by projecting them out by the Dirac equation itself: $(-i\gamma \cdot \partial + m)\psi = 0$. So there must be an intimate interplay between orbital and spin degrees of freedom.

There were two major attacks on this problem. To explain them I need to make brief mention of the operator formalism introduced by Fubini, Gordon, and Veneziano.[75] They noticed that the factorization of the open DRM could be described in terms of harmonic oscillator raising and lowering operators a_n^μ with commutation relations $[a_n^\mu, a_m^\mu] = \eta^{\mu\nu} n \delta_{n,-m}$, where η is the mostly positive Minkowski metric and $n = 0, \pm 1, \pm 2, \ldots$. Also $a_0^\mu = \sqrt{2\alpha'} p^\mu$ with p the energy momentum of the resonance. The mass squared of the resonance is then given by the operator $\alpha' m^2 = -1 + \sum_{n=1}^\infty a_{-n} \cdot a_n$, and the resonance states are in 1-1 correspondence with monomials of the a_{-n}^μ applied to a ground state $|0, p\rangle$. As Nambu, Nielsen, and Susskind more or less independently observed, the a's are just the normal mode operators of a "string" $x^\mu(\sigma, \tau)$, which, however, vibrates in space–time rather than space. Nambu[78] then postulated the correct dynamics for the string which explained how the ghostly oscillations in time x^0 were redundant.

But these stringy insights were not directly used in the construction of DRM with spin degrees of freedom. One approach by Bardakci and Halpern (BH)[88] was to add fermionic oscillators ψ_r^α where α is a Dirac spinor index. They initially assumed $r = \pm 1/2, \pm 3/2, \ldots$. The spinor indices are responsible for additional ghosts. To decouple them BH postulated Ward-like identities which added terms of the schematic spin-orbit form $\bar\psi\gamma^\mu\psi a_\mu$ to Virasoro operators. The cubic non-bilinear form of such a term made it difficult to work with, and progress on the study of these models was slow. The other approach initiated by Ramond[82] with important development by Neveu and Schwarz[83] (RNS) and me, introduced anticommuting Lorentz vectors b_n^μ, $n = 0, \pm 1, \ldots$ (R fermion sector) or b_r^μ, $r = \pm 1/2, \pm 3/2, \ldots$. (NS boson sector). These new operators contributed to the Virasoro generators in a bilinear way. The time components of the b's introduce more ghosts, but Ramond proposed Ward-like identities $F_n = 0$ where $F_n = \sum_k b_{-k} \cdot a_{k+n}$. These operators being only bilinear were much easier to work with than the BH setup. For this reason the new DRM amplitudes were the first to be constructed and analyzed in the RNS language.

The RNS amplitudes were shown to be ghost free for $D \leq 10$.[80,85] The leading open Regge trajectory still had unit intercept and the closed Regge trajectory still had intercept 2. But there were still tachyons, though these were on lower trajectories than in the bosonic model.

Interestingly, very early on (Spring, 1971) Mandelstam remarked to me that the open tachyon in the mesonic RNS amplitudes, the "pion", being in the odd G-parity sector, would decouple if the meson spectrum were restricted to even G-parity. I mentioned this little gem to my colleagues at CERN during my year-long tenure there. Later (1976) Gliozzi, Scherk, and Olive (GSO) observed[84] that the closed tachyon could also be decoupled if, in addition to the even G-parity restriction, noticed by Mandelstam, a Weyl restriction were made on the open fermions. After the even G-parity and GSO projections, the remaining resonance spectrum fell into multiplets of supersymmetry, giving birth to the superstring. The superstring amplitudes were Regge behaved and implied a ghost-free and tachyon free spectrum that included a massless spin 2 boson. I think it is important to note that this construction of superstring provided theoretical physicists with a fully consistent alternative paradigm for the description of physical phenomena, distinct from quantum field theory.

In addition to his G-parity insight for the RNS amplitudes, Mandelstam made substantial contributions to the BH approach to adding spin to DRM.[40,41] In the first of these papers, he went some distance to disentangling the K-degeneracy problem in the nonadditive BH models. These models supported two commuting Virasoro algebras, one of which is chosen to supply the Ward identities, and the other of which predicted an infinite degeneracy of the particle spectrum. Mandelstam showed that the K-degeneracy was a kind of gauge symmetry which can be handled by imposing a gauge condition. The second paper constructed a simple viable nonadditive model that seemed to reproduce many features of the RNS models, including a critical dimension of $D = 10$. He conjectured but did not definitively prove that they were in fact the same models. This result presaged later work by Green and Schwarz[89] who constructed a BH-style formulation of the superstring that was manifestly supersymmetric. Their spinor valued operators, however, were integer moded in contrast to the BH choice of half integer modes. The half integer choice would have violated supersymmetry.

6. Light-Cone Interacting String Formalism

Mandelstam's papers, which develop the light-cone interacting string formalism (Ref. 42 for the bosonic string and Ref. 43 for the Ramond–Neveu–Schwarz string), were essential for the future development of string theory. First of all they completed the program, initiated by Nambu's hypothesis[78] (see also Ref. 79) that the dual resonance model scattering amplitudes[73] were based on the dynamics of a relativistic string. The light-cone quantization of a single noninteracting relativistic string[86] had shown that the spectrum and degeneracies of excited resonances, implied by the pole singularities of dual resonance amplitudes, coincided (for the critical dimension $D = 26$) with the spectrum and degeneracies of the Nambu–Goto string. It was Mandelstam's work that definitively established that the scattering amplitudes themselves followed from string breaking and joining processes described by his path integral representation of the light cone quantum dynamics of the Nambu–Goto string. Secondly, by revealing the underlying physics of dual models, these papers laid the foundation for going beyond string perturbation theory (notably by summing planar interacting string diagrams). Here I will try to give a simplified overview of these papers, accenting these two aspects. I shall limit the detailed discussion to the bosonic string.[42] I shall only briefly mention his extension of the formalism to the RNS string,[43] which included, as a tour-de-force, the first calculation of amplitudes involving four or more external fermions.

6.1. *Interacting string path integral on the light-cone*

The operator quantization of the Nambu–Goto action in light-cone parametrization[86] is equivalent to path integration over the transverse string coordinates $\boldsymbol{x}(\sigma, \tau)$ defined on a rectangular domain $0 < \sigma < p^+ = (p^0 + p^z)/\sqrt{2}$, $0 < \tau = i(t + z)/\sqrt{2} < T$, with (imaginary time) action given by

$$S_{l.c.} = \frac{1}{2} \int_0^T d\tau \int_0^{p^+} d\sigma \left[\left(\frac{\partial \boldsymbol{x}}{\partial \tau}\right)^2 + T_0^2 \left(\frac{\partial \boldsymbol{x}}{\partial \sigma}\right)^2 \right].$$

Then Mandelstam's proposal for a typical contribution to the evolution of a system of strings initially at $\tau = 0$ and finally at $\tau = T$ is simply to alter the geometry of the (σ, τ) domain, as shown in Fig. 8. The horizontal lines in this diagram are internal boundaries where open strings end, so the diagram describes the time evolution of a system of open strings which break and rejoin from time to time.

For the open string in the critical dimension, the worldsheet path integral uses the light-cone action for the free open string. The complete contribution to the evolution amplitude is obtained by doing the path integral over $\boldsymbol{x}$ summing over all numbers of horizontal lines and integrating over their lengths and locations in σ, τ.

Fig. 8. Typical interacting string diagram with two incoming strings three outgoing strings and 7 loops.

As Mandelstam showed[42] the dependence of the diagram on the initial and final $x(\sigma)$ is obtained by solving a suitable Poisson equation via Neumann functions. But there is additional dependence on the P_i^+, τ_i, which characterize the geometry of the diagram, given by the measure factor, which is in turn related to the determinant of the Laplacian defined on the domain represented by the diagram. Since each component of x supplies an identical determinant factor this measure dependence must be of the form

$$\mathcal{M}(P_i^+, \tau_i) = [\mu(P_i^+, \tau_i)]^{D-2}. \qquad (15)$$

Thus it is evident that Lorentz invariance can be achieved for *at most* one value of D. In his pioneering papers on the subject,[42] Mandelstam inferred the measure factor μ by establishing first that his formalism was Lorentz covariant for $D = 26$, and then, setting $D = 26$, he obtained the measure by evaluating it in a special Lorentz frame that simplified the calculation. From the known D dependence, this determines the measure μ. In later work[66] he calculated the necessary determinants directly, as I shall briefly describe in the following.

6.2. *The 3 string vertex*

I shall explain the importance of the measure factor in the case of the 3 open bosonic string vertex represented in Fig. 9. Since the path integral automatically reproduces the properly normalized probability amplitude, Lorentz covariance requires an answer of the form:

$$\text{Vertex} = \frac{1}{\sqrt{P_1^+ P_2^+ (P_1^+ + P_2^+)}} F_{h_1 h_2 h_3}(m_1^2, m_2^2, m_3^2) \qquad (16)$$

where h_i, m_i are the helicities and masses of the three open string states. The factors of $\sqrt{P_i^+}$ are simply the light-cone analogs of the familiar $\sqrt{E_i}$ factors in relativistic probability amplitudes. They can only come from the measure factor of the path integral, and prescribe that under the scaling $P_i^+ \to \lambda P_i^+$, the measure factor should scale as $\lambda^{-3/2}$.

Fig. 9. Vertex diagram for three open strings.

By defining the path integral on a worldsheet lattice, a direct calculation shows that the measure factor scales as $\lambda^{-(D-2)/16}$ for bosonic string,[90] which is compatible with Lorentz invariance only if $D = 26$ the critical dimension. It is important to appreciate that this scaling behavior is entirely due to the sharp 360° corner in the worldsheet boundary where the two strings join. This can be inferred from an inspiring paper by M. Kac.[91] Kac analyzed the eigenspectrum of the Laplacian defined on a polygonal domain, bounded by straight line segments meeting at various angles θ_i. He proved the result

$$\text{Tr } e^{t\nabla^2/2} \sim \frac{\text{Area}}{2\pi t} - \frac{\text{Perimeter}}{4\sqrt{2\pi t}} + \sum_{\text{corners}} \frac{1}{24}\left(\frac{\pi}{\theta_i} - \frac{\theta_i}{\pi}\right) + o(1).$$

The light-cone vertex is a 360° corner, so putting $\theta = 2\pi$ gives

$$\frac{1}{24}\left(\frac{\pi}{\theta} - \frac{\theta}{\pi}\right) \to -\frac{1}{16} \qquad (17)$$

which agrees with the light-cone lattice result. McKean and Singer[92] generalized Kac's result to an arbitrary smooth geometry.

In Ref. 66, Mandelstam evaluated the determinants that enter the measure factor for his path integrals by exploiting the conformal transformation properties of the determinant of the Laplacian:

$$-\frac{1}{2}\delta(\ln(-\nabla^2)) = \frac{1}{24\pi}\int dA \, g^{ab}\frac{d\sigma}{dz_a}\frac{d\sigma}{dz_b} + \frac{1}{12\pi}\int d\ell \, k\sigma + \frac{1}{24\pi}\int dA R\sigma$$

$$+ \frac{1}{24}\sum_{\text{corners}}\left(\frac{\pi}{\theta_i} - \frac{\theta_i}{\pi}\right)\sigma(z_i) \qquad (18)$$

where the new metric is $e^{2\sigma}$ times the old metric. The terms involving the curvature k, R and corners refer to the *old metric*, which Mandelstam took to be the image of the interacting string diagram in the upper half (whole) z-plane for open strings (closed strings). The determinant of the Laplacian in the z-plane as a function of the radii of the semi-circles (circles) can be easily inferred in the limit where all but one of these radii are small and the last one is large.

The boundaries of the light-cone diagram at large initial and final times map to semi-circles (circles) in the upper half (whole) z-plane. The only corners in the z-plane are the 90° corners where the open string semi-circles meet the real axis. For closed string amplitudes the corner terms are absent all together. Mandelstam's evaluation neglected corner terms. However it correctly gave the vital P^+ dependence associated with the 360° corner at the interaction point in the light-cone diagram, because that corner is the image under the conformal transformation of a nonsingular point on the real axis of z-plane: The conformal transformation formula, with corner terms neglected, correctly reproduces Kac's result for those corners in the target manifold that are images of nonsingular points in the original manifold!

The neglect of the contributions from the 90° corners, which are present in both the target and original manifolds, is inconsequential for the calculation of the S-matrix. This is because those same corner contributions, which are multiplicative, are present in the free string propagator, which determines the normalization of the asymptotic states of the S-matrix. Even if they had been properly included, they would have nonetheless completely cancelled out of the S-matrix! The only corner that matters, the 360° one at the interaction point, is correctly treated in Mandelstam's evaluation.

6.3. *Scattering amplitudes for the superstring*

Before Mandelstam's application of his light-cone interacting string formalism to the RNS dual resonance models, there were already formulas for the scattering amplitudes involving any number of bosonic particles with zero or 2 fermionic particles. These amplitudes had been obtained using operator techniques rather than path integrals. The extension of these operator methods to amplitudes with more than two fermionic particles, however, was bogged down.

In this setting Mandelstam's light-cone path integral calculation of amplitudes with arbitrary numbers of fermions[43] had a stunning impact. The measure part of these calculations required the calculation of the determinant of the worldsheet Dirac operator rather than the worldsheet Laplacian. However, with the RNS boundary conditions the z-plane images of the joining point are nonsingular points, so the important part of the measure can be inferred again by conformal mapping, just as in the bosonic case. The boundary conditions on either leg of the 360° angle at the joining point are $S_1 = S_2$ on one side of the angle vertex but $S_1 = -S_2$ on the other side. Then because the worldsheet fermions have conformal weight $1/2$ the conformal transformation to the upper half plane which rotates the line on which $S_1 = -S_2$ by 180° flips that line's boundary condition to $S_1 = S_2$, so there is no discontinuity at the image of the interaction point. This enabled Mandelstam to write down the amplitude for any number of fermions effortlessly.

Applying the light-cone path integral methods to the Green–Schwarz light-cone formulation of the superstring is not so simple.[65,66] This is because the boundary conditions in this case are $S_1 = S_2$ for *all* boundaries (because the worldsheet fermion fields are integer moded on all external legs of a vertex function). Thus the image of the joining point on the upper half plane is the location of a discontinuity in the boundary conditions on the real axis of the upper half plane. This is a 180° "corner" that must be included in the McKean–Singer conformal transformation formula. Thus the conformal transformation is not the only source of P^+ factors: the z-plane determinant is no longer trivial. Since the superstring amplitudes can also be obtained in the RNS formulation it was possible to figure out the required z-plane determinant indirectly. It would certainly be interesting to fill this gap with a direct calculation, for example, along the lines used for Dirichlet–Neumann 180° corners in Ref. 93.

7. Quark Confinement

One of the frustrations of string theory was that, although it was discovered in the effort to understand the strong interactions, it fell short of reaching that goal. The string mass spectrum, inevitably including massless vector and massless tensor particles, was more suited to gauge and gravity theories! After the discovery of asymptotic freedom and charm in 1973 and 1974, overwhelming evidence accumulated that a non-Abelian gauge theory of quarks and gluons based on color SU(3) gauge group was the correct theory of strong interactions. For this choice to be valid, however, the hypothesis of quark confinement, that quarks and gluons are permanently trapped inside of hadrons must hold. Understanding the theoretical physics underlying quark confinement attracted Mandelstam's attention and dominated his work for the rest of the decade.

7.1. *Early papers on gauge theory and gravity*

Before describing Mandelstam's work on quark confinement, I would like to describe briefly four interesting papers he wrote on the fundamental formulation of gauge theories and quantum gravity.[10,11,25,26] Gauge theories and general relativity in their standard treatment use dynamical variables which are not directly observable: the vector potential A^μ in the first case and coordinate systems and metric $g_{\mu\nu}$ in the second. All four papers have the ambitious goal of setting up the fundamental quantum mechanics of such systems without the initial introduction of such quantities. The case of quantum electrodynamics is easiest to explain. The electromagnetic fields $F_{\mu\nu}$ have direct physical meaning and are included in the list of fundamental quantities.

In the standard formulation a charged field satisfies equations in which the vector potential, related to the fields by $F_{\mu\nu} = \partial_\mu A_\nu - \partial_\nu A_\mu$, makes a direct appearance $-(\partial - iQA)^2\phi + m^2\phi = 0$. Then quantization proceeds by promoting ϕ, A to operators subject to commutation relations. In Ref. 10, Mandelstam instead promotes $F_{\mu\nu}$ and a path dependent charged field $\phi(P_x)$ to operators. In the formulation with potentials this path dependent field would be

$$\phi(P_x) = \phi(x) \exp\left\{ iQ \int_{P_x} dx^\mu A_\mu(x) \right\}. \tag{19}$$

Mandelstam replaces such a relation with a rule giving the path dependence

$$\delta\phi(P_x) = iQ\delta\sigma^{\mu\nu} F_{\mu\nu} \phi(P_x) \tag{20}$$

where $\delta\sigma^{\mu\nu}$ is a surface element spanning the closed curve formed by δP and the original path. The homogeneous Maxwell equations guarantee that a global deformation of the path is independent of the surface chosen to span the change.

Clearly quantizing path dependent dynamical variables involves many complicating issues, and the papers which confront and deal with them are vintage Mandelstam. He systematically identifies each complication and through thorough and close reasoning shows how to handle it, The second paper[11] applies the same principles to the quantization of Einstein's theory of gravity. The fundamental variables are the curvature tensor and matter fields, *both* of which are given path dependence. The setup is very much more involved, but Mandelstam forthrightly deals with every obstacle in his path.

The two 1962 papers are mostly concerned with the fundamental setup of the new formalism without regard to the strength of interactions. The 1968 papers[25,26] are concerned with obtaining the Feynman rules of perturbation theory from Mandelstam's gauge or coordinate independent formalism in gauge theories and gravity respectively. In addition to electrodynamics, he also obtains the Feynman rules for non-Abelian gauge theory. His results are in accord with those of the standard procedures, including the rules for Feynman–Fadeev–Popov ghosts.

These four papers express a vision of quantum gauge theories and quantum gravity which is in principle more in tune with their classical analogues. A modern implementation of this vision can be found in Wilson's formulation of lattice gauge theory, in which the fundamental dynamical variables are group elements U_L (as opposed to Lie algebra elements) assigned to the links L of the lattice. Instead of the $-F^2$ term in the Lagrangian, the lattice theory uses the product of U's about each plaquette. Path dependent observables such as Wilson loops play a central role in extracting physical information from lattice gauge theory.

7.2. *Understanding QCD*

In the mid to late 1970s Mandelstam's research concentrated on investigating the physics of quark confinement in QCD. By that time Nielsen and Olesen,[87] motivated by the close connection between superconductivity and the Higgs mechanism in Abelian gauge theory, had found stable classical solutions of the coupled gauge and Higgs equations of motion in which a long magnetic flux tube centered on a line was stabilized by its interaction with the Higgs field. In the semi-classical approximation these classical solutions can be interpreted as new quantum states beyond the vacuum and gauge and Higgs particles.

In a superconductor such states exist because of the Meissner effect, in which the magnetic field $\boldsymbol{B}$ is expelled from the interior of a superconducting medium. If north and south magnetic monopoles were placed at finite separation in a superconductor that filled all of space, the magnetic flux from one would have to end on the other. But the Meissner effect would require the flux to be collimated in a tube sufficiently thin, so that the magnetic field would have the strength to destroy superconductivity inside the flux tube. This would cost an energy proportional to the separation R of the monopoles, i.e. there would be a linear confining potential energy between the monopoles. Nielsen and Olesen and also Nambu observed that if quarks were monopoles, this would provide a mechanism for quark confinement.

The reason why the vacuum of the Higgs mechanism is a superconductor can be understood in terms of the vacuum polarization tensor $R^{\mu\nu} = (q^\mu q^\nu - q^2 \eta^{\mu\nu})R(q^2)$. The q dependent dielectric parameter of the vacuum, thought of as a medium, is related to R by the formula $\epsilon(q^2) = 1 + R(q^2)$, with the static dielectric constant the limit $q \to 0$. The potential of the Higgs field is in the shape of a Mexican hat so that the phase of the Higgs field is a massless scalar, which produces a massless pole in $R(q^2) \sim K/q^2$. Thus $\epsilon(q^2) \to \infty$ as $q \to 0$. But in a Lorentz covariant gauge theory, $\epsilon(q^2)\mu(q^2) = 1$ with μ the magnetic permeability. It follows that $\mu(0) = 0$, so the Higgs vacuum is a perfect diamagnet, and so exhibits the Meissner effect.

In Refs. 46 and 47, Mandelstam studied this effect in the SU(2) non-Abelian gauge theory. There is of course the Dirac quantization condition that the monopole strength must be integer multiples of the reciprocal of non-Abelian charge. He showed that only monopoles of unit strength would produce a flux tube. (For SU(n) flux tubes would form between monopoles of strength $1, 2, \ldots, n-1$.) The flux is defined modulo n, and he showed for $n = 2$ that a two unit flux tube solution could be continuously deformed to the vacuum. As a model of confinement, the quantization condition on the monopole charge would conflict with asymptotic freedom, a demonstrable property of QCD: the magnetic charge would have to blow up at short distances.

Moreover asymptotic freedom for QCD implies that the dielectric parameter $\epsilon(q^2)$ increases as q^2 increases. To see this recall that the effective charge is $g^2(q^2) = g_0^2/(4\pi\epsilon(q^2))$. Asymptotic freedom means that $g^2(q^2)$ decreases with q^2, implying that $\epsilon(q^2)$ increases with q^2 or ϵ decreases as q^2 decreases. The static properties of the medium are given by the limit $q \to 0$. This is *not* indicative of the behavior of a diamagnetic medium — it is rather the behavior of a paramagnetic medium, since $\mu = 1/\epsilon$ increases as q^2 decreases. Quark confinement in QCD should correspond to a perfect dia-electric and, by Lorentz invariance, a perfect paramagnetic medium. So Mandelstam proposed replacing the Higgs field by a field which creates a magnetic monopole. He then proposed a model of the QCD vacuum that contained a condensate of this field. In other words his proposal is essentially the dual of a superconductor in which electric and magnetic quantities are interchanged. The question of whether such a vacuum ansatz minimized the energy was left unresolved in these papers.

7.2.1. *Electric and magnetic variables in gauge theory*

In his next paper on this subject[49] Mandelstam embarks on a thorough and systematic study of electric/magnetic duality in non-Abelian gauge theories. One has the usual magnetic vector potentials to start with, but Mandelstam constructed electric vector potentials in terms of them. In addition he considered operators that create loops of electric or loops of magnetic flux. The former are simply the standard Wilson loops, and the latter are related to loops considered previously by 't Hooft. He then used these constructions to delineate the possible phases of the gauge theory. Two of the possible phases are magnetic confinement a la Nielsen and Olesen and electric confinement a la Wilson. It was hoped that QCD picks the latter phase. Neither of these phases would support massless particles. A third phase with massless gauge particles could not be ruled out.

In four space–time dimensions, the explicit construction of functions of quantum field operators is fraught with the distracting presence of ultraviolet divergences. This clouds the details of the attempted construction of "electric" variables in terms of "magnetic" variables. Thus it is reassuring that Mandelstam was able to implement rigorously an explicit operator mapping of the quantum field operators in two popular toy models in two space–time dimensions. In Refs. 54, 55 he demonstrated the equivalence of the massive Thirring model and the sine–Gordon scalar field model, both in two space–time dimensions. The first model is a QFT of a Fermi field ψ with an interaction term quartic in the fields. In the massless case it had been exactly solved by Thirring. The second toy model is a QFT of a scalar field ϕ with an interaction potential $V(\phi) = (1 - \cos\phi)$. It was interesting because it admitted soliton solutions, which were exact static classical solutions with the property $\phi(x = \infty) = \phi(x = -\infty) + 2\pi$. Evidence for the equivalence of the two models had already been given by Coleman. But Mandelstam was able to construct the Thirring field ψ as an explicit operator function of the sine–Gordon field ϕ. Since the model is two-dimensional, he had complete control over ultraviolet divergences. An interesting feature of Mandelstam's mapping between the two theories is that the weak coupling regime of one maps onto the strong coupling regime of the other, and vice versa.

7.2.2. *An approximation scheme for QCD*

If all quarks are massless, there are no free parameters in QCD with the gauge group SU(3): It is classically scale invariant, but ultraviolet divergences break scale invariance so that the apparently free coupling parameter is exchanged for a single mass scale Λ. Alternatively the coupling depends on the momentum $\alpha_s(q^2)$, and Λ specifies the scale at which α_s has some standard value, e.g. $\alpha_s(\Lambda^2) = 1$. The physical analogy associating confinement with a dual Meissner effect is insightful, and Mandelstam's effort to implement it via monopole condensation was convincing in a qualitative way. But in the inspiring paper,[50] he takes a different tack. Instead of trying to construct an explicit model for the monopole condensate, he argues that it is enough if magnetic effects are sufficiently enhanced in the gluon propagator at low frequencies and wave numbers. By replacing the sum over Feynman diagrams with the set of Schwinger–Dyson equations for multigluon amplitudes, one can explore the effect of such enhancements on the physics of confinement.

For instance, a static confining linear potential in a Coulomb gauge would arise from the Fourier transform of a small q behavior $(\boldsymbol{q}^2)^{-2}$. In the context of the gluon propagator in a Lorentz covariant gauge this behavior would translate to $(q^2)^{-2}$. Mandelstam's strategy was to explore the consistency of such behavior with the Schwinger–Dyson equations, and if so to find (or at least prove the existence of) a solution that exhibits it. The full set of Schwinger–Dyson equations was too complicated, so Mandelstam truncated them to get a tractable equation. He was able to establish that the desired small q behavior was consistent with the truncated equations. Through a rough iterative numerical solution of the equations he obtained pretty convincing evidence that a suitable solution existed. Although the truncated S–D equations could not be achieved as the limit of a parameter of the theory (there are none!), it was clear what graphs had been omitted in the approximation. At least in principle, one could hope to restore them gradually to improve the solution. In the rest of the paper, however, he stuck with his original truncation together with well-described approximations for dealing with multigluon amplitudes and quark propagators. He closed with a discussion of a Bethe–Salpeter equation for the propagation of a quark–antiquark system using the ingredients just described. One output was linearly rising Regge trajectories. This was not surprising given the infrared behavior of the gluon propagator.

8. Closing Tribute

Stanley Mandelstam was a truly impressive theoretical physicist. His work has touched on nearly all of the significant problems in quantum field theory and elementary particles of his lifetime. Each of his published articles is a significant and inspiring achievement and a gold mine of closely reasoned theoretical physics. But Stanley's impact extended way beyond his published work. I am especially aware of his impact on the development of string theory, since that was a central focus of my early research. The discovery and development of dual resonance models, a.k.a string theory, was a spectacular shift from the existing theoretical paradigm of quantum field theory. Young particle theorists, who participated in these developments, were privileged to experience the thrill and excitement of creating something truly new, even though the new paradigm fell short of its original goal. But their work might never have been pursued without the support and encouragement of three masters in the older generation: Stanley Mandelstam, Yoichiro Nambu, and Sergio Fubini. Each, in his own way, made pioneering advances that enabled and guided the developments. These three men not only shared their considerable achievements and wisdom with us youngsters, but they also came down into the trenches and shared a good deal of the burden with us. As one of Stanley's Ph.D. students, I was particularly aware of the ways Stanley would inspire others to take an important step — always with a light and gentle touch, often with a one line provocative remark. He set me off on my first research project in just this way. I don't remember his exact words, but it was something like: "this question may well be soon resolved by others, but you might like to try to understand the (Bardakci–Mandelstam) Ward-identity in the operator formalism." It was a good problem. Stanley was the go-to person whenever a student or postdoc at Berkeley reached an impasse. He would usually know the resolution, but if not he would almost always suggest a new path which led to success. Every string theorist, young and old, owes a tremendous debt to Mandelstam, Nambu, and Fubini. They are all gone now, but their vision and spirit live on.

Acknowledgments

I would like to thank Lars Brink for encouraging me to write this essay on the work of my Ph.D. supervisor. I also thank Korkut Bardakci for helpful comments, especially for drawing my attention to Ref. 4.

References

Articles by Stanley Mandelstam

1. S. Mandelstam, Dynamical variables in the Bethe-Salpeter formalism, *Proc. Roy. Soc. Lond. A* **233**, 248 (1955), doi:10.1098/rspa.1955.0261.
   Mandelstam shows how to extract matrix elements of dynamical variables from solutions of the Bethe-Salpeter equations.
2. S. Mandelstam, Determination of the pion-nucleon scattering amplitude from dispersion relations and unitarity. General theory, *Phys. Rev.* **112**, 1344 (1958), doi:10.1103/PhysRev.112.1344.
   Mandelstam introduces double dispersion relations for 4 particle scattering amplitudes (2 in, 2 out), later known as the Mandelstam representation.
3. S. Mandelstam, Analytic properties of transition amplitudes in perturbation theory, *Phys. Rev.* **115**, 1741 (1959), doi:10.1103/PhysRev.115.1741; Construction of the perturbation series for transition amplitudes from their analyticity and unitarity properties, *Phys. Rev.* **115**, 1752 (1959), doi:10.1103/PhysRev.115.1741.
   The Mandelstam representation is refined into its modern form. s, t, u notation appears. In the second paper he shows how to reproduce the perturbation series for amplitudes, usually given as a sum of Feynman diagrams, as a consequaence of analyticity and unitarity.
4. S. Mandelstam, Some rigorous properties of transition amplitudes, *Nuovo Cim.* **15**, 658 (1960), doi:10.1007/BF02724997.
   Mandelstam rigorously proves that the domain of analyticity of the two particle scattering amplitudes in two variables contains a region bounded by $|stu| < a$, with a a constant depending on the particle masses. This is an extention of the Lehmann ellipse domain of analyticity.
5. G. F. Chew and S. Mandelstam, Theory of low-energy pion-pion interactions, *Phys. Rev.* **119**, 467 (1960), doi:10.1103/PhysRev.119.467; Theory of low-energy pion-pion

interactions II, *Nuovo Cim.* **19**, 752 (1961), doi:10.1007/BF02733371.
Chew and Mandelstam join forces to analyze pion scattering amplitudes.

6. S. Mandelstam, Unitarity condition below physical thresholds in the normal and anomalous cases, *Phys. Rev. Lett.* **4**, 84 (1960), doi:10.1103/PhysRevLett.4.84.

7. G. F. Chew, S. Mandelstam and H. P. Noyes, S-wave dominant solutions of the pion-pion integral equations, *Phys. Rev.* **119**, 478 (1960), doi:10.1103/PhysRev.119.478.

8. G. F. Chew, S. C. Frautschi and S. Mandelstam, Regge poles in π-π scattering, *Phys. Rev.* **126**, 1202 (1961), doi:10.1103/PhysRev.126.1202.
Frautschi joins the Chew-Mandelstam collaboration to bring Regge poles into S-matrix theory.

9. S. Mandelstam, J. E. Paton, R. F. Peierls and A. Q. Sarker, Isobar Approximation of production processes, *Annals Phys.* **18**, 198 (1962), doi:10.1016/0003-4916(62)90067-2.

10. S. Mandelstam, Quantum electrodynamics without potentials, *Annals Phys.* **19**, 1 (1962), doi:10.1016/0003-4916(62)90232-4.
Mandelstam formulates a version of QED without potentials, but with path dependent dynamical variables instead.

11. S. Mandelstam, Quantization of the gravitational field, *Annals Phys.* **19**, 25 (1962), doi:10.1016/0003-4916(62)90233-6.
Mandelstam attacks quantum gravity!

12. S. Mandelstam, An extension of the Regge formula, *Annals Phys.* **19**, 254 (1962).

13. S. Mandelstam, Regge poles as consequences of analyticity and unitarity, in *Proc. 11th Int. Conf. on High Energy Physics (ICHEP62)*, pp. 513–515.

14. S. Mandelstam, Pion and nucleon physics — theoretical, in *Proc. 11th Int. Conf. on High Energy Physics (ICHEP62)*, pp. 739–750.

15. S. Mandelstam, Regge poles as consequences of analyticity and unitarity, *Annals Phys.* **21**, 302 (1963), doi:10.1016/0003-4916(63)90110-6.

16. S. Mandelstam, Regge formalism for relativistic particles with spin, *Nuovo Cim.* **30**, 1113 (1963), doi:10.1007/BF02828820.
Paper on spin and Regge theory and implications for the Gribov Pomeranchuk phenomenon.

17. S. Mandelstam, Cuts in the angular momentum plane. 1, *Nuovo Cim.* **30**, 1127 (1963), doi:10.1007/BF02828821; S. Mandelstam, Cuts in the angular momentum plane. 2, *Nuovo Cim.* **30**, 1148 (1963), doi:10.1007/BF02828822.
Mandelstam's two papers on cuts in the angular momentum plane. The first shows AFS cuts absent (because they sum planar diagrams). The second shows that they are present in diagrams with third double spectral functions.

18. S. Mandelstam, Regge behavior of the spinor channel in the vector-spinor theory, in *Proc. 12th Int. Conf. on High Energy Physics (ICHEP64)*, pp. 373–375.

19. S. Mandelstam, Non-Regge terms in the vector channel of the vector-spinor theory, in *Proc. 12th Int. Conf. on High Energy Physics (ICHEP64)*, pp. 376–377.

20. R. Omnes and S. Mandelstam, On the extension of the non-relativistic three-body scattering amplitude to complex values of the total angular momentum, in *Proc. 12th Int. Conf. on High Energy Physics (ICHEP64)*, pp. 404–406.

21. S. Mandelstam, Non-Regge terms in the vector-spinor theory, *Phys. Rev.* **137**, B949 (1965), doi:10.1103/PhysRev.137.B949.
In vector spinor theory, a Kronecker delta in the J-plane is absent in the spinor channel but present in the vector channel. A Kronecker delta represents an "elementary" particle, i.e. one not on a Regge trajectory.

22. S. Mandelstam, Recent work on Regge poles and the three-body problem, November 1966, UCRL-17250.
23. S. Mandelstam, Dynamics based on indefinitely rising Regge trajectories, in *Int. Conf. on Particles and Fields*, Rochester, NY, USA, 28 August–1 September 1967, *Conf. Proc. C* **670828**, 605 (1967).
    Proposal first announced at a conference to seek dynamical models based on indefinitely rising Regge trajectories.
24. S. Mandelstam and L. L. Wang, Gribov–Pomeranchuk poles in scattering amplitudes, *Phys. Rev.* **160**, 1490 (1967), doi:10.1103/PhysRev.160.1490.
    Mandelstam and Wang (now Chau) clarify the Gribov-Pomeranchuk argument for fixed poles in J at nonsense points and show that the argument for an essential singularity fails.
25. S. Mandelstam, Feynman rules for electromagnetic and Yang-Mills fields from the gauge independent field theoretic formalism, *Phys. Rev.* **175**, 1580 (1968), doi:10.1103/PhysRev.175.1580.
    Derivation of Feynman rules for QED and nonabelian gauge theory from gauge independent path dependent formalism.
26. S. Mandelstam, Feynman rules for the gravitational field from the coordinate independent field theoretic formalism, *Phys. Rev.* **175**, 1604 (1968), doi:10.1103/PhysRev.175.1604.
    Derivation of Feynman rules for gravity from coordinate independent path dependent formalism.
27. S. Mandelstam, Dynamics based on rising Regge trajectories, *Phys. Rev.* **166**, 1539 (1968), doi:10.1103/PhysRev.166.1539.
    First publication proposing that the bootstrap be implemented assuming narrow (zero width) resonances and exactly linear Regge trajectories. Precedes Veneziano's discovery.
28. S. Mandelstam, Relations between partially conserved axial-vector current, axial-vector charge commutation relations, and conspiracy theory, *Phys. Rev.* **168**, 1884 (1968), doi:10.1103/PhysRev.168.1884.
    In this paper Mandelstam shows how to obtain some of the consequences of current algebra, for hadronic scattering amplitudes, from S-matrix theory and "conspiracy" relations among several Regge trajectories, one of which is the pion trajectory . . . .
29. S. Mandelstam, Rising Regge trajectories and dynamical calculations, *Comments Nucl. Part. Phys.* **3**, 65 (1969); S. Mandelstam, Rising Regge trajectories and dynamical calculations. 2, *Comments Nucl. Part. Phys.* **3**, 147 (1969).
    Two conference proceedings on rising trajectory dynamics.
30. S. Mandelstam, Veneziano formula with trajectories spaced by two units, *Phys. Rev. Lett.* **21**, 1724 (1968), doi:10.1103/PhysRevLett.21.1724.
    Post Veneziano, a proposal to remove odd trajectories from the four point function.
31. S. Mandelstam, Generalizations of the Veneziano and Virasoro models, *Phys. Rev.* **183**, 1374 (1969), doi:10.1103/PhysRev.183.1374.
    Here Stanley obtains a double integral representation for the Virasoro model and generalizes it in a single formula which interpolates between Veneziano and Virasoro.
32. K. Bardakci and S. Mandelstam, Analytic solution of the linear-trajectory bootstrap, *Phys. Rev.* **184**, 1640 (1969), doi:10.1103/PhysRev.184.1640.
    With Bardakci, Stanley factorizes the generalized Veneziano model on resonances in 1-1 correspondence with the excited states of a string. They also discover the first Ward-like identity in hind sight corresponding to $L_1 - L_0$.

33. S. Mandelstam, Currents in s-matrix theory, in *Topical Conference on Weak Interactions*, Geneva, Switzerland, 1969, *Conf. Proc. C* **690114**, 349 (1969).
34. S. Mandelstam, Relativistic quark model based on the Veneziano representation. i. Meson trajectories, *Phys. Rev.* **184**, 1625 (1969), doi:10.1103/PhysRev.184.1625.
    Starting with the n point dual resonance amplitude, Stanley adds spin and internal symmetry factors to give the amplitudes the character of a relativistic quark model.
35. S. Mandelstam, Multiparticle Veneziano formulas corresponding to minimal nonplanar Feynman diagrams, *Phys. Rev. D* **1**, 1720 (1970), doi:10.1103/PhysRevD.1.1720.
36. S. Mandelstam, Relativistic quark model based on the Veneziano representation. ii. General trajectories, *Phys. Rev. D* **1**, 1734 (1970), doi:10.1103/PhysRevD.1.1734.
37. S. Mandelstam, Relativistic quark model based on the Veneziano representation. iii. Baryon trajectories, *Phys. Rev. D* **1**, 1745 (1970), doi:10.1103/PhysRevD.1.1745.
    In the preceding three papers Stanley introduces new dual resonance amplitudes which have the character of scattering of multi-strand strings, called minimal nonplanar amplitudes. Then he uses them to generalize his quark model to amplitudes involving baryons, among others things.
38. S. Mandelstam, Feynman-like diagrams in dual-resonance models, *Comments Nucl. Part. Phys.* **4**, 95 (1970).
39. S. Mandelstam, Dispersion theory and the complex j-plane, Fields and Quanta 1 (1971) 295–344.
40. S. Mandelstam, K degeneracy in nonadditive dual resonance models, *Phys. Rev. D* **7**, 3763 (1973), doi:10.1103/PhysRevD.7.3763.
    This paper addresses a problem arising in the Bardakci–Halpern approach to spin in dual models which they dubbed nonadditive models.
41. S. Mandelstam, Simple nonadditive dual resonance model, *Phys. Rev. D* **7**, 3777 (1973), doi:10.1103/PhysRevD.7.3777.
    Mandelstam constructs a simple nonadditive model which resembles the RNS model and might be equivalent to it.
42. S. Mandelstam, Interacting string picture of dual resonance models, *Nucl. Phys. B* **64**, 205 (1973); Lorentz properties of the three-string vertex, *Nucl. Phys. B* **83**, 413 (1974).
    Introduction of the lightcone interacting string formalism which definitively showed that dual resonance amplitudes followed from string theory.
43. S. Mandelstam, Manifestly dual formulation of the Ramond model, *Phys. Lett. B* **46**, 447 (1973); Interacting string picture of the Neveu–Schwarz–Ramond model, *Nucl. Phys. B* **69**, 77 (1974).
    These papers calculated string amplitudes with more than two fermionic legs for the first time. A technical tour-de-force.
44. S. Mandelstam, Dual-resonance models, *Phys. Rep.* **13**, 259 (1974), doi:10.1016/0370-1573(74)90034-9.
    An important review article on dual resonance models.
45. S. Mandelstam, Recent work on string models, in *Proc. 17th Int. Conf. on High-Energy Physics (ICHEP 74)*, pp. I.260–264.
46. S. Mandelstam, Vortices and quark confinement in nonabelian gauge theories, *Phys. Lett. B* **53**, 476 (1975), doi:10.1016/0370-2693(75)90221-X.
    Mandelstam's first proposal of a promising quark confinement mechanism, elaborated in the following article.
47. S. Mandelstam, Vortices and quark confinement in nonabelian gauge theories, *Phys. Rep.* **23**, 245 (1976), doi:10.1016/0370-1573(76)90043-0.

48. S. Mandelstam, Introduction to quark confinement and the dynamics of strong interactions, UCB-PTH-78-5. Joseph H. Weis, Memorial Symposium on Strong Interactions, 30 November 1978, Seattle, Washington, CNUM: C78-11-30.
49. S. Mandelstam, Charge-monopole duality and the phases of nonabelian gauge theories, *Phys. Rev. D* **19**, 2391 (1979), doi:10.1103/PhysRevD.19.2391.
   A systematic analysis of the possible phases of nonabelian gauge theories.
50. S. Mandelstam, Approximation scheme for QCD, *Phys. Rev. D* **20**, 3223 (1979), doi:10.1103/PhysRevD.20.3223.
   Proposal of a practical scheme for approximate calculations of quark anti quark mesons in QCD.
51. S. Mandelstam, Review of recent results on QCD and confinement, UCB-PTH-79-9, in *9th Int. Symp. on Lepton and Photon Interactions at High Energy*, Batavia, Illinois, 23–29 August 1979, CNUM: C79-08-23.
52. S. Mandelstam, General introduction to confinement, *Phys. Rep.* **67**, 109 (1980), doi:10.1016/0370-1573(80)90083-6.
53. S. Mandelstam, The possible role of monopoles in the confinement mechanism, Trieste Monopole Mtg.1981:0289, (1982).
54. S. Mandelstam, Soliton operators for the quantized sine–Gordon equation, *Phys. Rev. D* **11**, 3026 (1975), doi:10.1103/PhysRevD.11.3026.
   A mathematical excursion: The construction of an explicit operator mapping between the massive Thirring model and the sine–Gordon scalar field theory, both in two spacetime dimensions.
55. S. Mandelstam, Soliton operators for the quantized sine–Gordon equation, *Phys. Rep.* **23**, 307 (1976), doi:10.1016/0370-1573(76)90052-1.
56. S. Mandelstam, Light cone superspace and the vanishing of the beta function for the $N = 4$ model, UCB-PTH-82-10, in *21st Int. Conf. on High Energy Physics (ICHEP 82)*; *J. Phys. Colloq.* **43**, 331 (1982), doi:10.1051/jphyscol:1982367.
57. S. Mandelstam, Light cone superspace and the ultraviolet finiteness of the $N = 4$ model, *Nucl. Phys. B* **213**, 149 (1983), doi:10.1016/0550-3213(83)90179-7.
   This important paper establishes that the $\mathcal{N} = 4$ supersymmetric nonabelian gauge theory is free of ultraviolet divergences.
58. S. Mandelstam, Covariant superspace with unconstrained fields, *Phys. Lett. B* **121**, 30 (1983), doi:10.1016/0370-2693(83)90195-8.
59. S. Mandelstam, *Workshop on Problems in Unification and Supergravity*, 13–16 January 1983, La Jolla, California, CNUM: C83-01-13; *AIP Conf. Proc.* **116**, 99 (1984), doi:10.1063/1.34597.
60. S. Mandelstam, Ultraviolet finiteness of the $N = 4$ model, *20th Annual Orbis Scientiae: Dedicated to P.A.M. Dirac's 80th Year*, Miami, Florida, 17–22 January 1983.
61. S. Mandelstam, The current theory of strong interactions and the problem of quark confinement, in *Old and New Questions in Physics, Cosmology, Philosophy, and Theoretical Biology*, A. Van Der Merwe (ed.), pp. 265–274.
62. S. Mandelstam, Composite vector meson and string models, Print-84-1018 (UC, BERKELEY). G. F. Chew Jubilee, Berkeley, California, 29 September 1984. CNUM: C84-09-29
63. J. Lepowsky, S. Mandelstam and I. Singer, in *Proc. Vertex Operators in Mathematics and Physics*, Berkeley, CA, USA, 10–17 November, 1983, doi:10.1007/978-1-4613-9550-8.

64. S. Mandelstam, Review of aspects of string theory, LPTENS 85/13, *Symp. on the Occasion of the Niels Bohr Centennial: Recent Developments in Quantum Field Theory*, Copenhagen, Denmark, 6–10 May 1985, CNUM: C85-05-06.
65. S. Mandelstam, Interacting string picture of the fermionic string, *Prog. Theor. Phys. Suppl.* **86**, 163 (1986), doi:10.1143/PTPS.86.163.
    This paper extends the interacting string formalism to the lightcone Green–Schwarz formulation of the superstring.
66. S. Mandelstam, The interacting string picture and functional integration, in *Unified String Theories*, eds. M. Green and D. Gross (World Scientific, 1986); *Proceedings, Unified String Theories* (Santa Barbara 1985), pp. 46–102 and Calif. Univ. Berkeley, UCB-PTH-85-47 (85,REC.NOV.) 58p.
67. S. Mandelstam, The N loop string amplitude, *2nd Nobel Symposium on Elementary Particle Physics: Unification of Fundamental Interactions*, Marstrand, Sweden, 2–7 June 1986, CNUM: C86-06-02.5; *Phys. Scripta T* **15**, 109 (1987), doi:10.1088/0031-8949/1987/T15/013.
68. S. Mandelstam, The n loop string amplitude: Explicit formulas, finiteness and absence of ambiguities, *Phys. Lett. B* **277**, 82 (1992), doi:10.1016/0370-2693(92)90961-3.
69. S. Mandelstam, Factorization in dual models and functional integration in string theory, in *The Birth of String Theory*, Andrea Cappelli *et al.* (eds.), pp. 294–311, arXiv:0811.1247 [hep-th].

Other References

70. G. 't Hooft, A planar diagram theory for strong interactions, *Nucl. Phys. B* **72**, 461 (1974).
71. R. Dolen, D. Horn and C. Schmid, *Phys. Rev. Lett.* **19**, 402 (1967); doi:10.1103/PhysRevLett.19.402; R. Dolen, D. Horn and C. Schmid, *Phys. Rev.* **166**, 1768 (1968), doi:10.1103/PhysRev.166.1768.
72. M. Ademollo, H. R. Rubinstein, G. Veneziano and M. A. Virasoro, Bootstraplike conditions from superconvergence, *Phys. Rev. Lett.* **19**, 1402 (1967); Bootstrap of meson trajectories from superconvergence, *Phys. Rev.* **176**, 1904 (1968).
73. G. Veneziano, Construction of a crossing-symmetric, Regge behaved amplitude for linearly rising trajectories, *Nuovo Cim. A* **57**, 190 (1968).
74. S. Fubini and G. Veneziano, Level structure of dual-resonance models, *Nuovo Cim. A* **64**, 811 (1969), doi:10.1007/BF02758835.
75. S. Fubini, D. Gordon and G. Veneziano, A general treatment of factorization in dual resonance models, *Phys. Lett. B* **29**, 679 (1969), doi:10.1016/0370-2693(69)90109-9.
76. M. S. Virasoro, Subsidiary conditions and ghosts in dual resonance models, *Phys. Rev. D* **1**, 2933 (1970), doi:10.1103/PhysRevD.1.2933.
77. C. B. Thorn, Linear dependences in the operator formalism of Fubini, Veneziano, and Gordon, *Phys. Rev. D* **1**, 1693 (1970), doi:10.1103/PhysRevD.1.1693.
78. Y. Nambu, Duality and hadrodynamics, Lectures at the Copenhagen Summer Symposium (1970).
79. T. Goto, Relativistic quantum mechanics of one-dimensional mechanical continuum and subsidiary condition of dual resonance model, *Prog. Theor. Phys.* **46**, 1560 (1971).
80. P. Goddard and C. B. Thorn, Compatibility of the dual Pomeron with unitarity and the absence of ghosts in the dual resonance model, *Phys. Lett. B* **40**, 235 (1972).

81. R. C. Brower, Spectrum generating algebra and no ghost theorem for the dual model, *Phys. Rev. D* **6**, 1655 (1972).
82. P. Ramond, Dual theory for free fermions, *Phys. Rev. D* **3**, 2415 (1971); C. B. Thorn, Embryonic dual model for pions and fermions, *Phys. Rev. D* **4**, 1112 (1971).
83. A. Neveu and J. H. Schwarz, Factorizable dual model of pions, *Nucl. Phys. B* **31**, 86 (1971); A. Neveu, J. H. Schwarz and C. B. Thorn, Reformulation of the dual pion model, *Phys. Lett. B* **35**, 529 (1971); A. Neveu and J. H. Schwarz, Quark model of dual pions, *Phys. Rev. D* **4**, 1109 (1971).
84. F. Gliozzi, J. Scherk and D. I. Olive, Supersymmetry, supergravity theories and the dual spinor model, *Nucl. Phys. B* **122**, 253 (1977).
85. E. Corrigan and P. Goddard, The absence of ghosts in the dual fermion model, *Nucl. Phys. B* **68**, 189 (1974).
86. P. Goddard, J. Goldstone, C. Rebbi and C. B. Thorn, Quantum dynamics of a massless relativistic string, *Nucl. Phys. B* **56**, 109 (1973).
87. H. B. Nielsen and P. Olesen, *Nucl. Phys. B* **61**, 45 (1973), doi:10.1016/0550-3213(73)90350-7.
88. K. Bardakci and M. B. Halpern, New dual quark models, *Phys. Rev. D* **3**, 2493 (1971).
89. M. B. Green and J. H. Schwarz, Supersymmetrical dual string theory, *Nucl. Phys. B* **181**, 502 (1981).
90. R. Giles and C. B. Thorn, A lattice approach to string theory, *Phys. Rev. D* **16**, 366 (1977).
91. M. Kac, Can one hear the shape of a drum?, *Am. Math. Mon.* **73**, 1 (1966).
92. H. P. McKean and I. M. Singer, Curvature and eigenvalues of the Laplacian, *J. Diff. Geom.* **1**, 43 (1967).
93. C. B. Thorn, Determinants for the lightcone worldsheet, *Phys. Rev. D* **86**, 066010 (2012), doi:10.1103/PhysRevD.86.066010, arXiv:1205.5815 [hep-th].

Scientific Biography of Stanley Mandelstam: 1981–2016

Nathan Berkovits

*ICTP South American Institute for Fundamental Research,
Instituto de Física Teórica, Universidade Estadual Paulista,
Rua Dr. Bento T. Ferraz 271, 01140-70, São Paulo, Brazil
nberkovi@ift.unesp.br*

In this contribution to the memorial volume, I will review Stanley Mandelstam's work after 1980 and concentrate on his powerful application of light-cone gauge methods in super-Yang–Mills theory and superstring theory.

1. Introduction

When I arrived in Berkeley as a graduate student in 1983, I of course had heard about Stanley's important work on S-matrix theory and was interested in working under his supervision. At Berkeley, the first step to being accepted by an advisor is to take a reading course with him, so I knocked on Stanley's door to see if this would be possible. When he asked what kind of research I was interested in, I remember replying that I was not interested in studying field theories like QCD because they seemed too complicated. And his response was "So you are interested in studying string theory?"

I had no idea what was string theory, but I of course said "Yes" in order to have the chance to study with him. At the time, there were only a few researchers actively working in string theory which included Stanley, John Schwarz together with Lars Brink and Michael Green, and a young group at Chicago consisting of Dan Friedan, Emil Martinec and Steve Shenker. As I later learned, Stanley seemed to thrive in this atmosphere where he could carefully work out his ideas without the pressure of publishing quickly in order to avoid being scooped. Although he was very interested in the comments of others about his work, he had a quiet self-confidence in his original ideas and did not seem to worry if others would follow up on his ideas or if his work would be adequately cited.

Soon after he accepted me for a reading course, Stanley went away for a year on sabbatical to Paris. During that year, I studied all the dual model and early string theory papers I could find and was very lucky to find Warren Siegel, a postdoc in Berkeley at the time and one of the few people interested in string theory. Moreover,

Warren had the patience to answer all my questions on string theory as well as teach me about supersymmetry. The papers of Stanley Mandelstam and Warren Siegel share the property that they are not easy to read, but once you understand them, contain extremely powerful ideas that will be useful for the rest of your career.

By the time Stanley returned from his sabbatical, I had learned how worldsheet supersymmetry could be used to simplify some of the light-cone expressions from his early papers on the Ramond–Neveu–Schwarz (RNS) formalism, and he fortunately accepted me as a PhD student. As mentioned in Charles Thorn's biography, Stanley had developed in the 1970s the "light-cone interacting string picture" of both the bosonic string[1] as well as the RNS spinning string.[2] The interacting string picture of the bosonic string was much simpler than that of the RNS string, however, the bosonic string was inconsistent because of tachyons. Although the naive RNS string also had a tachyon, one could perform a projection on the RNS string which removed the tachyon and left a space–time-supersymmetric spectrum. The resulting projected string theory was called the "superstring," and there were conjectures based on one-loop computations that scattering amplitudes of this superstring were finite to all orders in perturbation theory. Since the massless sector of the superstring spectrum includes gravity (together with its supersymmetric completion to "supergravity"), superstring theory had the potential of providing physicists with their holy grail of a consistent quantum theory of gravity.

Despite this exciting potential, there were only a few groups working in superstring theory in 1983. In the late 1970s and early 1980s, Michael Green and John Schwarz had developed a version of the superstring with manifest space–time supersymmetry but which was only defined in light-cone gauge.[3] Because of the manifest space–time supersymmetry, the Green–Schwarz formalism was much more powerful than the RNS formalism for computing scattering amplitudes of external fermions. As was mentioned in Thorn's biography, the computation of even 4-fermion tree-level scattering amplitudes using the RNS formalism was a Herculean task that Mandelstam was the first to compute.[2] But using the light-cone Green–Schwarz formalism, the 4-fermion amplitudes were no harder to compute than the 4-gluon amplitudes because of manifest space–time supersymmetry.

Using the light-cone formalism developed by Green and Schwarz, scattering amplitudes could only be computed if all but two of the external strings had infinitesimal width, which in light-cone gauge implied that all but two of the external states had infinitesimal momentum P^+ in the longitudinal light-cone direction. This restriction prevented Green and Schwarz from computing amplitudes with more than 4 external strings. However, Mandelstam's light-cone interacting string picture allows all external states to have finite P^+ momentum. At the time I began research with Stanley, he had succeeded in modifying his interacting string picture for the RNS formalism to an interacting string picture for the Green–Schwarz formalism, and was able for the first time to derive N-point massless tree amplitudes with manifest space–time supersymmetry for an arbitrary number N of external states.[4]

Although Mandelstam's remarkable supersymmetric expression for the N-point tree amplitude was argued to be Lorentz covariant, it has not yet been derived from a manifestly Lorentz-covariant formalism.

At this time, Green and Schwarz were developing a classical Lorentz-invariant and space–time supersymmetric worldsheet action for the superstring[5] which was closely related to a space–time supersymmetric action for the superparticle found by Brink and Schwarz[6] and further studied by Siegel.[7] This classical worldsheet action had a fermionic symmetry called kappa-symmetry which could be gauge-fixed in light-cone gauge to reduce the action to the previous light-cone formalism of Green and Schwarz. There was a great deal of effort to find alternative quantizations of this covariant Green–Schwarz action which preserved the manifest Lorentz covariance, but all such attempts at covariant quantization of the Green–Schwarz formalism were unsuccessful at computing scattering amplitudes.

On the other hand, the Chicago group of Friedan, Martinec and Shenker at this time were developing a new understanding of the covariant quantization of the RNS formalism in which super-Riemann surfaces with worldsheet supersymmetry played a crucial role.[8] The ghosts coming from gauge-fixing the worldsheet supersymmetry were necessary for constructing vertex operators for fermionic states, and by bosonizing the ghosts and combining them with the RNS fermionic matter variables into spin fields, the horrendously complicated computations of scattering amplitudes for external fermions in the RNS formalism could be simplified to a simple correlation function computation of these vertex operators.

Soon after these important developments involving covariant descriptions of both the Green–Schwarz and RNS formalisms, Green and Schwarz showed that in certain types of superstring theory, anomaly cancellation almost uniquely determines the choice of gauge group.[9] And because the anomaly-free gauge group was a promising candidate for Grand Unified Theories generalizing the standard model of particle physics, there was a huge explosion of interest in superstrings among physicists who interpreted the theory not just as a theory of quantum gravity, but as a theory which unified a certain supersymmetric generalization of the standard model with gravity. I do not remember Stanley being very enthusiastic about this new application of superstring theory to describe particle physics phenomenology, and his main interests continued to be studying and verifying the internal consistency of superstring theory as a quantum theory of gravity.

After the work of Green–Schwarz and Friedan–Martinec–Shenker, there were still many open questions concerning the consistency of superstring theory. In addition to the question of finiteness of scattering amplitudes which was closely related to space–time supersymmetry, one also needed to show for consistency that superstring scattering amplitudes are unitary and Lorentz-covariant. Using Mandelstam's light-cone interacting string picture of either the RNS or Green–Schwarz formalism, unitarity was manifest but Lorentz covariant was not manifest. And using the covariant RNS description of Friedan-Martinec-Shenker, Lorentz covariance was manifest but unitarity and space–time-supersymmetry were not manifest.

Mandelstam's goal over the next decades was to derive explicit expressions for superstring scattering amplitudes and prove that they had all of the desired properties of finiteness, unitarity and Lorentz covariance. Although he did not write many papers during this period, he supervised two PhD students after me, Sang-Jin Sin and Arjun Berera (both with contributions to this memorial volume), and all of our thesis topics were on the consistency properties of superstring amplitudes. During the late 1980s, these consistency properties became a hot topic because of a mathematical question related to potential ambiguities in defining the supermoduli space of super-Riemann surfaces when the surface is nearly degenerate.[10] His many insights on the light-cone RNS and Green–Schwarz formalisms led to worldsheet conformally invariant formalisms for the superstring with manifest spacetime supersymmetry where the consistency properties are easier to verify.[11] And very recently, an extensive discussion and proof of the consistency properties of perturbative superstring amplitudes using the covariant RNS formalism has appeared in the important papers of Edward Witten[12] and Ashoke Sen.[13]

In 1982, Mandelstam had been the first to prove the finiteness of $N = 4$ super-Yang–Mills theory[14] using light-cone gauge arguments, and although this work preceded my interaction with him, this important proof will be briefly reviewed in the next section since it was not included in Thorn's biography. (A more detailed discussion of light-cone $N = 4$ super-Yang–Mills is in Lars Brink's contribution to this volume.) And in the following sections, Mandelstam's explicit formulas for bosonic[15] and superstring amplitudes[16] will be reviewed and his arguments for the various consistency properties of superstring amplitudes including unitarity, Lorentz covariance, absence of ambiguities and finiteness will be discussed.

2. Finiteness of $N = 4$ Super-Yang–Mills Theory

The maximal number of supersymmetries consistent with Yang–Mills theory is 16, and in four dimensions, this maximally supersymmetric theory is called $N = 4$ super-Yang–Mills. Unlike ordinary four-dimensional Yang–Mills theory which is conformally invariant at the classical level, but has quantum divergences which require renormalization counterterms that break conformal invariance at the quantum level, it had been conjectured based on explicit loop computations that $N = 4$ super-Yang–Mills theory is finite to all orders and is therefore conformally invariant at both the classical and quantum level.

However, a proof of finiteness of $N = 4$ super-Yang–Mills was lacking. The most powerful methods for verifying the cancellations of quantum divergences in supersymmetric theories involve the extension of ordinary Feynman diagram rules to superspace Feynman diagram rules where the space–time field for the gluon, $A_m(x)$ with $m = 0$ to 3, is extended to a space–time superfield, $\mathcal{A}_M(x, \theta)$, describing both the bosonic gluon and its fermionic gluino superpartners where θ^α are anticommuting coordinates, α is a spinor index, and $M = (m, \mu)$ ranges over both space–time vector and space–time spinor indices. Under space–time supersymmetry, x^m and θ^α

transform in a simple manner, and the cancellation due to supersymmetry between the contributions of the bosonic and fermionic fields in the internal loops of the Feynman diagrams becomes manifest.

But since there are 16 θ coordinates, the superfield $\mathcal{A}_M(x,\theta)$ has 2^{16} component fields for each value of M. Therefore, the superfield $\mathcal{A}_M$ needs to be highly constrained if it is to describe only the physical degrees of freedom of $N=4$ super-Yang–Mills consisting of 8 bosons (one gluon and six scalars) and 8 fermions. There does not exist a Lorentz-covariant description of these constraints for $N=4$ super-Yang–Mills which can be applied offshell to compute superspace Feynman diagrams. However, as Mandelstam showed in Ref. 14, there is a lightcone description of $N=4$ super-Yang–Mills where the superfields are unconstrained and only contain the 8 bosonic and 8 fermionic physical degrees of freedom. Using this light-cone description of $N=4$ super-Yang–Mills, he was able to define superspace Feynman rules and prove the cancellation of all quantum divergences.

In this light-cone description, there are only four anticommuting coordinates θ^A for $A=1$ to 4, and the superfield $\Phi(x,\theta)$ can be expanded as

$$\Phi = i(2P^+)^{-1}V + (2P^+)^{-1}\theta^A \psi_A + \frac{i}{4}\theta^A \theta^B s_{AB}$$
$$+ \frac{1}{6}\epsilon_{ABCD}\theta^A \theta^B \theta^C \bar{\psi}^D + 2iP^+(\theta)^4 \bar{V}, \tag{1}$$

where $P^+ = P^0 + P^3$ is the momentum in the longitudinal light-cone direction, V and $\bar{V}$ are the positive and negative helicity components of the gluon, $s_{AB} = -s_{BA}$ are the 6 scalars, and ψ_A and $\bar{\psi}^A$ are the 8 propagating components of the gluino. The reality condition of the superfield Φ is a bit unusual and is

$$\bar{\Phi} = (2P^+)^{-2} D_1 D_2 D_3 D_4 \Phi, \tag{2}$$

where $D_A = \frac{\partial}{\partial \theta^A} + 2P^+ \theta^A$, however, it is otherwise unconstrained.

It is straightforward to construct the superspace action for $N=4$ super-Yang–Mills in terms of Φ as the integration of a superspace Lagrangian over both $\int d^4x$ and $\int d^4\theta$. And using the superspace Feynman rules derived from this action, Mandelstam was able to prove using standard power-counting arguments that quantum divergences are absent to all orders in perturbation theory, i.e. that the β function for the conformal anomaly vanishes to all orders. The only subtle point of the proof involved the use of power-counting arguments in light-cone gauge.

Power-counting arguments for loop momenta assume that one can analytically continue to imaginary values of P_0, but the presence in light-cone gauge of explicit factors of $(P^+)^{-1}$ prevents this naive analytic continuation. Using his unique insights about light-cone gauge, Mandelstam was able to avoid this problem by arguing that one can slightly modify the usual light-cone gauge choice such that the $(P^+)^{-1}$ poles are replaced by $(P^+ + \epsilon P^-)^{-1}$ poles where ϵ is an infinitesimal parameter. With this modification, there is no problem in analytically continuing to imaginary values of P_0 and the standard power-counting arguments are valid. This resolution is typical of Mandelstam's work in which he carefully analyzes every possible subtlety and does not publish his work until he has found solutions to every point. His solutions are usually ingenious with deep implications, but because they are unfamiliar to the general physics community, they often are accepted without being understood.

3. Bosonic String Amplitudes

As reviewed in Thorn's biography, Mandelstam developed in the 1970s the light-cone interacting string picture first for the bosonic string[1] and then for the RNS spinning string.[2] After his finiteness proof of $N = 4$ super-Yang–Mills, the main focus of Mandelstam's research was the explicit computation of multiloop superstring amplitudes. The first step was to extend the bosonic tree amplitude result to bosonic multiloop amplitudes[15] using his manifestly unitary light-cone string picture, and this will be reviewed here. The second step of generalizing the bosonic multiloop amplitudes to superstring multiloop amplitudes[16] will be reviewed in the following section.

3.1. *Bosonic tree amplitudes*

The starting point for bosonic tree amplitudes in Mandelstam's light-cone interacting string diagram is the conformal map from the complex plane to the closed string diagram

$$\rho(z) = \sum_{r=1}^{N} \alpha_r \log(z - z_r), \tag{3}$$

where $\alpha_r = P_r^+$ is the longitudinal light-cone component of the momentum of the rth external string, and the circumference of the closed string in light-cone gauge is $2\pi|\alpha_r|$. $\rho = \tau + i\sigma$ is a complex variable parametrizing the light-cone string diagram where τ is the light-cone time and σ is the position along the string, and if the rth external string is outgoing/incoming, α_r is positive/negative so that $\text{Re}(\rho) \to +\infty/-\infty$ when $z \to z_r$. Also, when z goes around the point z_r, $\rho \to \rho + 2\pi i \alpha_r$ as expected for a string of circumference $2\pi\alpha_r$.

For an N-point tree amplitude, one has $N - 2$ interaction points ρ_I where the strings split or join, and these points are located at the position $\rho_I \equiv \rho(z_I)$ where $\frac{\partial \rho}{\partial z}\big|_{z=z_I} = 0$, i.e. where $\sum_{r=1}^{N} \frac{\alpha_r}{z_I - z_r} = 0$. If $\rho_I \equiv \rho(z_I)$ are the $N - 2$ interaction points, the manifestly unitary expression for the N-point tree amplitude coming from performing the path integral over all light-cone interaction diagrams is

$$\mathcal{A} = g_s^{N-2} \prod_{I=2}^{N-2} \int d^2(\rho_I - \rho_1) |\Delta|^{-\frac{d-2}{2}} e^{-\sum_r P_r^- \tau(z_r)} \langle V_1(z_1) \cdots V_N(z_N) \rangle \tag{4}$$

$$= g_s^{N-2} \prod_{r=3}^{N-1} \int d^2 z_r \left|\det\left(\frac{d(\rho_I - \rho_1)}{dz_r}\right)\right|^2 |\Delta|^{-\frac{d-2}{2}}$$

$$\times e^{-\sum_r P_r^- \tau(z_r)} \langle V_1(z_1) \cdots V_N(z_N) \rangle, \tag{5}$$

where g_s is the string coupling constant, $|\Delta|$ is the determinant of the Laplacian for the string diagram, $e^{-\sum_r P_r^- \tau(z_r)}$ is the light-cone Wick-rotated version of the e^{iEt} time dependence of the quantum-mechanical wave-function, $\det\left(\frac{d(\rho_I - \rho_1)}{dz_r}\right)$ is the Jacobian factor obtained in transforming from the $N-3$ differences of the interaction point locations to $N-3$ of the z_r's, and $\langle V_1(z_1) \cdots V_N(z_N) \rangle$ is the correlation function on the complex plane of N light-cone vertex operators for the external states.

For example, if all external strings are tachyons, $V_r(z_r) = e^{iP_r^j x^j(z_r)}$ where $j = 1$ to 24 are the transverse directions and $\langle V_1(z_1) \cdots V_N(z_N) \rangle = \exp\left(\sum_{r,s} P_r^j P_s^j \log|z_r - z_s|\right)$. Using $\tau(z_r) = \sum_{s=1}^N \alpha_s \log|z_s - z_r|$ where a cutoff is implicitly included when $s = r$, one obtains the expression

$$e^{-\sum_r P_r^- \tau(z_r)} \langle V_1(z_1) \cdots V_N(z_N) \rangle = \exp\left(\sum_{r,s} P_r \cdot P_s \log|z_r - z_s|\right) \quad (6)$$

which is manifestly Lorentz invariant. However, since both $\det\left(\frac{d(\rho_I - \rho_1)}{dz_r}\right)$ and $|\Delta|^{-\frac{d-2}{2}}$ depend explicitly on α_r, the other terms in the amplitude are not manifestly Lorentz invariant.

An explicit computation of $\det\left(\frac{d(\rho_I - \rho_1)}{dz_r}\right)$ is extremely complicated since finding the zeros of $\sum_r \frac{\alpha_r}{z - z_r}$ involves finding the roots of a polynomial equation of order $N-2$ (analogous to solving the scattering equations $\sum_r \frac{P_r \cdot P_s}{z_s - z_r}$ of Cachazo–He–Yuan[17]). However, using techniques of McKean, Singer[18] and Alvarez,[19] Mandelstam was able to show that $|\Delta|$ is proportional to $\prod_{I=1}^{N-2} |c_I|^{\frac{1}{24}}$ where $\rho - \rho_I = \frac{1}{2}c_I(z - z_I)^2$ near the Ith interaction point. Mandestam furthermore showed that $\prod_{I=1}^{N-2} |c_I|^{-\frac{1}{2}} \det\left(\frac{d(\rho_I - \rho_1)}{dz_r}\right)$ has no singularites as a function of z_I, and therefore $\left|\det\left(\frac{d(\rho_I - \rho_1)}{dz_r}\right)\right|^2 |\Delta|^{-\frac{d-2}{2}}$ is a constant when $d = 26$ which can be computed to be proportional to $|(z_1 - z_2)(z_2 - z_N)(z_N - z_1)|^2$.

Finally, as mentioned in Thorn's biography, the divergent contribution in $|\Delta|$ near the ends of the strings cancels (when $d = 26$) the divergent contribution from $r = s$ in $\exp\left(\sum_{r,s} P_r \cdot P_s \log|z_r - z_s|\right)$ which comes from the tachyon mass. So the closed string tree-level scattering amplitude is

$$\mathcal{A} = g_s^{N-2} |z_1 - z_2|^2 |z_2 - z_N|^2 |z_N - z_1|^2$$
$$\times \prod_{r=3}^{N-1} \int d^2 z_r \exp\left(\sum_{r \neq s} P_r \cdot P_s \log|z_r - z_s|\right) \quad (7)$$

which of course agrees with the Lorentz-covariant computation in conformal gauge.

3.2. Bosonic multiloop amplitudes

To generalize the interacting string diagram to a higher genus surface for multiloop amplitude computations, one replaces the genus zero map $\rho(z) =$

$\sum_{r=1}^{N} \alpha_r \log(z - z_r)$ to the genus g map

$$\rho(z) = \sum_{r=1}^{N} \alpha_r \log \phi'(z, z_r) + \sum_{R=1}^{g} \alpha_R v_R(z). \tag{8}$$

where $\phi'(z, z')$ is an analytic function defined such that

$$N(z, z') = \log |\phi'(z, z')| - \frac{1}{2\pi} \operatorname{Re}(v_R(z) - v_R(z'))$$
$$\times [(\operatorname{Im} \tau)^{-1}]^{RS} \operatorname{Re}(v_S(z) - v_S(z')) \tag{9}$$

is the Green's function on the genus g surface, τ_{RS} is the period matrix, $\omega_R(z) = \frac{1}{2\pi i} dv_R(z)$ are the g single-valued holomorphic differentials satisfying $\oint_{A_R} \omega_S = \delta_{RS}$ and $\oint_{B_R} \omega_S = \tau_{RS}$ when integrated around the A-cycles and B-cycles, and the values of α_R in Eq. (8) corresponding to the P^+ momenta in the loops are chosen such that $\operatorname{Re}(\rho(z))$ is single-valued on the genus g surface.

In addition to integrating over the $2g - 3 + N$ differences of interaction points $\rho_I = \rho(z_I)$ where $\frac{\partial \rho}{\partial z}\big|_{z=z_I} = 0$, one also needs to integrate over the g internal string lengths α_R and twists θ_R which are determined by the changes in $\operatorname{Im}(\rho(z))$ when z goes around an A-cycle or B-cycle. So the manifestly unitary expression for the N-point g-loop amplitude is

$$\mathcal{A} = g_s^{N-2+2g} \prod_{I=2}^{2g-2+N} \int d^2(\rho - \rho_1) \prod_{R=1}^{g}$$
$$\times \int d\alpha_R \, d\theta_R |\Delta|^{-\frac{d-2}{2}} e^{-\sum_r P_r^- \tau(z_r)} \langle V_1(z_1) \cdots V_N(z_N) \rangle. \tag{10}$$

The moduli for the genus g Riemann surface with N punctures can be parametrized by the $N-3$ puncture locations $(z_3, \ldots, z_{N-1})$ together with the $3g$ complex Schottky variables (z_{1R}, z_{2R}, w_R) where the projective transformation

$$\frac{z' - z_{1R}}{z' - z_{2R}} = w_R \frac{z - z_{1R}}{z - z_{2R}} \tag{11}$$

is defined to transform the surface into itself. In terms of the puncture locations and Schottky variables, the amplitude can be expressed as

$$\mathcal{A} = g^{N-2+2g} \prod_{r=3}^{N-1} \int d^2 z_r \prod_{R=1}^{g} \int d^2 z_{1R} \, d^2 z_{2R} \, d^2 w_R \left| \det \left(\frac{d(\rho_I - \rho_1) d\alpha_R \, d\theta_R}{dz_r \, dz_{1R} \, dz_{2R} \, dw_R} \right) \right|^2$$
$$\times |\Delta|^{-\frac{d-2}{2}} e^{-\sum_r P_r^- \tau(z_r)} \langle V_1(z_1) \cdots V_N(z_N) \rangle. \tag{12}$$

As in the tree amplitude, the product of $\left| \det \left(\frac{d(\rho_I - \rho_1) d\alpha_R \, d\theta_R}{dz_r \, dz_{1R} \, dz_{2R} \, dw_R} \right) \right|^2$ and $|\Delta|^{-\frac{d-2}{2}}$ can be shown to be independent of α_r and was computed by Mandelstam in terms of the Schottky variables to be

$$\text{constant} \times \prod_S |w_S(z_{1S}-z_{2S})|^{-4} (\operatorname{Im}\tau)^{-13} {\prod}' |1-w|^{-24}, \qquad (13)$$

where $\prod'$ denotes the product over all conjugacy classes of projective transformations (excluding the identity transformation) of the genus g surface. So the final expression for the N-point g-loop tachyon scattering amplitude is

$$\mathcal{A} = g_s^{N-2+2g} \prod_{r=3}^{N-1} \int d^2 z_r \prod_{R=1}^{g} \int d^2 z_{1R}\, d^2 z_{2R}\, d^2 w_R$$

$$\times \prod_S |w_S(z_{1S}-z_{2S})|^{-4} (\operatorname{Im}\tau)^{-13} {\prod}' |1-w|^{-24} \exp\left(\sum_{r \neq s} k_r \cdot k_s N(z_r, z_s)\right). \qquad (14)$$

As shown by D'Hoker and Giddings,[20] this manifestly unitary computation in light-cone gauge agrees with the manifestly Lorentz covariant computation using the Polyakov approach in conformal gauge. The singularity properties of this g-loop amplitude can be studied by considering the limit when either two interaction points approach each other or when one of the internal string lengths α_R goes to zero. In addition to the expected singularities coming from physical poles, there is also an unwanted singularity coming from the dilaton tadpole when a massless scalar is emitted into the vacuum. Because of this dilaton tadpole divergence (as well as the tachyonic states), these bosonic string scattering amplitudes are inconsistent and Mandelstam focused most of his attention after 1980 on scattering amplitudes of the space–time supersymmetric version of string theory called the superstring.

4. Superstring Amplitudes

Although the light-cone interacting string picture for the RNS spinning string had been constructed by Mandelstam in the 1970s, there were two features that made this RNS interacting string picture more complicated than for the bosonic string. Fortunately, both of these features were simplified by developments in the 1980s. After describing these simplifications in the next two subsections, it will be explained how Mandelstam used these simpifications to generalize his bosonic multiloop expressions to explicit multiloop formulas for the superstring and verify their finiteness properties.[16]

4.1. *Super-worldsheets*

The first complicating feature in the light-cone RNS interacting string picture was the presence of interaction point operators[2,21] which are necessary for Lorentz invariance and are inserted at each joining or splitting point ρ_I where $\frac{\partial \rho}{\partial z} = 0$.

These light-cone interaction point operators $\mathcal{O}(\rho_I)$ play a role similar to the Friedan–Martinec–Shenker picture-changing operators $e^\phi(\partial x \cdot \psi + \cdots)$ in RNS superconformal gauge computations[8] and take the form

$$\mathcal{O}(\rho_I) = \lim_{\rho \to \rho_I} (\rho - \rho_I)^{\frac{3}{4}} \partial_\rho x^j(\rho) \psi^j(\rho), \tag{15}$$

where the factor of $(\rho - \rho_I)^{\frac{3}{4}}$ is needed since $\partial_\rho x^j(\rho) = \left(\frac{\partial \rho}{\partial z}\right)^{-1} \partial_z x^j(z)$ diverges as $(\rho - \rho_I)^{-\frac{1}{2}}$ and $\psi^j(\rho) = \left(\frac{\partial \rho}{\partial z}\right)^{-\frac{1}{2}} \psi^j(z)$ diverges as $(\rho - \rho_I)^{-\frac{1}{4}}$ near the interaction points ρ_I where $\frac{\partial \rho}{\partial z} = 0$.

However, unlike the Friedan–Martinec–Shenker picture-changing operators that can be inserted anywhere on the string worldsheet, unitarity requires that the $\mathcal{O}(\rho_I)$ operators are located at the points where $\frac{\partial \rho}{\partial z} = 0$. Since $\mathcal{O}(\rho_I)$ has a singular OPE with $\mathcal{O}(\rho_J)$ when two interaction point locations ρ_I and ρ_J coincide, one needs to also include contact term interactions in the light-cone prescription which remove this divergence from colliding interaction-point operators. The presence of these contact terms was first pointed out by Greensite and Klinkhammer[22] who argued that the light-cone superstring field theory action must have quartic interaction terms in order to guarantee positivity.

As was shown by Berkovits,[23] the complicating feature of interaction-point operator insertions can be avoided by replacing the conformal map $\rho(z) = \sum_r \alpha_r \log(z - z_r)$ from the complex plane to an ordinary worldsheet with the superconformal map

$$\rho(z, \kappa) = \sum_r \alpha_r \log(z - z_r - \kappa \kappa_r), \tag{16}$$

$$\phi(z, \kappa) = \left(\frac{\partial \rho}{\partial z}\right)^{-\frac{1}{2}} D_\kappa \rho(z, \kappa)$$

$$= \left(\sum_r \frac{\alpha_r}{z - z_r - \kappa \kappa_r}\right)^{-\frac{1}{2}} \sum_s \frac{\alpha_s(\kappa - \kappa_s)}{z - z_s} \tag{17}$$

from the $N=1$ super-plane parametrized by (z, κ) to an $N=1$ super-worldsheet parametrized by (ρ, ϕ) where κ_r are anticommuting parameters and $D_\kappa \equiv \frac{\partial}{\partial \kappa} + \kappa \frac{\partial}{\partial z}$.

Defining $\phi_I = \lim_{\rho \to \rho_I} (\rho - \rho_I)^{\frac{1}{4}} \phi(\rho)|_{\kappa=0}$ to be the fermionic modulus associated with the interaction point ρ_I, the super-worldsheet action depends on ϕ_I as $S_0 + \sum_I \phi_I \mathcal{O}(\rho_I)$ where S_0 is the action on an ordinary worldsheet. So integrating the exponential of the super-worldsheet action over the fermionic moduli ϕ_I pulls down the interaction-point operator insertions $\mathcal{O}(\rho_I)$. Furthermore, the contact term insertions when interaction points collide is automatically included in this super-worldsheet prescription as can be seen by performing a change of variables from ϕ_I to other moduli which are non-singular when the interaction point operators collide. When expressed in terms of the non-singular moduli, the amplitude has no divergences when interaction points collide. So the contact term insertions are automatically generated by the surface term which arises when changing variables from the ϕ_I moduli to the non-singular moduli.

4.2. Space–time supersymmetry

A second complicating feature of the RNS interacting string picture is the nature of Ramond vertex operators which change the periodicity conditions of the 8 light-cone RNS fermions ψ^j. As in the covariant prescription of Friedan–Martinec–Shenker, Ramond vertex operators in light-cone gauge can be described by spin fields constructed by bosonizing ψ^j. However, this bosonization procedure breaks the manifest worldsheet supersymmetry and reduces the manifest light-cone $SO(8)$ symmetry down to $U(4)$. Moreover, the presence of Ramond states implies that loop amplitudes in superstring computations are unitary only after summing over all possible spin structures of ψ^j. Since unphysical divergences only cancel after summing over spin structures, it is difficult to use the light-cone RNS formalism to prove finiteness of superstring multiloop amplitudes.

As was explained in the introduction, Green and Schwarz developed an alternative formalism for the superstring in light-cone gauge in which space–time supersymmetry is manifest.[3] Unlike the RNS formalism where the fermionic variables ψ^j are $SO(8)$ vectors with either periodic or anti-periodic boundary conditions, the fermionic variables in the Green–Schwarz formalism, θ^α for $\alpha = 1$ to 8, are $SO(8)$ spinors which are always periodic on the worldsheet. So multiloop scattering amplitudes using the Green–Schwarz formalism do not require summing over spin structures, and cancellation of divergences due to space–time supersymmetry are much easier to verify. Furthermore, after bosonizing the RNS ψ^j and Green–Schwarz θ^α fermionic variables, there is a simple field redefinition between the resulting bosons which allows a formal equivalence proof of the two formalisms.[24]

The light-cone worldsheet action for the left-moving fermionic Green–Schwarz variables is

$$\int d\sigma \, d\tau \, \theta^\alpha \left(\frac{\partial}{\partial \tau} + \frac{\partial}{\partial \sigma} \right) \theta^\alpha \tag{18}$$

and since $\theta^\alpha(\rho)$ carries conformal weight $\frac{1}{2}$ and is periodic on the string worldsheet, $\theta^\alpha(z) = \left(\frac{\partial \rho}{\partial z}\right)^{\frac{1}{2}} \theta^\alpha(\rho)$ has square-root cuts at the external string locations z_r where $\frac{\partial \rho}{\partial z} \to \infty$ and at the interaction points where $\frac{\partial \rho}{\partial z} \to 0$. To decide if $\theta^\alpha(z)$ has square-root poles or square-root zeros at these points, one needs to break the manifest $SO(8)$ to $U(4)$ and split θ^α into (θ^A, θ_A) where $A = 1$ to 4 and (θ^A, θ_A) transform in the (fundamental, anti-fundamental) representation of $U(4)$. One can then choose boundary conditions so that $\theta^A(z)$ has square-root poles at z_r and square-root zeros at z_I, whereas $\theta_A(z)$ has square-root zeros at z_r and square-root poles at z_I. This splitting breaks the $SO(8)$ vector x^j into (x^L, x^{AB}, x^R) where (x^L, x^R) are $SO(6)$ singlets and $x^{AB} = -x^{BA}$ is an $SO(6)$ vector.

As in the light-cone RNS formalism, the light-cone Green–Schwarz formalism requires the insertion of operators $\mathcal{O}(\rho_I)$ at the interaction points $\rho_I = \rho(z_I)$ where $\frac{\partial \rho}{\partial z} = 0$. Mandelstam showed that the cubic string vertex of the RNS light-cone formalism is reproduced by the light-cone Green–Schwarz formalism if one defines the light-cone Green–Schwarz interaction-point operator as[25,4]

$$\mathcal{O}(\rho_I) = \lim_{\rho \to \rho_I} \left[\partial_\rho x^L (\rho - \rho_I)^{\frac{1}{2}} + \frac{1}{2} \partial_\rho x^{AB} \theta_A \theta_B (\rho - \rho_I)^{\frac{3}{2}} \right.$$
$$\left. + \frac{1}{24} \partial_\rho x^R \epsilon^{ABCD} \theta_A \theta_B \theta_C \theta_D (\rho - \rho_I)^{\frac{5}{2}} \right], \tag{19}$$

where the factors of $(\rho - \rho_I)$ are necessary since both $\partial_\rho x^j$ and θ_A diverge as $(\rho - \rho_I)^{-\frac{1}{2}}$ when $\rho \to \rho_I$.

The unusual boundary conditions of (θ^A, θ_A) make it difficult to construct a worldsheet supersymmetric action which reproduces these Green–Schwarz interaction point operators after integration over fermionic super-moduli. For this reason, most of Mandelstam's computations of superstring amplitudes used the light-cone RNS formalism, and the light-cone Green–Schwarz formalism was used only to analyze the finiteness properties due to the absence of tachyons and dilaton tadpoles. Nevertheless, starting from the light-cone Green–Schwarz formalism, Mandelstam was able to deduce a remarkable formula in Ref. 4 for the superstring tree amplitude with an arbitrary number of external massless states. His formula only had manifest $U(4)$ symmetry, but it was argued to be both Lorentz-covariant and space–time supersymmetric where 8 of the 16 space–time supersymmetries are manifest and are generated by $q_A = \sum_{r=1}^N \frac{\partial}{\partial \theta_r^A}$ and $q^A = \sum_{r=1}^N \alpha_r \theta_r^A$.

4.3. RNS tree amplitudes

After replacing the conformal map of $\rho = \sum \alpha_r \log(z - z_r)$ with the superconformal map $\rho = \sum \alpha_r \log(z - z_r - \kappa \kappa_r)$ and integrating over the $N - 2$ fermionic moduli ϕ_I, the steps for computing the bosonic string tree amplitude easily generalize to the RNS superstring tree amplitude with external Neveu–Schwarz states. When $d = 10$, the contribution from the super-determinant of the Laplacian $|\Delta|^{-\frac{d-2}{2}}$ cancels the super-Jacobian $\left|\text{sdet}\left(\frac{d(\rho_I - \rho_1)d\phi_I}{dz_r \, d\kappa_r}\right)\right|^2$ coming from the change of variables where (ρ_I, ϕ_I) are the super-moduli of the interaction points in the light-cone superworldsheet and (z_r, κ_r) are variables in the complex super-plane which are non-singular when two interaction points collide.

The resulting formula for the N-point tachyon tree-level scattering amplitude is

$$\mathcal{A} = g_s^{N-2} \int d^2\phi_1 \prod_{I=2}^{N-2} \int d^2(\rho_I - \rho_1) \int d^2\phi_I |\Delta|^{-\frac{d-2}{2}}$$
$$\times e^{-\sum_r P_r^- \tau(z_r, \kappa_r)} \left\langle \prod_{r=1}^N V_r(z_r, \kappa_r) \right\rangle$$
$$= g_s^{N-2} \int d^2\kappa_2 \prod_{r=3}^{N-1} \int d^2 z_r \int d^2 \kappa_r \left|\text{sdet}\left(\frac{d(\rho_I - \rho_1)d\phi_I}{dz_r \, d\kappa_r}\right)\right|^2 |\Delta|^{-\frac{d-2}{2}}$$

$$\times e^{-\sum_{r,s} P_r^- P_s^+ \log|z_r - z_s - \kappa_r \kappa_s|} \left\langle \prod_{r=1}^{N} e^{iP_r^j X^j(z_r, \kappa_r)} \right\rangle$$

$$= g_s^{N-2} |z_1 - z_2|^2 |z_N - z_2|^2 \int d^2\kappa_2 \prod_{r=3}^{N-1} \int d^2 z_r \int d^2 \kappa_r$$

$$\times \exp\left(\sum_{r \neq s} P_r \cdot P_s \log|z_r - z_s - \kappa_r \kappa_s| \right), \tag{20}$$

where

$$V_r(z_r, \kappa_r) = e^{iP_r^j X^j(z_r, \kappa_r)} \tag{21}$$

is the tachyon vertex operator, and this formula is easily generalized to N-point tree-level gravity scattering amplitudes by replacing the tachyon vertex operator with the graviton vertex operator

$$V_r(z_r, \kappa_r) = g_{jk} D_\kappa X^j \bar{D}_{\bar{\kappa}} X^k e^{iP_r^j X^j(z_r, \kappa_r)}. \tag{22}$$

However, for the tree-level scattering of Ramond states using this light-cone super-worldsheet formalism, the expressions are more complicated[26] because of the anti-periodic boundary conditions of the fermionic variables in the complex super-plane.

4.4. RNS multiloop amplitudes

The prescription of (20) generalizes in an obvious way to g-loop amplitudes by considering light-cone super-worldsheets with $2g + N - 2$ interaction points, and in Ref. 27, it was shown by Aoki, D'Hoker and Phong that this manifestly unitary light-cone prescription of Mandelstam agrees with the manifestly Lorentz-covariant Polyakov prescription in superconformal gauge. To obtain explicit expressions for the superstring multiloop amplitudes, the main complication is to compute the product

$$\left| \text{sdet} \left(\frac{d(\rho_I - \rho_1) d\phi_I \, d\alpha_R \, d\theta_R}{dz_r \, d\kappa_r \, dz_{1R} \, dz_{2R} \, dw_R \, d\kappa_{1R} \, d\kappa_{2R}} \right) \right|^2 |\Delta|^{-\frac{d-2}{2}} \tag{23}$$

on the higher-genus super-Riemann surface where α_R and θ_R are the internal string lengths and twists, and $(z_{1R}, z_{2R}, w_R, \kappa_{1R}, \kappa_{2R})$ are the super-Schottky variables.

The product of (23) was computed indirectly by Mandelstam by requiring that it has the correct analyticity properties. In his construction, Mandelstam found it convenient to define a modified string diagram to describe the super-Riemann surface using a superconformal map $\rho(z, \kappa)$ such that the odd $\frac{1}{2}$-form $\omega(z, \kappa) \equiv D_\kappa \rho(z, \kappa)$ has fixed A-periods and is holomorphic instead of meromorphic, i.e. the string diagram has no punctures corresponding to the external states. This does not affect the computation of (23) since its independence on the puncture locations for tree amplitudes implies that it is also independent of the puncture locations for multiloop amplitudes.

For even spin structures in which the fermionic variables have no zero modes, the N-point g-loop tachyon amplitude is

$$\mathcal{A} = g_s^{N-2+2g} \prod_{r=1}^{N} \int d^2 z_r \int d^2 \kappa_r \int d^{6g-6} \nu$$

$$\times \int d^{4g-4} \chi\, M(\nu, \chi) \exp\left(\sum_{r \neq s} k_r \cdot k_s N(z_r, \kappa_r; z_s, \kappa_s)\right), \qquad (24)$$

where (ν, χ) are the super-Teichmuller parameters, N is the Green's function on the super-Riemann surface,

$$M(\nu, \chi) = \text{constant} \times (\text{Im}\, \tau)^{-5} \mathcal{N} \bar{\mathcal{N}}, \qquad (25)$$

and the formulas for $\mathcal{N}$ and $\bar{\mathcal{N}}$ are manifestly Lorentz-invariant and are explicitly constructed in Eq. (11b) of Ref. 16 in terms of the holomorphic $\frac{1}{2}$-form $\omega = D_\kappa \rho$ together with Θ functions and prime forms on the higher-genus surface.

For odd spin structures, ψ^j has a fermionic zero mode and one can construct an even holomorphic $\frac{1}{2}$-form ω_E with no A-periods, i.e. $\oint_{A_R} dz\, \omega_E = 0$. Integration over the fermionic zero modes of ψ^j produces the term

$$\prod_{j=1}^{8} \left(\sum_{r=1}^{N} P_r^j W(z_r, \kappa_r) \right), \qquad (26)$$

where $W(z_r, \kappa_r) = \int^{z_r, \kappa_r} dz\, d\kappa\, \omega_E(z, \kappa)$. Although (26) is only $SO(8)$ covariant, Mandelstam showed that after allowing the odd $\frac{1}{2}$-form $\omega = D_\kappa \rho$ to have an unphysical pole at the location (z_0, κ_0) which can be chosen arbitrarily, and inserting a delta function to restrict the residue of this pole, (26) is covariantized to

$$\prod_{j=0}^{9} \left(\sum_{r=1}^{N} P_r^j W(z_r, \kappa_r) \right). \qquad (27)$$

The factor of $\sum_{r=1}^{N} P_r^+ W(z_r, \kappa_r)$ in (27) comes from the delta function insertion which imposes the usual light-cone gauge-fixing condition $\psi^+ = 0$, and the factor of $\sum_{r=1}^{N} P_r^- W(z_r, \kappa_r)$ in (27) comes from integrating over the constant fermionic modulus ϕ_c which appears in the superconformal map as $\rho(z, \kappa) = \int^{z, \kappa} \omega + \phi_c W(z, \kappa)$ and contributes to the amplitude through the $\exp(\sum_{r=1}^{N} P_r^- \tau(z_r, \kappa_r))$ term in the wave-function of the external states. Besides the factor of (27), the remaining terms in $\mathcal{N}$ for the odd spin structures are the same as in Eq. (11b) of Ref. 16 except that the factor of $(\Theta(0))^4$ is absent and the term $(\omega_E(z_0, \kappa_0))^{-2}$ is included so that the amplitude is independent of the location of (z_0, κ_0).

4.5. *Absence of ambiguities*

The explicit multiloop superstring amplitude expressions constructed by Mandelstam are manifestly Lorentz covariant and are derived from the unitary light-cone prescription. However, it was realized in the late 1980s that there is an apparent ambiguity in superstring amplitudes coming from the definition of the boundary of super-moduli space.[10] To regularize the logarithmic divergence of superstring amplitudes before summing over spin structures, one needs to introduce a cutoff ϵ near the boundary of super-moduli space. But shifting ϵ by a nilpotent quantity constructed out of a product of odd moduli will change the answer since

$$\int_{\epsilon+\kappa_1\kappa_2} dz\, f(z) = \int_\epsilon dz\, f(z) - \kappa_1\kappa_2 f(\epsilon). \tag{28}$$

So how should define the nilpotent contribution to the cutoff ϵ?

The resolution found by Mandelstam for this ambiguity is that unitary requires that the bosonic ρ_I variables in the (ρ_I, ϕ_I) light-cone supersheet supermoduli are ordinary complex numbers with no nilpotent part. This follows from requiring equivalence with the manifestly unitary component prescription in light-cone gauge where integration over the ϕ_I moduli in the super-worldsheet formalism replaces insertions of the interaction-point operators of (15) in the component formalism. So when expressed in terms of the (ρ_I, ϕ_I) moduli, the cutoff ϵ in the ρ_I moduli should be an ordinary complex number.

More explicitly, consider the degeneration of a Riemann surface of genus $g_1 + g_2$ into two surfaces of genus g_1 and g_2. Define $\rho_A - \rho_B$ to be the distance between two interaction points which are both on the surface of genus g_1. Then if the ratios of the other distances between interaction points on the surface of genus g_1 are held fixed, $\rho_A - \rho_B$ goes to zero as the surface degenerates. So one can introduce a cutoff $\rho_A - \rho_B \geq \epsilon$ to regularize the resulting divergence, and one should require that ϵ is an ordinary complex number and is the same for all spin structures. After summing over spin structures, the logarithmic divergences will cancel and one can take the limit where ϵ goes to zero.

4.6. *Finiteness*

To verify the finiteness of multiloop superstring amplitudes, the only dangerous divergence comes from the dilaton tadpole and occurs when either the Riemann surface degenerates into two surfaces or when the locations of all external states coincide. All other divergences are either physical poles, or can be removed by mass renormalization (e.g. the divergences when all but one of the external states coincide) or by integrating over real time instead of Wick-rotated Euclidean time (e.g. the divergences coming from the propagation of tachyons in loops before summing over spin structures). Note that the contact term divergences from colliding interaction point operators in light-cone gauge have already been eliminated by using the super-worldsheet formalism.

If space–time supersymmetry is unbroken and one has a manifestly super-Poincaré covariant string theory, the one-point function of the dilaton must vanish since the dilaton vertex operator V_{dilaton} is the supersymmetric variation of the dilatino vertex operator $V^\alpha_{\text{dilatino}}$, i.e. $q_\alpha V^\alpha_{\text{dilatino}} = V_{\text{dilaton}}$ where q_α is the space–time supersymmetry generator. So if space–time supersymmetry is unbroken, one can pull the contour integral of q_α off the dilatino vertex operator to prove that the one-point function $\langle V_{\text{dilaton}} \rangle = 0$.

Using the RNS formalism, the space–time supersymmetry generator q_α is only well-defined after summing over spin structures. But q_α is well-defined in the light-cone Green–Schwarz formalism, which has been shown to be equivalent to the light-cone RNS formalism through a field redefinition which relates both the quadratic kinetic term and the cubic three-string vertex. The only subtlety in proving absence of the dilaton tadpole in the light-cone Green–Schwarz formalism is that the vertex operator V_{dilaton} is at zero momentum, so the length $P_r^+ = \alpha_r$ of the string is zero.

To argue that there are no subtleties when the length of the string is zero, Mandelstam used that space–time supersymmetry in the light-cone Green–Schwarz formalism implies that the closed superstring scattering amplitude is multiplied by the factor $\delta^4\bigl(\sum_{r=1}^N \alpha_r \theta_r^A\bigr)\delta^4\bigl(\sum_{r=1}^N \alpha_r \bar\theta_r^A\bigr)$ where $A = 1$ to 4 is a $U(4)$ index and θ_r^A and $\bar\theta_r^A$ are the left and right-moving light-cone fermionic variables for the rth external string of length α_r. (For the dilaton one-point function, $N=1$ so there is only one α.) Since there is an $\alpha^{-\frac{1}{2}}$ normalization in θ^A (note that θ^A transforms as $(P^+)^{-\frac{1}{2}}$ under the Lorentz transformation generated by M_{+-}) and since the rest of the amplitude can be shown to be regular when $\alpha \to 0$, the amplitude behaves like $\alpha^8 (\alpha^{-\frac{1}{2}})^8$ which goes to zero when $\alpha \to 0$. So in light-cone gauge, space–time supersymmetry together with the α dependence of the scattering amplitude implies the vanishing of the dilaton tadpole and the resulting finiteness of multiloop superstring amplitudes.

References

1. S. Mandelstam, Interacting string picture of dual resonance models, *Nucl. Phys. B* **64**, 205 (1973).
2. S. Mandelstam, Manifestly dual formulation of the Ramond model, *Phys. Lett. B* **46**, 447 (1973); S. Mandelstam, Interacting-string picture of the Neveu–Schwarz model, *Nucl. Phys. B* **69**, 71–106 (1974).
3. M. B. Green and J. Schwarz, Supersymmetrical dual string theory, *Nucl. Phys. B* **181**, 502–530 (1981).
4. S. Mandelstam, Interacting string picture of the fermionic string, *Prog. Theor. Phys. Suppl.* **86**, 163 (1986).
5. M. B. Green and J. H. Schwarz, Covariant description of superstrings, *Phys. Lett. B* **136**, 367 (1984).
6. L. Brink and J. H. Schwarz, Quantum superspace, *Phys. Lett. B* **100**, 310 (1981).
7. W. Siegel, Hidden local supersymmetry in the supersymmetric particle action, *Phys. Lett. B* **128**, 397 (1983).
8. D. Friedan, E. Martinec and S. Shenker, Conformal invariance, supersymmetry and string theory, *Nucl. Phys. B* **271**, 93 (1986).

9. M. B. Green and J. H. Schwarz, Anomaly cancellation in supersymmetric $D = 10$ gauge theory and superstring theory, *Phys. Lett. B* **149**, 117 (1984).
10. J. J. Atick, J. M. Rabin and A. Sen, An ambiguity in fermionic string perturbation theory, *Nucl. Phys. B* **299**, 279 (1988).
11. N. Berkovits, Covariant quantization of the Green–Schwarz superstring in a Calabi–Yau background, *Nucl. Phys. B* **431**, 258 (1994), arXiv:hep-th/9404162; N. Berkovits, Super-Poincaré covariant quantization of the superstring, *J. High Energy Phys.* **0004**, 018 (2000), arXiv:hep-th/0001035.
12. E. Witten, Superstring perturbation theory revisited, arXiv:1209.5461 [hep-th]; E. Witten, More on superstring perturbation theory: An overview of superstring perturbation theory via super-Riemann surfaces, arXiv:1304.2832 [hep-th].
13. A. Sen and E. Witten, Filling the gaps with PCO's, *J. High Energy Phys.* **1509**, 004 (2015), arXiv:1504.00609 [hep-th]; A. Sen, Ultraviolet and Infrared divergences in superstring theory, arXiv:1512.00026 [hep-th]; A. Sen, Unitarity of superstring field theory, *J. High Energy Phys.* **1612**, 004 (2016), arXiv:1607.08244 [hep-th].
14. S. Mandelstam, Light cone superspace and the ultraviolet finiteness of the $N = 4$ model, *Nucl. Phys. B* **213**, 149 (1983).
15. S. Mandelstam, The interacting string picture and functional integration, in *Proceedings, Unified String Theories*, Santa Barbara, 1985, pp. 46–102; S. Mandelstam, The N loop string amplitude, *Phys. Scripta T* **15**, 109 (1987).
16. S. Mandelstam, The n-loop string amplitude: Explicit formulas, finiteness and absence of ambiguities, *Phys. Lett. B* **277**, 82–88 (1992).
17. F. Cachazo, S. He and E. Y. Yuan, Scattering of massless particles in arbitrary dimensions, *Phys. Rev. Lett.* **113**, 171601 (2014), arXiv:1307.2199 [hep-th].
18. H. P. McKean and I. M. Singer, Curvature and eigenvalues of the Laplacian, *J. Diff. Geom.* **1**, 43 (1967).
19. O. Alvarez, Theory of strings with boundaries: Fluctuations, topology, and quantum geometry, *Nucl. Phys. B* **216**, 125 (1983).
20. E. D'Hoker and S. B. Giddings, Unitary of the closed bosonic Polyakov string, *Nucl. Phys. B* **291**, 90 (1987).
21. S. Mandelstam, Lorentz properties of the three-string vertex, *Nucl. Phys. B* **83**, 413–439 (1974).
22. J. Greensite and F. R. Klinkhamer, New interactions for superstrings, *Nucl. Phys. B* **281**, 269 (1987).
23. N. Berkovits, Supersheet functional integration and the interacting Neveu–Schwarz string, *Nucl. Phys. B* **304**, 537 (1988).
24. E. Witten, $D = 10$ superstring theory, in *Fourth Workshop on Grand Unification* (Birkhauser, 1983), p. 395.
25. M. B. Green and J. H. Schwarz, Superstring field theory, *Nucl. Phys. B* **243**, 475 (1984).
26. N. Berkovits, Supersheet functional integration and the calculation of NSR scattering amplitudes involving arbitrarily many external Ramond strings, *Nucl. Phys. B* **331**, 659 (1990).
27. K. Aoki, E. D'Hoker and D. H. Phong, Unitarity of closed superstring perturbation theory, *Nucl. Phys. B* **342**, 149 (1990).

Stanley Mandelstam, 1928–2016:
Brief Biography and Selected Publications with Commentary[1]

Ling-Lie Chau (喬玲麗)*

University of California, Davis, CA 95616, USA
chau@physics.ucdavis.edu

The enduring influences of Stanley Mandelstam's publications are deep and diverse. They affect almost all the current major research efforts in theoretical and mathematical physics that try to deepen our understanding of the physical universe. Reviewing Stanley's accomplishments offers a rare opportunity for everyone interested, experts as well as nonexperts, to gain a perspective about the current status of theoretical and mathematical physics and what to look for in the future. This paper presents a brief biography of Stanley and a selection of his publications, grouped together according to subject matters, with commentary.[2]

Fig. 1. Stanley Mandelstam at the 1961 Solvay Conference.[3]

---

*Professor Emerita in Physics and GGAM (Graduate Group of Applied Mathematics), UC Davis.
[1]A contribution to the *"Memorial Volume for Stanley Mandelstam,"* editors N. Berkovits, L. Brink, L. L. Chau, K. K. Phua and C. Thorn (World Scientific Publishing, to be published in 2017). It is referred to as "the *Memorial Volume*" in the rest of the paper.
[2]For a very short highlight of this paper, see author's Physics Today Obituary for Stanley Mandelstam, May issue 2017. About the genesis of these two papers see the Acknowledgments in this paper.
[3]The photo is cropped from the wall-size photo of the 1961 Solvay conference on the 4th floor of the UC Davis Physics Department. (The author had it installed in the late 1980s.)

Contents

Stanley Mandelstam, 1928–2016:
  Brief Biography and Selected Publications with Commentary[1]   53
1. Brief Biography   54
2. Selected Subjects of Publications and Commentary   56
3. A Perspective   62
4. Selected Publications   62
Acknowledgments   66
Attachment: Photos of Stanley with colleagues, students, and friends   68

1. Brief Biography

Stanley Mandelstam was born in 1928, in Johannesburg, South Africa, and was the elder brother to a sister. His father was a grocer who had recently emigrated from Latvia. His mother was an elementary school teacher, born in South Africa to parents from Latvia.

He obtained a B.Sc. in chemical engineering in 1952 from University of Witwatersrand (or Wits), Johannesburg. By then, he had manifested his talent in mathematical physics with the publishing of the book [A], "*Variational Principles in Dynamics and Quantum Theory*," with Wolfgang Yourgrau.

Subsequently Stanley switched to the study of theoretical and mathematical physics — his true passion, on which he worked for the rest of his live. He obtained a B.A. in two years at Trinity College, Cambridge in 1954. In just another two years he received his Ph.D. from the University of Birmingham in 1956, under the direction of Rudolf E. Peierls, who had brought prominence and high visibility to theoretical physics there. Stanley's thesis work was published in two papers [B] in the *Proceedings of the Royal Society of London A*. By then, Stanley had developed a solid mastery of quantum field theory (QFT), including its use in calculating S-matrices. He would put these to brilliant uses in his work throughout his life. His writings demonstrate his attention to the close relevance between theory and experiments.

After continuing his research at Birmingham for another year, in 1957 he moved to New York City, hired as a Boese fellow, a research position in the Department of Physics at Columbia University 1957–58. He published the 1958 paper [C1], in which he pioneered the double dispersion relations for scattering amplitudes (elements of an S-matrix), now called the Mandelstam representation. It was a daring leap from the insights he had gained from perturbative QFT results and his advanced knowledge about functions of more-than-one complex variables.[4] His presentation of the paper at the 1958 American Physical Society (APS) Washington DC meeting caught the attention of Geoffrey (Geoff) Chew. The two met and had a discussion right after Stanley's talk. At the end of their discuussion, Chew made an offer on the spot to Stanley to go to UC Berkeley as a researcher. Stanley immediately agreed.

---
[4]A subject of mathematics that is still rarely taught to graduate students in physics.

He had two productive years, 1958–60, at UC Berkeley. He consolidated the Mandelstam representation into its final form using the Mandelstam variables, and published several single-author papers. He worked with Chew on three published papers to implement the Mandelstam representations into the S-matrix approach that Chew had developed earlier with collaborators. He initiated the idea of using Regge poles to treat the high energy behaviors in S-matrices among colleagues at Berkeley.

Then, he was hired back to Birmingham as a professor in 1960. Stanley's research output continued to be spectacular. He wrote an extensive review to summarize his work done at Berkeley, "Dispersion Relations in Strong-Coupling Physics." He published a paper with Chew and Steven Frautschi incorporating the Regge pole idea into the S-Matrix approach. Moreover, he published a paper with Ronald F. Peierls[5] and collaborators at Birmingham, applying the S-matrix approach. Some of these papers are listed in [C1] and [C2]. Stanley further worked by himself not only delving deeper into the complex plane of angular momentum with Regge poles and cuts and publishing many papers, e.g., the 1962 paper in [C5], but also striking out to pioneer two far-reaching new directions in research. He formulated quantum gauge theories in terms of gauge-invariant path-dependent fields (for Maxwell theory in 1962, and then also for Yang–Mills theory in 1968 when he derived the Feynman rules) [C3]. In the same journals, together with the two papers on gauge theories, were his two papers on quantum general relativity: the 1962 one giving the formulation in terms of coordinate-independent path-dependent fields, and the 1968 one deriving the Feynman rules [C4]. The importance of these two breakthroughs are discussed in (S3) and (S4) below. We call these formulations the Mandelstam path-field formulations, for gauge theories and for general relativity.

Thus in 1958–62, he made breakthroughs on four major subjects in theoretical physics, listed and commented upon below in (S1–S4), which correspond to the selected papers [C1–C4], and paved the ground work for the next breakthrough, subject (S5) corresponding to [C5]. These years can be called

"**Stanley Mandelstam's miracle years, 1958–1962**."

Following the 1958–60 publications [C1, C2], Stanley was invited to participate in the prestigious 1961 Solvay conference, celebrating the 50th anniversary of the famous inaugural 1911 Solvay conference on physics,[6] where he gave a talk and published the 1961 paper listed in [C1]. He was one of the youngest participants and appeared in the photo with so many distinguished theoretical physicists. In 1962 he was elected to become a Fellow of the Royal Society, a great honor for any physicist and even more so for a young 6-year post-Ph.D. physicist.

---

[5]Ronald F. Peierls (1935–2003) was the son of Sir. Rudolf E. Peirels. He later went to Institute for Advanced Study (IAS), Princeton NJ, 1961–62, and then to Brookhaven National Laboratory (BNL), NY, for the rest of his life. The author was a colleague of his and coauthored with him at BNL during 1969–86. He has been dearly missed. See more in the reference at (C2)

[6]M. Curie and A. Einstein attended the 1911 Solvay conference and appeared in that iconic conference photo with other distinguished physicists.

In 1963 Stanley returned to UC Berkeley as a professor in the Department of Physics.[7] He continued to produce ground-breaking work, e.g., those publications after 1963 in [C5–C9]), and make contributions to the subjects (S1–S9) listed below with comments. There were the development of the precursor, in the S-matrix approach culminated in 1968, for the eventual discovery of string theory;[8] the elucidation of mechanisms for quark confinement in quantum chromodynamics (QCD) in 1975-79; the nonperturbative constructions of the bosonization (or fermionization) in $(1+1)$-dimension QFTs in 1975; the proof of the perturbative UV finiteness and $\beta = 0$, in any gauge to all orders, of $N = 4$ supersymmetric Yang-Mills theory (SYM) in 1983; and from 1968 onward, many important contributions to the development of string theory that culminated in 1992 with the first proof of the pertubative UV finiteness of string theory, so string theory can be considered as a contender for the theory of quantum gravity.

In 1994 he became Professor Emeritus.[9] He continued to do research, use his office in the Department, and live in the same apartment he rented since 1980 in Berkeley (by choice, he was always a renter) until his death on June 11, 2016, age eighty-seven.

With his quiet, always polite, attentive, and kind ways, Stanley won the respect and love from his colleagues and friends. He is deeply missed.[10]

2. Selected Subjects of Publications and Commentary

Stanley made long-lasting major contributions to theoretical physics, which are here organized into the following nine diverse subjects (S1–S9), corresponding to the nine groups of selected publications [C1–C9], in addition to [A, B].

---

[7]I took Stanley's first QFT two-semester course at Berkeley, Fall 1963 and Spring 1964. There was no textbook. He taught completely from memory and wrote on the blackboard. So it was a special event when one day he brought with him a little strip of paper to class, which had the Klein-Nishina formula on it. The course was unusual and impressive. I had no difficulty writing down notes and studying them. Regrettably, he did not teach his path-field formulations, nor do the current QFT text books! For the anecdote of how Stanley and I coauthored a paper, see the reference at [C5]

[8]The term "string theory" is used to include "bosonic string theory" and "superstring theory," and when specification is needed one of the two latter terms is used.

[9]In addition to being honored by being elected a Fellow of the Royal Society, 1962, Stanley received the Dirac Medal and Prize from the International Centre for Theoretical Physics, 1991; became a Fellow of the American Academy of Arts and Sciences (AAAS), 1992; and received the Dannie Heineman Prize for Mathematical Physics, APS, 1992.

[10]See the web page of the Department of Physics, UC Berkeley, http://physics.berkeley.edu/remembering-stanley-mandelstam; the author's Physics Today Obituary for Stanley Mandelstam, May issue 2017; and this *Memorial Volume*. Also see the extensive biographic writings about Stanley in the *Memorial Volume*: S. Lee, "Stanley Mandelstam: The early years at a "Most Stimulating Theoretical Group," in Birmingham 1954–57; C. Thorn, "Scientific Biography of Stanley Mandelstam, Part I: 1955–1980", and N. Berkovits, "Scientific Biography of Stanley Mandelstam:1981–2016." Thorn's paper in the *Memorial Volume* also gives an almost-exhaustive list of Stanley's publications. Hopefully, by including the several additions selected in this paper, his list will be the exactly-exhaustive list of Stanley's publications.

(S1) The Mandelstam representations in the Mandelstam variables for S-matrices [C1]: The Mandelstam representations are double dispersion relations of S-matrices. The Mandelstam variables are the Lorentz-transformation invariant variables of all possible combinations of the $(3+1)$-dimension momenta of particles involved in the scattering. They form the fundamental framework and strategy for studying S-matrices. They in turn can help the development of quantum theories in which scattering amplitudes can be calculated.[11]

(S2) The S-matrix approach [C2]: In the 1959–61 papers, working with Chew,[12] and then Chew and Frautschi,[13] using the Mandelstam representations and the Regge pole[14] idea, he helped to develop the S-matrix approach for strong interactions. So what Chew had developed with his earlier collaborators: giving the interpretation of particles being poles in the energy-complex-plane and the idea of bootstrap (or nuclear democracy, or duality, or crossing symmetry) got important extensions. Chew advocated it as an alternative to QFT,[15] commonly called the S-matrix theory.[16]

Now QCD (quantum chromodynamics), a QFT, has been established as the theory for strong interactions. Due to its asymptotic freedom, the coupling strength is small at high energies, so perturbative calculations apply. QCD in high energies has been actively researched using both the perturbative QFT and the S-Matrix approaches, as Stanley had always practiced since the beginning of his career.

(S3) The Mandestam path-field formulation for quantum gauge theories and Feynman rules [C3]: His 1962 paper of [C3] used gauge-invariant path-dependent fields for quantum electrodynamics (QED, Maxwell gauge fields interacting with matter fields). Stanley was the first to give such a formulation after the 1959 Aharonov–Bohm (or AB) paper[17] showing the AB effects. Stanley considered the AB effect a call from nature. His formulation gave the precise answer.

---

[11] That was what Stanley did in his 1959 paper with the title, "Construction of the Perturbation Series for Transition Amplitudes from their Analyticity and Unitarity Properties," selected in [C1]. This strategy is still being popularly used.

[12] G. F. Chew, "Recollections of Stanley Mandelstam," a contribution to the *Memorial Volume*.

[13] S. Frautschi, "Stanley Mandelstam and my postdoctoral years at Berkeley," a contribution to the *Memorial Volume*. The paper gave a truly moving story about Stanley, and credited Stanley for introducing the Regge pole idea for obtaining high energy behaviors of S-matrices, which even led to Frautschi's being hired by Caltech because of Frautschi's good work on it.

[14] T. Regge, "Introduction to Complex Orbital Momenta," *Nuovo Cim.* **14**, 951 (1959); "Bound States, Shadow States and Mandelstam Representation," *Nuovo Cim.* **18**, 948 (1960).

[15] G. F. Chew, "*The Analytic S Matrix*," (Benjamin, New York, 1966).

[16] Here the term "S-matrix theory" is used by the author to mean a stand-alone theory; otherwise the term "S-matrix approach" is used, because S-matrices can be and ought to be studied in any theory for particles.

[17] Y. Aharonov and D. Bohm, "Significance of electromagnetic potentials in quantum theory," *Phys. Rev.* **115**, 485–491 (1959). Stanley did not refer to this AB paper in the References of his 1962 paper in [C3], but in the text he emphasized the important implications of the AB effects.

Then in the 1968 paper of [C3], he extended the formulation to include the Yang–Mills gauge fields interacting with matter fields, and in addition developed Feynman rules for perturbative calculations. We call this formulation the Mandelstam path-field formulation for quantum gauge theories.

Now experiments have established that the Standard Model of particle physics are QFTs of Yang–Mills fields interacting with matter fields: QCD for strong interactions and EWT (electroweak theory) for electroweak interactions. The Mandelstam path-field formulation has been actively used.[18] In addition, it has been cited in lattice gauge theory research,[19] as well as in loop quantum gravity research.[20]

In 1974, C. N. Yang[21] gave what he called the integral formalism for gauge theories (Maxwell and Yang–Mills). It involves path-dependent integrations. He discussed its advantages, gave geometric understandings to it, and made contrast to the differential formalism in which he and Mills constructed the Yang–Mills theory.[22] Using his integral formalism, Yang constructed a new theory for gravity.[23] In 1975, T. T. Wu and C. N. Yang[24] made the connection of the integral formalism to the mathematics of fiber bundles. The two papers received large number of citations. Yang's integral formalism has deepened and widened the mathematical-physics perspective for physics and for the Mandelstam path-field formulation.

In 1974 K. Wilson[25] gave the formulation of the Lattice gauge theory for computation. It naturally embodied the ideas of Mandelstam and Yang for continuous gauge theories: being an integral formulation that is path-dependent and gauge-invariant. The lattice gauge theory computations have been hugely successful in producing results that agree with experiments.[26]

---

[18] J. Terning, "Gauging non-local Lagrangians," *Phys. Rev. D* **44**, 887–897 (1991); C. Csaki, C. Grojean and J. Terning, "Alternatives to an elementary Higgs," *Rev. Mod. Phys.* **88**, 045001 (2016).

[19] M. Creutz, "Gauge fixing, the transfer matrix, and confinement on a lattice," *Phys. Rev. D* **15**, 1128–1136 (1977).

[20] C. Rowelli, "Ashtekar formulation of general relativity and loop-space non-perturbative quantum gravity: A report," *Class. Quant. Grav.* **8**, 1613–1676 (1991).

[21] C. N. Yang, "Integral formalism for gauge fields," *Phys. Rev. Lett.* **33**, 445–447 (1974) [Erratum: *ibid.* **35**, 1748 (1975).

[22] C. N. Yang and R. Mills, "Isotopic spin conservation and a generalized gauge invariance," *Phys. Rev.* **95**, 631–631 (1954); "Conservation of isotopic spin and isotopic gauge invariance," *Phys. Rev.* **96**, 191–195 (1954).

[23] See a more extensive discussion by C. N. Yang, "Gauge fields," pp. 487–561, in the *Proceedings of the Sixth Hawaii Topical Conference in Particle Physics*, Honolulu (University of Hawaii Press, 1976).

[24] T. T. Wu and C. N. Yang, "Concept of nonintegrable phase factors and global formulation of gauge fields," *Phys. Rev. D* **12**, 3845–3857 (1975).

[25] K. Wilson, "Confinement of quarks," *Phys. Rev. D* **10**, 2445 (1974).

[26] M. Creutz, "The lattice and quantized Yang–Mills theory," pp. 41–52 in the *Proceedings of Conference on 60 Years of the Yang–Mills Theory* (World Scientific, May 2016).

Now the terms "Wilson-loop" fields/formulations are almost universally used even when discussing continuous (non-lattice) gauge theories. The term "loop" is often misused, because the paths in the path-dependent gauge field associated with matter fields are open paths, not loops (closed paths), as made clear in Stanley's 1962 and 1968 papers of [C3].

(S4) The Mandelstam path-field formulation for quantum general relativity and the Feynman rules [C4]: In the same year as he pioneered his path-field formulation for quantum gauge theories, Stanley impressively also pioneered his path-field formulation for quantum general relativity, using coordinate-independent (diffeomorphism-invariant) path-dependent fields. These are Stanley's precise response to the call from the basic principle of Einstein's derivation of his general relativity theory.[27] Bryce S. DeWitt, in his highly cited 1967 paper,[28] referred to this 1962 Mandelstam paper of [C4] to be *"the most beautiful attempt at such a language."* However, DeWitt deemed the Mandelstam path-field formulation for quantum general relativity impractical. Nevertheless Stanley pushed on. Later in his 1968 paper of [C4], which referenced this 1967 paper by DeWitt, he derived the Feynman rules in his path-field formulation for quantum general relativity.

The ultraviolet divergences in perturbative QFT for general relativity called for further developments. One being actively pursued, as a competitor to superstring theory, is the loop quantum gravity approach,[29] which embodies the idea of using coordinate-independent (diffeomorphism-invariant) and path-dependent fields, as Mandelstam had practiced in his 1962 paper of [C4].

The idea of using path-fields, instead of local-fields, has also recently been applied in superstring theory researches,[30] through the AdS/CFT (gravity/conformal-field-theory) correspondence.[31]

(S5) The precursor for the discovery of string theory [C5]:[32] In the 1962–1963 papers [C5], two from several, Stanley published the results of his extensive work to extend the Regge-pole idea to the general analyticity properties of the complex-plane of angular momentum. In his 1967 and 1968 papers of [C5], using

---

[27]The manifestation of Mandelstam's coordinate-independent (diffeomorphism-invariant) path-dependent fields in Aharonov–Bohm-type experiments have recently been discussed in the literature.
[28]B. S. DeWitt, "Quantum theory of gravity. II. The manifestly covariant theory," *Phys. Rev.* **162**, 1195–1239 (1967).
[29]See the review paper on loop quantum gravity by C. Rovelli given in an earlier reference in the footnote format.
[30]B. Czech, L. Lamprou, S. McCandlish, S. B. Mosk and J. Sully, "A stereoscopic look into the bulk," *J. High Energy Phys.* **1607**, 129 (2016); arXiv:1604.03110, and references therein.
[31]I. R. Klebanov and J. M. Maldacena, "Solving quantum field theories via curved spacetimes," *Physics Today*, January 2009, pp. 28–33.
[32]See the moving tribute to Stanley, with expertise details, by P. Goddard, "The guiding influence of Stanley Mandelstam, from S-matrix theory to string theory," a contribution to the *Memorial Volume*.

the approximation of straight-line Regge trajectories, including taking the limit to infinity, he gave a theoretical model of S-matrices, later called the dual-resonance model. Amazingly, about a year later (according to the submissions dates of the papers), the Veneziano amplitude[33] became an explicit example of the model, and triggered the eventual discovery of string theory![34]

(S6) The elucidation of mechanisms for quark confinement in QCD [C6]: He took a leading role in elucidating the non-perturbative mechanisms for quark confinement and the phases in QCD. See the many publications selected in [C6]. These are highly influential and quoted papers, considering QCD now is the theory for strong interactions.

(S7) Non-perturbative constructions of the bosonization (or fermionization) in $(1+1)$-dimension QFTs [C7]: Stanley showed non-perturbatively by explicit construction, inspired by the perturbative results of S. Coleman, that operators for the creation and annihilation of quantum sine-Gordon solitons satisfy the anticommutation relations and field equations of the massive Thirring model. Thus he pioneered the non-perturbative operator transformation, now called bosonization or fermionization, that relate a $(1+1)$-dimension bosonic field theory to a $(1+1)$-dimension fermionic field theory. These papers are highly influential in condensed matter and mathematical physics.

In 1984, E. Witten generalized the nonperturbative construction to non-Abelian bosonization in $(1+1)$-dimension and showed that any fermi theory in $(1+1)$-dimension is equivalent to a local bose theory which manifestly possesses all the symmetries of the fermi theory.[35] This paper has revealed a new horizon for mathematical physics and for string theory.

(S8) The proof of perturbative ultraviolet finiteness and $\beta = 0$, in any gauge to all orders, of $N = 4$ SYM [C8]: Stanley alone, as was the case through most of his life, reached the finishing line first in the distribution and submission for publication of the paper that gave the proof in August 1982.

About two months later, in November 1982, the paper of L. Brink, O. Lindgren and B. E. W. Nilsson, who had been independently pursuing the same problem, was submitted for publication.[36] It reached the same result. In this paper they

---

[33] G. Veneziano, "Construction of a crossing-symmetric, Regge-behaved amplitude for linearly rising trajectories," *Nuovo Cim. A* **57**, 190–7. [ricevuto (received) il 29 Luglio (July) 1968, while the 1968 [C5] paper of Mandelstam was received 5 September 1967].

[34] David Gross called Stanley "The Godfather of String Theory" when he paid tribute to Stanley as the concluding speaker at Strings 2016, Beijing, China, August 1–5, 2016.

[35] E. Witten, "Non-Abelian bosonization in two dimensions," *Comm. Math. Phys.* **92**, 455–472 (1984).

[36] L. Brink, O. Lindgren, and B. E. W. Nilsson, "The ultraviolet finiteness of the $N = 4$ Yang–Mills theory," *Phys. Lett. B* **123**, 323–328 (1983).

referenced Mandelstam's preprint: Berkeley preprint, UCB-PTH-82/15 (August 1982), and commented on the differences in their perturbative approaches.[37]

Later it was reasoned that this result of ultraviolet finiteness and $\beta = 0$, in any gauge to to all oders, for $N = 4$ SYM holds even in the presence of the nonperturbative effects of instantons, by T. Banks and N. Seiberg in 1986 and again by N. Seiberg in 1988.[38]

Being a QFT from superstring theory in some limit and being a popular model for studying the AdS/CFT correspondences from superstring theory, $N = 4$ SYM has been actively studied for learning quantum gravity that is on the other side of the AdS/CFT correspondence. The result of its being ultraviolet finite has helped to give guidance and checks for other studies, e.g. calculating the S-matrices.

Interestingly, there is the Leibbrandt–Mandelstam prescription, which was developed out of the methods given in Stanley's 1983 paper of [C8], used in the computation of quark and gluon distribution functions for very large nuclei![39]

These accomplishments, commented in (S1–S8) coresponding to the selected papers [C5–C9]), have established Stanley as a supreme creative master of QFT (perturbative as well as non-perturbative) and of the S-matrix approach.

(S9) Many important publications on string theory and eventually the first long-waited proof of pertubative ultraviolet finiteness of superstring theory so it can be considered as a contender for being the theory of quantum gravity [C9]: Since 1986 Stanley had made many important contributions to the development of string theory.[40] Ultimately he derived explicit formulas for all n-loop superstring amplitudes and showed their ultraviolet finiteness and the absence of ambiguities in his 1992 paper of [C9]! Stanley worked years toward finding out the result, and produced Ph.D. students working on the subject. In his 1992 paper of [C9], he acknowledged three for helpful discussions (in alphabetical order with others): A. Berera, N. Berkovits, and S. J. Sin.[41] He also referred to two papers by Berkovits. Later Berkovits showed the perturbative ultraviolet finiteness of superstring theory in different ways.[42]

---

[37] See also, L. Brink, "Stanley Mandelstam and me and life on the light-cone," a contribution to the *Memorial Volume*.
[38] T. Banks and N. Seiberg, "Non-perturbative infinities," *Nucl. Phys.* B **273**, 157–164 (1986); N. Seiberg, "Supersymmetry and nonperturbative beta functions," *Phys. Lett.* B **206**, 75–80 (1988).
[39] L. McLerran and R. Venugopalan, "Computing quark and gluon distribution functions for very large nuclei," *Phys. Rev.* D **49**, 2233–2241 (1994).
[40] J. Schwarz, "Reminiscences of Stanley Mandelstam," J. Polchinski (Ph.D. 1980, advisor Stanley), "Grad school with Stanley Mandelstam", and papers by Thorn and by Berkovits referenced before, all contributed to the *Memorial Volume*.
[41] Each of all three, A. Berera (1992 Ph.D.), N. Berkovits (1988 Ph.D.) and S. J. Sin (1989 Ph.D.), have contributed a paper to the *Memorial Volume*.
[42] N. Berkovits, "Finiteness and unitarity of Lorentz-covariant Green–Schwarz superstring amplitudes," *Nucl. Phys.* B **408**, 43–61 (1993); "Multiloop amplitudes and vanishing theorems using the pure spinor formalism for the superstring," *JHEP* **0409**, 047 (2004); See also Berkovits' paper contributed to the *Memorial Volume* referenced before.

So, Stanley was also a supreme master of string theory according his work described in (S5,S9) coresponding to the selected papers [C5,C9]).

3. A Perspective

Interestingly, despite its origin being motivated by the studies of strong interactions,[43] superstring theory now is mainly studied as a framework for quantum gravity because of its perturbative ultraviolet finiteness.[44] The experimentally established Standard Model of particle physics, which includes EWT for electroweak interactions[45] and QCD for strong interactions[46] are all regular QFTs. It has no need of the two important necessary ingredients for superstring theory: supersymmetry and more-than-(3+1) extra-dimensions. The Standard Model is not yet embraced in the superstring theory framework,[47] nor have the supersymmetry and extra-dimensions yet been seen experimentally. Superstring theory advocates very much hope and strive for that some limit and/or duality of superstring theory might lead to the inclusion of the Standard Model, as well as to the discovery of the ultimate unification of all interactions, described in one mathematical framework. To all these and more, Stanley had made important contributions, as elaborated in (S1–S9) of the previous section.[48]

Stanley was a supreme master of QFT, the S-matrix approach, and string theory. His influences on theoretical and mathematical physics are deep and diverse, in almost all the current major research efforts trying to deepen our understanding of the physical universe. Stanley's influences live on.

4. Selected Publications

The selections are organized according to subjects given in the previous section and in chronological order of the first publication in each subject, except [C9] which has the latest selected publication of 1992.

[A] **W. Yourgrau and S. Mandelstam**
"*Variational Principles in Dynamics and Quantum Theory,*" 1st edition (1952); 2nd edition (1961); 3rd edition, (Dover Publications Inc. NY 1968).[49]

---

[43]To them Stanley made important contributions (S1, S2, S5).
[44]That was proven first by Stanley in 1992 (S9).
[45]To them Stanley's path-field formulation has been applied for more efficient calculations (S3).
[46]Its quark confinement mechanisms Stanley had elucidated (S6).
[47]This was very much in Stanley's mind, see his 1985 paper in [C9].
[48] "*Stanley Mandelstam's nine dragons in theoretical physics,*" we call his nine subjects of achievements.
[49]Wolfgang Yourgrau obtained Ph.D. in physics from Humboldt University, Berlin, Germany in 1932 (while in Berlin he had studied under A. Einstein) and was forced by the Nazi's to "wander" around the world and happened to overlap with Stanley's study at Wits in South Africa!
The 3rd edition has the prefaces of earlier editions. They gave acknowledgements to Werner Heisenberg, Erwin Schwinger, L. de Broglie, M. Born, etc.! Also the authors' institutions and titles in the prefaces gave information of their careers 1952-1968. The third edition was published in 1968 after Stanley had settled down for good as a professor in Physics at UC Berkeley and Wolfgand as a professor in History and Philosophy of Science at the University of Denver, coincidently both in 1963.

[B] **S. Mandelstam**
"Dynamical variables in the Bethe-Salpeter formalism," *Proceedings of the Royal Society of London A* **233**, 246–266 (1955);
"Uniqueness of solutions of the Bethe-Salpeter equation for scattering," *ibid.* **237**, 496–516 (1956).[50]

[C1] **S. Mandelstam**
"Determination of the pion-nucleon scattering amplitude from dispersion relations and unitarity: general theory," *Phys. Rev.* **112**, 1344–1360 (1958);
"Analytic properties of transition amplitudes in perturbation theory," *Phys. Rev.* **115**, 1741–1751 (1959);
"Construction of the perturbation series for transition amplitudes from their analyticity and unitarity properties," *Phys. Rev.* **115**, 1752 (1959);
"Unitarity condition below physical thresholds in the normal and anomalous cases," *Phys. Rev. Lett.* **4**, 84 (1960);
"Some rigorous properties of transition amplitudes," *Nuovo Cim.* **15**, 658 (1960);
"Two-dimensional representations of scattering amplitudes and their applications," pp. 209–233, in "*Quantum Theory of Fields*," *Proceedings of the 1961 Twelfth Solvay Conference on Physics*, Chair: Sir Lawrence Bragg (Cambridge), October, 1961, University of Brussels, Belgium, Eds. R. Stoops, (Interscience Publishers, a division of John Wiley & Sons, Inc., New York, 1961).

[C2] **G. F. Chew and S. Mandelstam**
"Theory of low-energy pion pion interactions," *Phys. Rev.* **119**, 467–477 (1960);
"Theory of low-energy pion pion interactions II," *Nuovo Cim.* **19**, 752 (1961);
S. Mandelstam
"Dispersion relations in strong-coupling physics," in *Reports on Progress in Physics XXV*, pp. 99–162 (1962), ed. A. C. Stickland; an extensive review referencing earlier papers (up to and without the next paper) and Regge paper;
G. F. Chew, S. C. Frautschi and S. Mandelstam
"Regge poles in pi pi scattering," *Phys. Rev.* **126**, 1202–1208 (1962);
S. Mandelstam, J. E. Paton, Ronald F. Peierls, and A. Q. Sarker.[51]
"Isobar approximation of production processes," *Annals Phys.* **18**, 198–225 (1962).

---

[50] Stanley's Ph.D. thesis publications, both communicated by Rudolf Peierls, his thesis advisor.
[51] Ronald and I coauthored papers. The one with T. L. Trueman we are most proud of, "Estimates of production cross sections and distributions for W bosons and hadrons jets in high energy pp and $p\bar{p}$ collisions," *Phys. Rev. D* **16**, 1397 (1977). Our results for the quantities in the title agreed with the 1983 experiments, whose observation of W^+, W^- and Z^0 earned the 1984 Nobel Prize in Physics, http://www.nobelprize.org/nobel_prizes/physics/laureates/1984/.

[C3] **S. Mandelstam**
"Quantum electrodynamics without potentials," *Annals Phys.* **19**, 1–24 (1962);
"Feynman rules for electromagnetic and Yang–Mills fields from the gauge independent field theoretic formalism," *Phys. Rev.* **175**, 1580–1623 (1968).

[C4] **S. Mandelstam**
"Quantization of the gravitational field," *Annals Phys.* **19**, 25–66 (1962);
"Feynman rules for the gravitational field from the coordinate independent field theoretic formalism," *Phys. Rev.* **175**, 1604–1623 (1968).

[C5] **S. Mandelstam**
"An extension of the Regge formula," *Annals Phys.* **19**, 254–261 (1962);
"Regge poles as consequences of analyticity and unitarity," *Annals Phys.* **21**, 302–343 (1963);

S. Mandelstam and L. L. Wang (now Chau)[52]
"Gribov-Pomeranchuk Poles in Scattering Amylitudes," *Phys. Rev.* **160**, 1490–1493 (1967);

S. Mandelstam
"Dynamics based on indefinitely rising Regge trajectories," pp. 604–615, in the *Proceedings of the 1967 Rochester International Conference on Particles and Fields*, University of Rochester, Rochester, New York, August 28–September 1, 1967, eds. R. Hagen, G. Guralnik and V. S. Mathur (Interscience Publishers, 1967);
"Dynamics based on rising Regge trajectories," *Phys. Rev.* **166**, 1539–1552 (1968). As mentioned on the page of (S5), this paper was received on September 5, 1967, while the 1968 Veneziano-amplitude paper, *Nuovo Cim. A* **57**, 190–7 (1968), was received on July 29, 1968, and did refer to this paper of Mandelstam in Ref. 2.

---

[52] I included this paper here so I can tell the story how it came about. I got my Ph.D. in 1966 under the guidance of Geoffrey Chew. My thesis was essentially a clipping together of three published papers with a covering page. For family reasons, I delayed my graduation and then stayed on working at Berkeley until Fall 1967 (before I went to Institute for Advanced Study at Princeton). During that time, one day I saw the announcement of a talk to be given by Stanley with a title and abstract that seemed like something that I had been working on. I went to the talk, and indeed it was. After the talk, I went to his office and told him that. Further I told him some results that he did not cover in his talk. He was surprised, and he walked to the window and thought. After a few minutes, he came back from the window and said, "You are right. I did not know that." A few days later, he looked me up in my office and handed me a manuscript and said, "Your name should be on this paper." I reviewed it and agreed with the content, which had also included the part that he said he did not know. So I happily agreed to have my name on it and suggested a few minor revisions. That was how our co-authoered paper came about. So we did not really collaborated on that paper. It is interesting to note that it was received by Physical Review on March 20, 1967. Now looking back on his publications, he had very few co-authored papers, totaly only seven: one with Yourgrau at Wit in South Africa, one with Ronald F. Peierls and collaboratories at University of Birmingham, four with Chew and collaborators at Berkeley, and the one with me!

[C6] **S. Mandelstam**
"Vortices and quark confinement in non-Abelian gauge theories," *Phys. Lett. B* **53**, 476–478 (1975);
"Vortices and quark confinement in non-Abelian gauge theories," in pp. 245–249 of *Phys. Rept.* Vol. 23, Issue 3 (1976), which is the *Proceeding of the Meeting on "Extended systems in field theory,"* Ecole Normale Supérieure, Paris, June 16–21, 1975, eds. J. L. Gervais and A. Neveu;
"Charge-monopole duality and the phases of non-Abelian gauge theories," *Phys. Rev. D* **19**, 2391–2401 (1979);
"Approximation scheme for QCD," *Phys. Rev. D* **20**, 3223 (1979);
"General introduction to confinement," pp. 109–121, *Phys. Rept.* Vol. 67, Issue 1 (1980), which is the *Proceedings of Les Houches Winter Advanced Study Institute on "Common trends in particle and condensed matter physics,"* February 1980, eds: E. Brezin, J.-L. Gervais and G. Toulouse.

[C7] **S. Mandelstam**
"Soliton operators for the quantized sine–Gordon equation," *Phys. Rev. D* **11**, 3026–3030 (1975);
"Soliton operators for the quantized sine–Gordon equation," pp. 307-313 of *Phys. Rept.* Vol. 23, Issue 3 (1976), which is the *Proceeding of the Meeting on Extended Systems in Field Theory*, Ecole Normale Supérieure, Paris, June 16–21, 1975, eds. J. L. Gervais and A. Neveu.

[C8] **S. Mandelstam**
"Light cone superspace and the ultraviolet finiteness of the $N = 4$ model," *Nucl. Phys. B* **213**, 149–168 (1983);
"Ultraviolet finiteness of the $N = 4$ model," pp. 167–177, in the book *"High-Energy Physics,"* of the series Studies in the Natural Sciences, Volume 20, Authors: Behram Kursunoglu, Editors: Stephan L. Mintz, Arnold Perlmutter, (Springer 1985); ISBN: 978-1-4684-8850-0; In Honor of P. A. M. Dirac in his 80th Year, 17–22 Jan 1983. Miami, Florida.

[C9] **S. Mandelstam**
"Veneziano formula with trajectories spaced by two units," *Phys. Rev. Lett.* **21**, 1724 (1968);
"Generalizations of the Veneziano and Virasoro models," *Phys. Rev.* **183**, 1374 (1969);
"Manifestly dual formulation of the Ramond model," *Phys. Lett. B* **46**, 447 (1973);
"Interacting string picture of dual resonance models," *Nucl. Phys. B* **64**, 205 (1973);
"Interacting string picture of the Neveu–Schwarz–Ramond model," *Nucl. Phys. B* **69**, 77 (1974);

"Dual-resonance models," *Phys. Rept.* **13**, 259–353 (1974);

"Dual-resonance models," pp. 593–637, in *Les Houches, June Institute on "Structural Analysis of Collision Amplitudes,"* June 2–27, 1975, Amsterdam, (North-Holland Pub. Co., New York: American Elsevier Pub. Co., 1976), eds. Roger Balian and Daniel Iagolnitzer;

"Introduction to Strings Model and Vertex Operators," pp. 15–35, in *Proceedings on Vertex Operators in Mathematics and Physics*, eds. P. J. Lepowsky, S. Mandelstam and I. Singer, Berkeley, CA, USA, 1983 (Springer-Verlag, 1985);[53]

"Composite Vector Mesons and String Models," pp. 97–105, in *"A Passion for Physics, Essays in Honor of Geoffrey Chew,"* for his sixtieth anniversary, Berkeley, 5 June 1984, eds. C. DeTar, J. Finkelstein, and C. I. Tan (World Scientic, Singapore, 1985);[54]

"Interacting string picture of the fermionic string," *Prog. Theor. Phys. Suppl.* **86**, 163 (1986);

"The n loop string amplitude: Explicit formulas, finiteness and absence of ambiguities," *Phys. Lett. B* **277**, 82–88 (1992).

Acknowledgments

I would like thank Professor Kok Khoo Phua and Professor Lars Brink for inviting me to participate in organizing the *"Memorial Volume for Stanley Mandelstam"* (the *Memorial Volume*) and to contribute a paper. My original plan was to write an article with the title "Endearing Memories of Stanley Mandelstam" about anecdotes and photos of Stanley to show his humor and good-heartedness, and to let the scientific side of Stanley be written up by those contributors to the *Memorial Volume* who are experts/practitioners of string theory — the theory to which Stanley laid the precursor in the 1960s and made concerted efforts to contribute till the end of his life.[55]

Then came the invitation from Physics Today (represented by Ms. Gayle Parraway), asking me to write an Obituary for Stanley Mandelstam (the Obituary). It was a surprise to me. After much hesitation and thought, I decided to take

---

[53] I was at the wonderful conference and presented the paper " Supersymmetric Yang-Mills Fields as An Integrable System and Connections with Other Non-linear Systems."

[54] I was at the wonderful celebration and presented the paper "Comments on Heavy Quark Decays and CP Violation".

[55] I have never being a practitioner of string theory, however chance has made me encounter string theory ever since it was still in its 1968 embryonic stage. [That episode was documented in the book *A Brief History of String Theory, from Dual Models to M-Theory* (Springer, Heidelberg, 2014) by D. Rickles. My name was Ling-Lie Wang, as was correctly written, but I was at Institute for Advanced Study, Princeton, not Princeton University as was incorrectly witten.] I wrote about this on the occasion of celebrating John Schwarz's 75th birthday in 2016, with the title "My Encounters with String Theory," for chatting with friends. Though not a practitioner, I have tried to be an alert observer on its development and owned almost all string theory books and even studied some of them.

the challenge. While researching and organizing thoughts for writing the Obituary, to my amazement, I discovered many new insights about Stanley's accomplishments and long lasting influences through his publications. Their depth and immensity are truly impressive. I deeply regret that I did not fully appreciated what he had accomplished when he was still living. Telling the story properly compelled me to write this paper. The Obituary has a limit of less than 850 words, in which I could give only the highlights but could not offer any details supported by Stanley's publications or references to other literature. To my pleasant surprise, it has turned out to be beneficial to work both of these two manuscripts. The contributed papers to the *Memorial Volume* have been very helpful to me in writing this paper. Also I am happy to have become better at contributing to the *Memorial Volume* because of the research and thoughts devoted for writing this paper. In our dedication to celebrate Stanley's life for the good of our physics community, my hope is that publishing this paper in the *Memorial Volume* as well as the Obituary in *Physics Today* will serve readers better with their differing focus and depth. I owe sincere thanks to Physics Today for inviting me to write an Obituary for Stanley Mandelstam.

I would also like to give my heartfelt gratitude to the Department of Physics of UC Berkeley, in particular the Director of the Berkeley Center for Theoretical Physics (BCTP) Professor Yasunori Normura, Managers Ms. Eleanor Crump and Mr. Brian Underwood, for their most helpful assistance in providing information and records related to Stanley: photos, especially the photo used as the cover of the *Memorial Volume*, the list of Stanley's Ph.D. students, the exact catalog of Stanley's collection of 232 physics and mathematics books, and requesting fee-waivers for reprinting some of Stanley's papers in this *Memorial Volume*. I am touched by their gracious offer to me to have all Stanly's books! I gratefully accepted the three that Stanley co-authored and the one that has Stanley's paper. Besides keeping them as mementos, I have put them to good use for writing this paper, as listed in Selected Publications [A, C2, C9].

Now the paper is done, I give my heartfelt thanks to Professors Steven C. Frautschi, Yasunori Nomura, David Pines, John H. Schwarz, Nathan Seiberg, and John Terning for reviewing this manuscript and making helpful and encouraging comments, and especially to Steven for saying "Your mention of my Memorial Volume contribution in the reference at (S2) is accurate and appropriate.", to David for saying "You make a very convincing case for Stanley having played a key role in theoretical physics, from the early days of QFT to string theory," and "You are clearly the right person to write it." and to John H. for saying "Everything you wrote is factually correct."; and to Dr. Richard Breedon, Ms. Eleanor Crump, and Mr. Weiben Wang for very helpful editorial comments.

Of course, any errors in this paper are mine alone.

In looking forward to the final production of the *Memorial Volume* by the World Scientific, I give my sincere appreciation to Mr. Chee Hok Lim of World Scientific for his always outstanding performance in the production process.

Attachment: Photos of Stanley with colleagues, students, and friends

Fig. 2. The wall size photo of the Solvay Conference on Physics 1961 (which celebrated the 50th anniversary of the 1911 inaugural one), (The author had it installed on the 4th floor of the Department of Physics, UC Davis. in the late 1980s.)
First row, seated, left to right: S. Tomonaga, W. Heitler, Y. Nambu, N. Bohr, E. Perrin, J. R. Oppenheimer, Sir W. L. Bragg, C. Møller, C. J. Goter, H. Yukawa, R. E. Peierls, H. A. Bethe;
Second row: I. Prigogine, A. Pais, A. Salam, W. Heisenberg, F. J. Dyson, R. D. Feynman, L. Rosenfeld, P. A. M. Dirac, L. van Hove, O. Klein;
Third row: A. S. Whightman, S. Mandelstam, G. F. Chew, M. L. Goldberger, G. C. Wick, M. Gell-Mann, G. Källén, E. Wigner, G. Wentzel, J. Schwinger, M. Cini.

Fig. 3. Stanley Fest celebrating his 80th birthday, KITP, UC Santa Barbara, February 13, 2009.

Fig. 4. After-conference dinner, 13th February 2009.

Fig. 5. Stanley (middle) with four of his UC Berkeley Ph.D. students, 13th February 2009, From left to right: J. Polchinski (Ph.D. 1980), C. Thorn (Ph.D. 1971), S. J. Sin (Ph.D. 1989), N. Berkovits (Ph.D. 1988).

Fig. 6. Stanley (5th from the left) with six of his UC Berkeley QFT class students, 13th February 2009, N. Berkovits, J. H. Schwarz, C. Thorn, S. J. Sin, D. J. Gross, L. L. Chau (QFT classes, Fall 1963 & Spring 1964). (Chau, Gross, and Schwarz all got Ph.D. in 1996, with G. F. Chew as adviser, at UC Berkeley.)

Fig. 7. S. Mandelstam and G. F. Chew at SFMOMA, San Francisco, 2010, during one of the outings organized by L. L. Chau.

Fig. 8. Annual New Year's Eve Dinner, 31st December 2007, since 1999,
From left to right: G. F. Chew, L. L. Chau, S. Mandelstam,
Sadly now without Stanley from 2016 onward.

Stanley was an exceptional scholarly gentleman (君子). His spirit lives on.

Stanley Mandelstam: The Early Years at a "Most Stimulating Theoretical Group"

Sabine Lee
*Department of History, University Birmingham,
Edgbaston, Birmingham, B15 2TT, UK
s.lee@bham.ac.uk*

After a childhood spent in the small town of Glencoe in the Natal Midlands of South Africa, and schooling in the neighbouring town of Dundee, Stanley Mandelstam went to read chemical engineering at Witwatersrand, Johannesburg completing his study with a B.Sc. in 1952. This subject appears to have been a choice of reason initiated by his mother who had encouraged Stanley to opt for a vocational degree.[1] It was not until he moved to Cambridge that he found what his nephew Ian Abramson would later call his 'first love', namely mathematical physics. This would remain his personal vocation and all his future studies and work were in this field. Mandelstam completed a physics degree at Cambridge in 1954[2] and then moved to the University of Birmingham which, at the time had what was later described as 'the most stimulating theoretical group in the world' around Rudolf Peierls.[3] It is his time at Birmingham between 1954 and 1957 and his short spell as Professor of Mathematical Physics there between 1960 and 1963 as well as the interim three years at Columbia and Berkeley that this brief paper will focus on.

When Rudolf Peierls took up his newly-created Chair in Mathematical Physics at the University of Birmingham in 1937, the university had no theoretical physics to speak of. Less than two decades later, Birmingham was considered as THE go-to place to study and research the subject. It was a vibrant place led by 'Prof' as Rudof Peierls was affectionately referred to by his postgraduate students and junior colleagues. Most likely it was this vibrancy and the unusual concentration of impressive talent at Birmingham that attracted the young South African to join the group after his stint at Cambridge. By 1954, when Mandelstam considered his options for graduate study, the Department of Mathematical Physics, as it was known then, had firmly established itself in the international teaching and research landscape. And in fact, the names of physicists moving in and out of the department in that period read like a who is who in physics: Freeman Dyson, Sam

Edwards, Julian Schwinger, Dick Dalitz, Nina Byers, Gerry Brown, Ed Salpeter, James Langer, Brian Flowers, John Bell, Paul Matthews, Denys Wilkinson, Elliot Lieb to name but a few.

If Mandelstam was impressed with Prof and Birmingham, the latter was similarly impressed with Mandelstam. He described him as 'very bright'[4] and 'promising'[5] while at the same time being 'charming' and 'educated'. Prof. was very keen to foster his career by enabling him to participate in the lively research exchange that had been developed between Peierls and some of his American colleagues, most notably Hans Bethe at Cornell, but also Freeman Dyson at Princeton or Robert Serber at Columbia. A letter from Peierls to Robert Serber, in which he summarized Mandelstam's remarkable achievements, demonstrates clearly that Peierls, who was not known for being impressed easily, found Mandelstam's work extraordinary. Not only had he completed the theoretical physics course at Cambridge in two years, not a mean feat in itself for someone from an engineering rather than physics background; he had also completed the academic requirements for a Ph.D. at Birmingham in only two years. Mandelstam had done so by publishing two important papers in formal field theory, which dealt with the nature of the solutions to the Bethe–Salpeter equation. The first of these papers, "Dynamical variables in the Bethe–Salpeter formalism",[6] was published within a year of commencing his research, and by Prof's own admission, he would have awarded him a PhD for this achievement only, had the regulations permitted this.[7] The second paper, an extension of his earlier work, was published shortly afterwards,[8] and his doctoral duely awarded.

That Peierls held him in high regard also beyond the formal field theoretical work, is clear in his comments about Mandelstam's subsequent research, a more phenomenological approach of interpretation of high energy experiments. This work had been facilitated by a UK government grant relating to data collection and interpretation of experiments at the high energy lab at the University of Cambridge, specifically the meson production in p–p collisions combining phase-space arguments with the picture of the isobaric states and linking the results with experimental work at Birmingham.[9]

The glowing reference provided by Peierls in the above-cited letter to Robert Serber had the desired effect, and in late 1957, Mandelstam joined Serber's department at Columbia as a Boese fellow, before — a year later — moving on to Berkeley. Berkeley would become his home for most of the the subsequent half century first as Professor of Physics and upon his retirement in 1994 as emeritus.

From Columbia, he engaged in a regular and intensive exchange of letters with Prof. The tone and substance of these letters bears witness to both the scientific and the non-scientific interests and accomplishments of both scholars. Shortly after arriving in New York, Mandelstam reported back about the many 'distractions' that the city that never sleeps had on offer — very much in contrast to Birmingham which still suffered from post-war austerity. He talked of concerts, theatre, and art galleries which clearly fascinated him. Yet, first and foremost, his correspondence — not surprisingly — comprised the newest developments in physics.

The most exciting research, Mandelstam reported, took place in Jack Steinberger's bubble chamber group.[10] The activities of the bubble chamber program at Columbia were concentrated around the Nevis Laboratories at the Physics Department, though many early pictures analysed there had been taken at the Brookhaven National Laboratory, initially at the Cosmotron (1952–1966) and later (from 1960, i.e. after Mandelstam's time at Columbia) at the Alternating Gradient Synchrotron. The program had been initiated by Jack Steinberger in 1956 — just prior to Mandelstam's arrival in the US, and the excitement among the young scientist about the work of this group was palpable. And the enthusiasm would later be shared by the Nobel Committee which would award Steinberger, Leon M. Lederman and Melvin Schwartz the 1988 Nobel Prize for their development of the neutrino beam method and their demonstration of the doublet structure of the leptons through the discovery of the muon neutrino[11] based on the work at Columbia in the early 1960s.

Mandelstam himself was working on dispersion relations. The purpose of the research was to find a relativistic analogue to the methods developed by Geoffrey Chew, Frances Low and George and Freda Salzman[12] in order to calculate the pion–nucleon scattering amplitude in terms of two coupling constants only. As the usual dispersion relations by themselves were not sufficient he assumed a representation which exhibited the analytic properties of the scattering amplitude as a function of the energy and the momentum transfer. Requiring unitarity conditions for the two reactions $\pi + N \to \pi + N$ and $N + \bar{N} \to 2\pi$ he approximated those by neglecting states with more than two particles.[13] Some years later, Peierls would comment on this work as 'outstanding', describing its effect as so profound that 'it is no exaggeration to say that no paper is being written now on the theory of high energy physics which does not in some way rely on Mandelstam's paper'. Recalling Geoffrey Chew's own assessment of Mandelstam, he reported to John Cockroft, who was trying to entice Mandelstam to take up a position in Cambridge, that the former regarded Mandelstam as the best young man who had been in Berkeley in the time he himself had been there. Given those accolades, it is hardly surprising that Columbia, Birmingham, Cambridge, Berkeley and also Stanford were fighting over Mandelstam in the late 1950s.[14] As we all know, Berkeley would eventually win the day, but for a short period between 1960 and 1963 Mandelstam returned to Birmingham to take up a professorship in Mathematical Physics.

Steven Frautschi, in this volume, relates interesting aspects of the relationship within the research group at Berkeley and also about the workings of Anglo-Saxon research groups more generally — a non-hierarchical collegiality where distinction, if at all existent, was not based on seniority but on scientific standing (which are not at all synonymous). He comments on an encounter of Stanley Mandelstam and Geoffrey Chew and himself with Wolfgang Heisenberg, who mistook Mandelstam's comparatively young age for a student role in the scientific hierarchy. Frautschi's uncompromising denial of Mandelstam's role as a junior figure says a lot about the former's respect of the latter. Of course many physicists — even of the Germanic tradition, would agree that the most inspired contributions often come from exceptional young scholars, anyway. This is possibly best captured in Pauli's infamous statement about physics and age when he first met Viktor Weisskopf and greeted him with: 'Ach, so young and already so unknown'.[15] Even though Heisenberg may not have known or recognised Mandelstam in their meeting in the late 1950s, Mandelstam never suffered the fate of being young **and** unknown, because some of his outstanding contributions had indeed already emerged in the early stages of his career.

During Mandelstam's second stay at Birmingham, now as Professor of Mathematical Physics — a mere six years after completing his PhD — the collaboration with colleagues from Berkeley continued to influence his work, as is evident in his paper on the theory of low-energy pion–pion interaction with Chew[16] and his work on Regge Poles.[17]

In another respect, both Chew's and Peierls' influence on Mandelstam could be felt. He was operating firmly within his mentors' tradition of enthusiastically engaging in teaching and taking pride and joy in bringing the subject to life for his students. Despite his already considerable scientific accomplishments he did so with great humility. This was captured poignantly by one student's remark that Mandelstam was "the only person I know who does not refer to the Mandelstam Variables by their so-designated name".[18]

Mandelstam also found time to put pen to paper to write not only for the upper echelons of the physics profession but also for an interested student audience. Not unlike Prof's 'Surprises in Theoretical Physics' more than two decades later, in 1955 Mandelstam and Wolfgang Yourgrau wrote 'Variational Principles in Dynamic and Quantum Theory'. In this survey, they examined the relationship to dynamic and quantum theory by foregrounding the historical and theoretical developments of the concepts and thereby elucidating the development of quantum mechanics in what a reviewer described as remarkably lucid. This unique combination of intellect and humility, scientific creativity and lucidity proved to be a most powerful toolset accounting for the remarkable achievements while at Birmingham and in the subsequent half century at Berkeley.

References

1. Remembering Stanley Mandelstam, http://physics.berkeley.edu/news-events/news/20160629/remembering-stanley-mandelstam-1928-2016 (accessed 6.10.2016).
2. Most sources state that he completed a B.A. Peierls mentions to colleagues that it was a part III in physics, the degree that would be the standard entry route for prospective PhD students at Birmigham in this field. Letter Rudolf Peierls to John Cockroft, 10 August 1959, in S. Lee, *Sir Rudolf Peierls. Selected and Scientific Correspondence*, Vol. 2 (World Scientific, 2009) (hereinafter *Peierls Correspondence*), letter 676.

3. G. E. Brown, Fly with eagles, *Annu. Rev. Nucl. Part. Sci.* **51**, 1 (2001).
4. Letter Rudolf Peierls to R. H. Dalitz, 4 April 1955, *Peierls Correspondence*, letter 610.
5. Rudolf Peierls to Robert Serber, 6 December 1956, *Peierls Correspondence*, letter 642.
6. S. Mandelstam, Dynamical variables in the Bethe–Salpeter formalism, in *Proceedings of the Royal Society of London A: Mathematical, Physical and Engineering Sciences*, Vol. 233, No. 1193 (The Royal Society, 1955).
7. Rudolf Peierls to Robert Serber, 6 December 1956, *Peierls Correspondence*, letter 642.
8. S. Mandelstam, Uniqueness of solutions of the Bethe–Salpeter equation for scattering, in *Proceedings of the Royal Society of London A: Mathematical, Physical and Engineering Sciences*, Vol. 237, No. 1211 (The Royal Society, 1956).
9. Rudolf Peierls to John Cockroft, 10 August 1959, *Peierls Correspondence*, letter 676.
10. M. Chretien, J. Leitner, N. P. Samios, M. Schwartz and J. Steinberger, π–p elastic scattering at 1.44 Bev, *Phys. Rev.* **108**, 383 (1957).
11. G. Danby, J. M. Gaillard, K. Goulianos, L. M. Lederman, N. Mistry, M. Schwartz and J. Steinberger, Observation of high-energy neutrino reactions and the existence of two kinds of neutrinos, *Phys. Rev. Lett.* **9**, 36 (1962).
12. G. F. Chew, M. L. Goldberger, F. E. Low and Y. Nambu, Relativistic dispersion relation approach to photomeson production, *Phys. Rev.* **106**, 1345 (1957); G. Salzman and F. Salzman, Solutions of the static theory integral equations for pion–nucleon scattering in the one-meson approximation, *Phys. Rev.* **108**, 1619 (1957).
13. S. Mandelstam, Determination of the pion–nucleon scattering amplitude from dispersion relations and unitarity. General theory, *Phys. Rev.* **112**, 1344 (1958).
14. See Gerry Brown to Rudolf Peierls, 8 September 1959, *Peierls Correspondence*, letter 678.
15. L. M. Lederman and D. Teresi, *The God Particle: If the Universe is the Answer, What is the Question?* (Houghton Mifflin Harcourt, 1993), p. 182.
16. G. F. Chew and S. Mandelstam, Theory of the low-energy pion–pion interaction, *Phys. Rev.* **119**, 467 (1960).
17. G. F. Chew, S. C. Frautschi and S. Mandelstam, Regge poles in π–π scattering, *Phys. Rev.* **126**, 1202 (1962).
18. Remembering Stanley Mandelstam, http://physics.berkeley.edu/news-events/news/20160629/remembering-stanley-mandelstam-1928-2016 (accessed 6.10.2016).

The Guiding Influence of Stanley Mandelstam, from S-Matrix Theory to String Theory

Peter Goddard

School of Natural Sciences, Institute for Advanced Study,
Princeton, NJ 08540, USA
pgoddard@ias.edu

The guiding influence of some of Stanley Mandelstam's key contributions to the development of theoretical high energy physics is discussed, from the motivation for the study of the analytic properties of the scattering matrix through to dual resonance models and their evolution into string theory.

1. The Mandelstam Representation

When I began research on the theory of the strong interactions in Cambridge in 1967, the focuses of study were the Regge theory of the high energy behavior of scattering amplitudes, and the properties of these amplitudes as analytic functions of complex variables. Most prominent amongst the names conjured with in these subjects was that of Stanley Mandelstam: the complex variables that the scattering amplitudes depended on were the *Mandelstam variables*; the complex space they varied over was the *Mandelstam diagram*; and the proposal of the *Mandelstam representation* had provided the inspiration for much of the study of the analytic properties of scattering amplitudes that was then in full spate.

Entering the field then, one was not readily conscious of the fact that this conceptual framework had its origins less than ten years earlier, in 1958–59, in a seminal series of papers[1–3] by Mandelstam. The book, *The Analytic S-Matrix*, by Eden, Landshoff, Olive and Polkinghorne,[4] published in 1966, the bible for research students in Cambridge at the time, begins with the slightly arch sentence, "One of the most important discoveries in elementary particle physics has been that of the complex plane." The ideas of analyticity in energy had been around for some time, used, for example, to derive dispersion relations, essentially by application of Cauchy's theorem (as reviewed by Mandelstam in Ref. 5). The real breakthrough that Mandelstam made, which underlay and motivated the developments described in that still relevant book, was to show how scattering amplitudes could and should be thought of as functions of *more than one* complex variable.

Like many deep insights, Mandelstam's perception that scattering amplitudes could be regarded as analytic functions of momentum invariants, although so absorbed into the conceptual framework of particle physics that it is taken for granted now, initially was difficult for some to accept, as Geoffrey Chew recounts.[6] Marvin Goldberger listed among his excuses for failing to understand the significance of what Mandelstam was doing at that time that he had "never understood a word that Stanley says on any subject. He is almost always right, has fantastic understanding, intuition, and mathematical power, but to me he is far from lucid in his presentation of his wisdom."[7] Certainly, unwittingly on his part, Mandelstam's formidable technical skills could make his papers a steep uphill journey for others who were less gifted in this regard.

Mandelstam[1] considered the two-to-two scattering amplitude, A, as an analytic function of s and t, the square of center-of-mass momentum and the momentum transfer invariant, respectively; so these and, more generally, the square of the sum of any subset of the momenta in an N-particle scattering process became known as the Mandelstam variables. (See Ref. 8 for an excellent detailed account of Mandelstam's research in the period discussed in this article.)

Viewing the scattering amplitude for pion–nucleon scattering, $A(s,t)$, as an analytic function of s and t, with assumptions about the singularity structure of $A(s,t)$, which could be established in low order perturbation theory in quantum field theory (QFT), Mandelstam could apply Cauchy's theorem twice to obtain a 'double dispersion relation', and thus established a new representation of the scattering amplitude, which became known as the *Mandelstam representation*. Although Chew was skeptical at first, Mandelstam quickly convinced him that this was the right way to think of the two-to-two scattering amplitude, effectively showing how to analytically continue the amplitude in energy and angle, something that Chew had in fact been trying to do for some time.[9]

Chew immediately recruited Mandelstam to Berkeley, and, apart from a period 1960–63 as professor back in Birmingham, where he had taken his PhD in the group of Rudolf Peierls, he remained in Berkeley for the rest of his life. Before graduate studies in Birmingham, and following his first degree, a BSc from Witwatersrand, Mandelstam took a BA from Trinity College, Cambridge in 1954, sharing the university prize for the best performance in applied mathematics and theoretical physics in the final degree examination with Jeffrey Goldstone, also at Trinity. At that time, Paul Dirac held the Lucasian Professorship in Cambridge, as he still did when I was a graduate student. When, in 1967, a chair assigned to theoretical particle physics was established, I believe an attempt was made to attract Mandelstam back to Cambridge, but he could not be moved from Berkeley.

2. *S*-Matrix Theory

Mandelstam and Chew began collaborating, applying Mandelstam's approach to the study of pion–pion scattering.[10,11] In this context, the idea of 'crossing', whereby the analytic continuation of the amplitude $A(s,t)$ for the process $A + B \to C + D$, from the physical region, in which $s > 0$, $t < 0$, to a region in which $s < 0$, $t > 0$, gives the amplitude for the process $A + \bar{C} \to \bar{B} + D$, where $\bar{B}$ is the antiparticle of B, could actually be used in dynamical calculations, relating bound states in one channel to forces in the other. The development of these ideas led Chew to formulate his 'bootstrap hypothesis'. This was the proposition that the requirements of analyticity, which was related to causality, and the requirement of unitarity, which seemed fundamental to the probabilistic interpretation of quantum theory, perhaps together with some suitable asymptotic assumptions at high energy, should determine uniquely the scattering amplitudes, i.e. the S-matrix. Along with this went Chew's principle of 'nuclear democracy', in which no particles would be more fundamental than any other but all were in some sense bound states of one another.

Mandelstam's calculations in perturbative QFT showed that the amplitudes had poles and branch points. That these singularities are present could be seen to be a necessary consequence of unitarity. Omnès and Froissart, in their 1963 text book *Mandelstam Theory and Regge Poles*,[12] set out what they called the assumption that the scattering amplitude only have the singularities required by unitarity 'The Mandelstam Hypothesis'. The hope of Chew's bootstrap approach, a hope that became almost a tenet of faith, was that the scattering amplitudes would be determined uniquely if this 'Mandelstam Hypothesis' held.

Chew was a powerful and charismatic evangelist for Mandelstam's conceptual framework. Polkinghorne recalled it being said in the early 1960s, with a different religious metaphor, that "There is no God but Mandelstam and Chew is his prophet."[13] But it seems that, in this context, God himself was an agnostic, for Mandelstam never fully subscribed to the bootstrap. As Chew put it, "Stanley never endorsed it but, being a very mild person, he did not fight it, and the term also appeared in one of our papers.... He feels a need for something fundamental... he always believed he had firm ground under his feet."[9] Reading Mandelstam's papers one certainly feels the firm ground under one's feet, and for him at that time, firm ground meant calculations based on perturbative QFT.

At the beginning of the 1960s, Mandelstam's work presented two great theoretical challenges, different in direction: the first, from the point of view of QFT, was to prove the Mandelstam representation, order by order in perturbation theory; and, the second, from the S-matrix point of view, was to determine the singularity structure of the scattering amplitudes implied by unitarity. Both of these challenges are discussed in the book of Eden *et al.*[4] In 1961, Landshoff, Polkinghorne and Taylor[14] and, independently, Eden[15] published proofs of the Mandelstam representation in perturbation theory, but, unfortunately, later that year the four authors found that their arguments were incomplete, because there are terms in perturbation theory that possess isolated real singularities, anodes, to which are attached complex singularities which vitiate the proof of the Mandelstam representation.[16] While, a proof is still lacking, these investigations did lead to a deeper understanding of the singularities of Feynman integrals.

The program of determining the singularity structure of the S-matrix from unitarity was carried forward most extensively by David Olive, as described in Chapter 4 of Ref. 4, and in the book, *The Analytic S Matrix*, by Chew,[17] published in 1966 at almost exactly the same time as Ref. 4 (but lacking the hyphen), which relies heavily on Olive's work in its early chapters. Later, Chew saw this book as presenting the culmination of his bootstrap ideas, but with some disappointment because, in trying to make as complete as possible a dynamical analysis of pion–pion system within this framework, they had not been able to go beyond two-particle branch cuts in the analytic structure of the scattering amplitudes, in order to take account of singularities corresponding to multi-particle states.[9]

With colleagues in Cambridge, Olive developed his methods further to describe in some generality the singularities of the S-matrix at real points of the physical region of the Mandelstam variables and the corresponding discontinuities.[18] However, little progress has been made in describing singularities, real or complex, outside the physical region, which would be necessary in order to discuss the validity of the Mandelstam representation.

3. Regge Theory

As Chew explains in his book, for the bootstrap principle conceivably to lead to a unique possible S-matrix, some asymptotic conditions on the amplitudes for large values of s, t, etc., must be imposed, just as for an analytic function of a single complex variable, z, to be constrained severely, an asymptotic condition on the function is required, such as being bounded by a power of z for $|z| \to \infty$. Fortuitously, in 1959, just before Chew and Mandelstam's study of pion–pion scattering, work of Tullio Regge,[19,20] aimed at proving the Mandelstam representation for potential scattering, led to developments that provided what just was needed here and, indeed, reshaped much of theoretical research on strong interaction physics throughout the 1960s.

Regge showed that the amplitude, a_J, for the non-relativistic scattering of a particle off a potential at a specific angular momentum, J, could be continued away from integer J to complex values, so as to yield an analytic function of J, and this might have poles at $J = \alpha_E$, depending on the energy, E; these became known as Regge poles. For a given angular momentum, J, $J = \alpha_E$ determines the energy at which a bound state or resonance occurs. Thus, the curve, $J = \alpha_E$, known as a Regge trajectory, relates resonances with different angular momenta, or spin, corresponding to different integer values. If one considers analytically continuing the scattering amplitude, $a(E,\theta)$, at scattering angle θ for fixed energy, E, to the unphysical region of large $z = \cos\theta$, it behaves like z^{α_E} as $z \to \infty$, where α_E is the Regge trajectory for which the real part of α_E is largest for the given value of E, called the leading Regge trajectory. [Here, for simplicity, we shall just refer to meson resonances with integer spin.]

It was Mandelstam who first argued for the importance of Regge theory for high energy particle physics, through unpublished discussions with Chew, Frautschi and others, for which he never sought credit.[21,22] He saw that Regge theory applied to relativistic scattering would provide the appropriate boundary condition for the S-matrix.[23] As Elliot Leader later put it, Regge's great imaginative leap, of introducing complex angular momentum in non-relativistic quantum mechanics, might have ended in oblivion if Mandelstam had not demonstrated its striking consequences in high-energy scattering processes[24] (quoted in Ref. 25).

In the relativistic context, the energy E can be replaced by s and $z = \cos\theta$ by t, at fixed s. Then a Regge trajectory takes the form $J = \alpha(s)$, and corresponds to a sequence of resonances at mass squared s given by integral J, and determines a contribution $\beta(s)t^{\alpha(s)}$ to the behavior of the amplitude in the limit $t \to \infty$ at fixed s. The difference in the relativistic case is that this unphysical limit for the original process, $A + B \to C + D$ (the s-channel), is a physical limit, the high energy limit, for the 'crossed channel' $A + \bar{C} \to \bar{B} + D$ (the t-channel). Resonances in one channel are related to high energy behavior in the crossed channel.

Stimulated by Mandelstam, Chew and Frautschi[26] proposed that the known strongly interacting particles lay on nearly straight and parallel Regge trajectories and then Mandelstam collaborated with them on developing a calculational procedure for analyzing the interrelation between Regge poles, bound states, resonances and high energy behavior, arguing that Regge poles and behavior were general phenomena for strongly interacting particles.[27]

In his book,[17] Chew supplemented the assumption that the S-matrix should be as analytic as possible in the Mandelstam variables, consistent with unitarity, with an assumption of suitable analyticity in J, including that all particles should lie on Regge trajectories, and allowing for the possibility that there might be singularities other than poles in the complex J plane. In 1962, Amati, Fubini and Stanghellini (AFS)[28] proposed a mechanism to show that there should be cuts in the J plane based on perturbative QFT, but, soon after, Mandelstam[29] demonstrated both that the AFS cuts were in fact cancelled by other contributions and that an adaptation of their mechanism would produce a cut that would actually be present in the J plane. In so doing, he took the major step in establishing the nature of the singularity structure in the complex angular momentum plane.

4. Dual Models and String Theory

While a great deal of effort in strong interaction phenomenology from the mid-1960s onwards was devoted to understanding scattering data in terms of sums of Regge pole, and, where necessary, Regge cut, contributions at high energy and in terms of sums of resonance contributions at lower energies, the question arose as to whether these two sorts of description should be added together to get a more accurate parameterization of the scattering amplitudes or whether they should be regarded as alternative descriptions of the data, equivalent, or dual, in some sense. The former procedure, in which Regge contributions should be added to resonance contributions, was known as 'interference', and the latter, in which they were equivalent to one another, and which the description at high energy in terms of Regge poles could be determined from the parametrization of lower energy data in terms of resonances,[30] was known as 'duality'.

For a time, it seemed difficult to see analytically how a description of the scattering amplitude as a sum of resonance poles could be equivalent to an asymptotic expansion in terms of Regge poles, and there were suspicions that it might be mathematically impossible for this to be exactly the case, until, in the summer of 1968, Gabriele Veneziano produced, almost as a sort of *deus ex machina*, what very quickly became his famous eponymous formula,[31] $A(s,t) = B(-\alpha(s), -\alpha(t))$, where B is the classical Euler Beta function and the Regge trajectory is linear, $\alpha(s) = \alpha_0 + \alpha' s$. This is symmetric in s and t, meromorphic in s with poles whose residues are polynomials in t (and *vice versa*); it can be written as a sum over these poles or, equivalently, an asymptotic series of Regge pole contributions, realizing duality in a mathematical precise form. The resonances in one channel are precisely dual to the Regge poles in the other.

The Veneziano formula, describing two-to-two scattering, was generalized quickly to processes involving an arbitrary number of particles, by the construction of N-particle amplitudes which were also meromorphic and possessing Regge behavior, in an appropriate sense, for large values of Mandelstam variables. It was also generalized by finding other formulae for two particle scattering amplitudes, which possessed similar properties to the Veneziano formula. At first, these proposals, known as dual models, were either directed at providing sufficient flexibility in the amplitudes to enable them to be used to fit experimental data or to provide theoretical laboratories within which to study the form Regge theory might take for multi-particle processes.

However, soon it was appreciated that the Veneziano amplitude could be taken more seriously, in a sense, as being the starting point for a perturbative expansion for the scattering amplitudes, consistent with unitarity at least as a formal power series, in a similar way to perturbative QFT. Indeed, S-matrix theory, which in Chew's formulation of the bootstrap had offered the hope of determining the S-matrix uniquely, could also more modestly and more concretely provide the conceptual framework within which dual models could be viewed on comparable terms with perturbative QFT, as providing a model theory formally satisfying the basic postulates. For Mandelstam, who had been reluctant to give up the crutch that (perturbative) QFT provided to S-matrix theory, this gave an appealing alternative way to keep one's feet on solid ground. He later commented, after dual models had metamorphosed into string theory, "The string model originated as a model for the S-matrix, and it may well not have been discovered if S-matrix theory had not been vigorously pursued at the time."[32]

The Veneziano amplitude, extended to N-particle processes, is the starting point for a perturbative expansion for an S-matrix; higher order contributions need to be added, containing threshold cuts and the other singularities, corresponding to loop contributions in QFT. But, in order for it to be an appropriate starting point, even before considering whether the higher order contributions can be defined consistently, there are conditions that the N-particle Veneziano amplitude needs to satisfy, as a consequence of unitarity. In particular, the residue at each pole in a Mandelstam variable needs to factorize in the sense of being a sum over intermediate states of products of two amplitudes for fewer particles. In fact, this process, if it can be done consistently, determines the particle spectrum of the theory.

Mandelstam's early work had set much of the stage for the introduction of dual models, and, more recently, he had provided some of the immediate motivation[33] for Veneziano's breakthrough. He was quick to make a fundamental contribution to their development by establishing, with Bardacki[34] the factorization of the generalized Veneziano model, simultaneously with work of Fubini and Veneziano,[35] just nine months after that breakthrough. Their results demonstrated an unanticipated exponential growth in the degeneracy of states with mass, typical of a vibrating extended one-dimensional medium, as was pointed out over the following year by Nambu, Nielsen and Susskind,[36–38] describing it variously as a rubber band or string. At first, the main impact of this description was through the analogue approach introduced by Nielsen,[37] because it suggested appropriate mathematical techniques for calculating loop contributions, but it was unclear at the time whether it should be regarded merely as an analogy and calculational guide or as a deeper physical description.

The essential aspect of a relativistic string theory, absent in those earliest analogies, that enables the string to be more than only a qualitative correspondence, is that its (dynamically significant) oscillations should only be transverse. This requires the action for the string to depend just on the surface it traces out in space–time, rather than the particular way it is parametrized. In 1970, Nambu,[39] and then Goto,[40] wrote down such a reparametrization invariant action. After the structure of the physical states of the dual model, implied by factorization, was completely understood[41,42] in 1972, the way to quantize systematically the Nambu–Goto action, using light cone coordinates[43] became clear.

This showed definitively that the spectrum of the Veneziano dual model was precisely that of the quantized Nambu–Goto string, but it remained to show that the scattering amplitudes, defining the model, actually followed from the Nambu–Goto action. At the time, this seemed to present a formidable technical challenge, but, within nine months, developing a path integral approach, Mandelstam established that the scattering amplitudes for the Nambu–Goto string, as specified directly by their action, integrated over histories in which strings are allowed to split and join, are precisely the Veneziano model amplitudes.[44] In a series of papers,[44–47] he demonstrated the great power of path integral techniques for calculating amplitudes in string theory, although he remained largely alone in using them until Polyakov's seminal work in 1981.[48] In particular, Mandelstam was immediately able to calculate the amplitude for fermion–fermion scattering in the dual model of Neveu, Schwarz and Ramond,[49,50] a task that had proved much more difficult using previous techniques.[51,52] (For a more detailed account see Ref. 8 and Mandelstam's own recollections Ref. 53.)

By this time, 1974, string theory had emerged from its origin in dual models, a metamorphosis in which Mandelstam had played a key role. He wrote a review article,[54] which provided a definitive account of the theory at this stage of its development, before his own interest shifted away from string theory for some years.

5. Epilogue

I first met Stanley Mandelstam when he visited CERN in July 1971, and gave a seminar, entitled *Dual resonance models with quarks*, on his program to build a dual model including quarks as a basis for describing hadrons.[55] Although more than 45 years ago, I remember clearly meeting someone who had previously seemed almost mythical to me. The impression that was created by his combination of quiet modesty, formidable intellectual strength and lively wit, and his approachability did not diminish at all his status as one of my personal heroes.

I had come to CERN the previous year as a postdoctoral fellow, after completing my doctorate in Cambridge on S-matrix analyticity and Regge theory. I had intended to carry on with this work but, inspired by lectures by David Olive, just before I left Cambridge, and attracted by the lively group of young theoreticians at CERN working on dual models, which had begun to cluster around Daniele Amati, my interest had shifted to that area. I felt, as Mandelstam later said,[32] that dual models could best be understood at the time as providing an alternative to perturbative QFT as a model for the S-matrix, and one that had the advantage for describing strong interactions of possessing Regge behavior at finite order in the perturbative expansion.

Mandelstam, along with Amati, Fubini, Nambu and (quietly) Goldstone, was one of the relatively few senior figures in theoretical physics then working on the theory of dual models and lending it their support. The disapproval of many leading physicists, at CERN and elsewhere, gave the research in the area the extra frisson of forbidden fruit. At least in retrospect, "to be young was very heaven" then, but the difficulty of getting a permanent post if one worked on string theory was a large part of the reason that interest in the subject largely diminished for a decade after 1974 (see contributions in Ref. 56 for accounts of the period).

In his writings, Stanley Mandelstam was as he was in person: he spoke when he had something to say, and then one should listen. He advanced our understanding in many areas, including: S-matrix theory; Regge theory; dual models and string theory; and solitons and monopoles in QFT, always keeping his feet on the ground, making contributions that were both solid and seminal. Perhaps, if he had had less formidable technical skills, some of his work might have been more easily accessible, but the steep climb was easily justified by the wider and deeper view that one obtained. Although he wrote less than 50 original articles, and would not have stood out exceptionally by the contemporary dull citation metrics, which are now widely and lazily used, his influence remains profound, pervasive and enduring.

References

1. S. Mandelstam , Determination of the pion-nucleon scattering amplitude from dispersion relations and unitarity. General theory, *Phys. Rev.* **112**, 1344 (1958).
2. S. Mandelstam, Analytic properties of transition amplitudes in perturbation theory, *Phys. Rev.* **115**, 1741 (1959).
3. S. Mandelstam, Construction of the perturbation series for transition amplitudes from their analyticity and unitarity properties, *Phys. Rev.* **115**, 1752 (1959).
4. R. J. Eden, P. V. Landshoff, D. I. Olive and J. C. Polkinghorne, *The Analytic S-Matrix* (Cambridge University Press, 1966).
5. S. Mandelstam, Dispersion relations in strong-coupling physics, *Rep. Prog. Phys.* **25**, 99 (1962).
6. G. F. Chew, Recollections of Stanley Mandelstam, *Int. J. Mod. Phys. A* **32**, 1740001 (2017).
7. M. L. Goldberger, Fifteen years in the life of dispersion theory, in *Subnuclear Phenomena, 1969 International School of Physics "Ettore Majorana"*, ed. A. Zichichi (Academic Press, New York, 1970), p. 685.
8. C. B. Thorn, Scientific biography of Stanley Mandelstam: 1955–1980, *Int. J. Mod. Phys. A* **32**, 1740010 (2017).
9. F. Capra, Bootstrap physics: A conversation with Geoffrey Chew, in *A Passion for Physics, Essays in Honor of Geoffrey Chew*, eds. C. DeTar *et al.* (World Scientific, Singapore, 1985), p. 247.
10. G. F. Chew and S. Mandelstam, Theory of low-energy pion pion interactions, *Phys. Rev.* **119**, 467 (1960).
11. G. F. Chew and S. Mandelstam, Theory of low-energy pion pion interactions II, *Nuovo Cim.* **19**, 752 (1961).
12. R. Omnès and M. Froissart, *Mandelstam Theory and Regge Poles: An Introduction for Experimentalists* (Benjamin, New York, 1963), p. 80.
13. J. C. Polkinghorne, Salesman of ideas, in *A Passion for Physics, Essays in Honor of Geoffrey Chew*, eds. C. DeTar *et al.* (World Scientific, Singapore, 1985), p. 23.
14. P. V. Landshoff, J. C. Polkinghorne and J. C. Taylor, A proof of the Mandelstam representation in perturbation theory, *Nuovo Cim.* **19**, 939 (1961).
15. R. J. Eden, Proof of the Mandelstam representation for every order in perturbation theory, *Phys. Rev.* **121**, 1567 (1961).
16. R. J. Eden, P. V. Landshoff, J. C. Polkinghorne and J. C. Taylor, Acnodes and cusps on Landau curves, *J. Math. Phys.* **2**, 656 (1961).
17. G. F. Chew, *The Analytic S Matrix* (Benjamin, New York, 1966).
18. M. J. W. Bloxham, D. I. Olive and J. C. Polkinghorne, S-matrix singularity structure in the physical region: I. Properties of multiple integrals; II. Unitarity integrals; III. General discussion of simple Landau singularities, *J. Math. Phys.* **2**, 494, 545, 553 (1969).
19. T. Regge, Introduction to complex orbital momenta, *Nuovo Cim.* **14**, 951 (1959).

20. T. Regge, Bound states, shadow states and Mandelstam representation, *Nuovo Cim.* **18**, 948 (1960).
21. S. C. Frautschi, *Regge Poles and S-Matrix Theory* (Benjamin, New York, 1963), p. 144.
22. S. C. Frautschi, Stanley Mandelstam and my postdoctoral years at Berkeley, *Int. J. Mod. Phys. A* **32**, 1740004 (2017).
23. S. C. Frautschi, My experiences with the S-matrix program, in *A Passion for Physics, Essays in Honor of Geoffrey Chew*, eds. C. DeTar et al. (World Scientific, Singapore, 1985), p. 44.
24. E. Leader, Why has Regge pole theory survived?, *Nature* **271**, 213 (1978).
25. D. Rickles, *A Brief History of String Theory, from Dual Models to M-Theory* (Springer, Heidelberg, 2014).
26. G. F. Chew and S. C. Frautschi, Regge trajectories and the principle of maximum strength for strong interactions, *Phys. Rev. Lett.* **8**, 41 (1962).
27. G. F. Chew, S. C. Frautschi and S. Mandelstam, Regge poles in π–π scattering, *Phys. Rev.* **126**, 1202 (1961).
28. D. Amati, S. Fubini and A. Stanghellini, Asymptotic properties of scattering and multiple production, *Phys. Lett.* **1**, 29 (1962).
29. S. Mandelstam, Cuts in the angular-momentum plane I, II, *Nuovo Cim.* **30**, 1128, 1147 (1963).
30. R. Dolen, D. Horn and C. Schmidt, Finite energy sum rules and their application to π–N charge exchange, *Phys. Rev.* **166**, 1768 (1968).
31. G. Veneziano, Construction of a crossing-symmetric, Regge behaved amplitude for linearly rising trajectories, *Nuovo Cim. A* **57**, 190 (1968).
32. S. Mandelstam, Composite vector mesons and string models, in *A Passion for Physics, Essays in Honor of Geoffrey Chew*, eds. C. DeTar et al. (World Scientific, Singapore, 1985), p. 97.
33. S. Mandelstam, Dynamics based on rising Regge trajectories, *Phys. Rev.* **166**, 1539 (1968).
34. K. Bardakci and S. Mandelstam, Analytic solution of the linear-trajectory bootstrap, *Phys. Rev.* **184**, 1640 (1969).
35. S. Fubini and G. Veneziano, Level structure of dual-resonance models, *Nuovo Cim. A* **64**, 811 (1969).
36. Y. Nambu, Quark model and the factorization of the Veneziano amplitude, in *Proceedings of the International Conference on Symmetries and Quark Models held at Wayne State University*, June 18–20, 1969, ed. R. Chand (Gordon and Breach, New York, 1970), p. 269, reprinted in *Broken Symmetry, Selected Papers of Y. Nambu*, eds. T. Eguchi and K. Nishijima (World Scientific, Singapore, 1995), pp. 258–277.
37. H. B. Nielsen, An almost physical interpretation of the integrand of the n-point Veneziano model, paper submitted to the *15th International Conference on High Energy Physics*, 1970, and Nordita preprint (1969).
38. L. Susskind, Dual-symmetric theory of hadrons I, *Nuovo Cim. A* **59**, 457 (1970).
39. Y. Nambu, *Duality and Hadrodynamics* (Notes prepared for the Copenhagen High Energy Symposium, unpublished, 1970), published in *Broken Symmetry, Selected Papers of Y. Nambu*, eds. T. Eguchi and K. Nishijima (World Scientific, Singapore, 1995), p. 280.
40. T. Goto, Relativistic quantum mechanics of a one-dimensional continuum and subsidiary condition of dual resonance model, *Prog. Theor. Phys.* **46**, 1560 (1971).
41. R. C. Brower, Spectrum-generating algebra and no-ghost theorem in the dual model, *Phys. Rev. D* **6**, 1655 (1972).

42. P. Goddard and C. B. Thorn, Compatibility of the dual pomeron with unitarity and the absence of ghosts in the dual resonance model, *Phys. Lett. B* **40**, 235 (1972).
43. P. Goddard, J. Goldstone, C. Rebbi and C. B. Thorn, Quantum dynamics of a massless relativistic string, *Nucl. Phys. B* **56**, 109 (1973).
44. S. Mandelstam, Interacting string picture of dual resonance models, *Nucl. Phys. B* **64**, 205 (1973).
45. S. Mandelstam, Manifestly dual formulation of the Ramond model, *Phys. Lett. B* **46**, 447 (1973).
46. S. Mandelstam, Lorentz properties of the three-string vertex, *Nucl. Phys. B* **83**, 413 (1974).
47. S. Mandelstam, Interacting string picture of the Neveu–Schwarz–Ramond models, *Nucl. Phys.* **69**, 77 (1974).
48. A. M. Polyakov, Quantum geometry of bosonic strings, *Phys. Lett. B* **103**, 207 (1981).
49. P. Ramond, Dual theory for free fermions, *Phys. Rev. D* **3**, 2415 (1971).
50. A. Neveu and J. H. Schwarz, Factorizable dual model of pions, *Nucl. Phys. B* **31**, 86 (1971).
51. J. H. Schwarz and C. C. Wu, Evaluation of dual fermion amplitudes, *Phys. Lett. B* **47**, 453 (1973).
52. E. F. Corrigan, P. Goddard, D. I. Olive and R. A. Smith, Evaluation of the scattering amplitude for four dual fermions, *Nucl. Phys. B* **67**, 477 (1973).
53. S. Mandelstam, Factorization in dual models and functional integration in string theory, in *The Birth of String Theory*, eds. A. Cappelli *et al.* (Cambridge University Press, 2012), p. 294.
54. S. Mandelstam, Dual resonance models, *Phys. Rep.* **13**, 259 (1974).
55. S. Mandelstam, Relativistic quark model based on the Veneziano representation. I meson trajectories, II general trajectories, III baryon trajectories, *Phys. Rev.* **184**, 1625 (1969); *Phys. Rev. D* **1**, 1734 (1970); *Phys. Rev. D* **1**, 1745 (1970).
56. A. Cappelli, E. Castellani, F. Colomo and P. Di Vecchia (eds.), *The Birth of String Theory* (Cambridge University Press, 2012).

Remembering Stanley: From a Source of Inspiration to a Fair Strong Competitor

G. Veneziano

Theory Division, CERN, CH-1211 Geneva 23, Switzerland
Collège de France, 11 place M. Berthelot, 75005 Paris, France

I will recall my first indirect encounters with Stanley Mandelstam through courses and readings, the way he deeply inspired my research, how he supported a new scientific program in the middle of much skepticism, and how we became healthy and fair competitors.

1. First Indirect Encounters

How did I first encounter his name? Like for most theorists, I am pretty sure, when I first heard about the Mandelstam variables s, t, u. That was, I guess, during my 4th year course on Quantum Field Theory given by Professor Raoul Gatto at the University of Florence around 1963.

More important was the next (indirect) encounter. While working at my M.Sc. Thesis, a year or so later, Gatto asked me to study a review article by Daniele Amati and Sergio Fubini, two Italian physicists I did not personally know at the time (who could have imagined then that I would have written many papers with both of them during my career!). It dealt with a number of theoretical approaches to pion–nucleon scattering and contained, if I remember well, a section dealing with the high energy behavior of that process according to the recently formulated (~ 1959) Regge theory of complex angular momentum.

The original approach by Tullio Regge (incidentally, another great, recently disappeared physicist) would look at that process from the direct (s)-channel perspective. It would describe the baryonic bound states and resonances of different spin appearing in that channel as belonging to a bunch of "Regge trajectories" $\alpha_i(s)$ but this would have nothing to do, in principle, with the high-energy limit of pion–nucleon scattering. However, G. F. Chew and S. Mandelstam had made the crucial observation that bosonic Regge poles in the crossed (t)-channel (nucleon–antinucleon annihilation into two pions) would determine the high energy, small momentum transfer limit of the process through their trajectories $\alpha_j(t)$.

This observation marked the beginning of an entire domain of research in the physics of strong interactions which, with varying amounts of appreciation by the scientific community, has been going on till now. In any case it made on me, as a student, an enormous impression. Think of it: the same Regge poles, originally introduced to interpolate between states of different angular momentum in the physical (i.e. $t > 0$) region in t-channel, when extrapolated to its unphysical (i.e. $t < 0$) region would describe the asymptotic behavior of the physical process occurring in the crossed s-channel. It must have looked like magic to me and would have shaped the course of my research for years to come.

I did also hear, of course, of Mandelstam's double dispersion relations. They looked very deep and extremely promising to me at the time, but somehow they would not have affected as much my later research.

2. An Important Input to My Research

A couple of year later I was a graduate student at the Weizmann Institute in Israel. My official thesis advisor was H. J. Lipkin but, actually, I was working under the guidance of the newly arrived Hector Rubinstein. In the winter of 1967 Miguel Virasoro had joined the two of us and we had started thinking about constraints on scattering amplitudes coming from Regge and Chew–Mandelstam theory, i.e. both from contributions of s-channel resonances and from those of t-channel Regge poles. The connection between the two was very intriguing and it was made even more so by two kinds of considerations:

- For sometime Chew had been pushing a "Regge bootstrap" program based on what he called "Nuclear Democracy," the assumption that *all* strongly interacting particles (hadrons) lie on Regge trajectories (for positive values of their argument) and that *all* asymptotic behaviors are determined by *the same* Regge trajectories (for negative values of their argument). Chew's idea was that this assumption, together with some very general properties (such as analyticity, crossing and unitarity), would completely fix the strong-interaction S-matrix.
- More recently, Stanley had stressed (e.g. at his Summer Lectures in Tokyo, 1966) the potential importance of an amazing property of the experimentally observed Regge trajectories: their linearity ($\alpha_i(t) \sim a_i + \alpha'_i t$) as well as their universal slope ($\alpha'_i \sim \alpha'$) both, of course, up to small deviations and/or within experimental uncertainty. This was very different from what Regge had found in his pioneering work based on nonrelativistic potential scattering. In that case, trajectories would start from a negative integer for large negative t, would rise for a while going through a small number of resonances for small positive t, and, finally, would turn over and fall back to a negative integer for large positive t. By contrast, in strong interaction physics they looked like straight lines all the way as if they wanted to go through resonances of arbitrarily high mass and angular momentum.

It took us however till the summer of 1967 before we could put together these different theoretical and experimental hints. As far as I am concerned the "revelation" happened in Erice when Murray Gell-Mann mentioned a paper by R. Dolen, D. Horn and C. Schmid (DHS) that had just come out of the Caltech theory group. The paper had to do with pion–nucleon charge exchange ($\pi^- p \to \pi^0 n$) where DHS observed that the contribution to the scattering amplitude coming from s-channel baryonic resonances was well represented, on average, by the contribution coming from the t-channel bosonic Regge trajectories. The two contributions were "Dual" to each other (this property became thus known as DHS duality) so that adding them would give the wrong scattering amplitude (by about a factor 2). Gell-Mann stressed that this observation could lead to a new and easier bootstrap than the one advocated by Chew.

Now things started to fit nicely together. Mandelstam's linear trajectories would support resonances all the way to the high energies needed to apply the predictions of (the Chew–Mandelstam version of) Regge theory. For theoretical simplicity and elegance we needed a process in which s- and t-channel shared the same Regge trajectories. Therefore, instead of looking at pion–nucleon scattering, the three of us, together with Marco Ademollo, then visiting Harvard, picked up a more convenient process: $\pi\pi \to \pi\omega$ which allows Regge trajectories with the quantum numbers of the ρ-meson in all three channels. And the first results were quite encouraging....

I will skip the rest of that story: it led, eventually, to my 1968 paper describing that reaction in terms of an Euler Beta-function, marking the beginning of what became known as the "Dual Resonance Model." We shall therefore jump to the fall of 1968.

3. A Rare and Precious Supporter

After the Vienna Conference (September 1968) I moved to the brand new CTP (Center for Theoretical Physics) at MIT for my first (and, as it turned out, last!) postdoc. Sergio Fubini, who had joined MIT a year earlier and was the senior scientist responsible for that first job of mine, had been an enthusiastic supporter of the new development. However, most of the establishment did not share that enthusiasm: at best there was indifference towards the DRM, at worse there was loud criticism. There were reasons for that attitude, of course: the DRM was neither pure S-matrix theory a la Chew (the unitarity constraint, for instance, was notably absent) nor was it like any quantum field theory (QFT), starting with a classical Lagrangian and proceeding through a well defined set of quantization recipes. Thus, while a whole new young generation was excited by the novelty, most of the establishment was diffident about something they could hardly classify within their own know-how.

Just a handful of senior physicists were encouraging this new line of research. One was Gell-Mann, a friend and admirer of Fubini: he has been always supportive of the DRM (and of its later string-theory incarnation), albeit without getting much involved personally. Steven Weinberg (also at MIT at the time) was particularly interested in the relationships with Current Algebra. In general, the staff at MIT was rather sympathetic. Chew was also supportive of the DRM, probably seeing in it an explicit realization of his bootstrap dream. But three senior physicists stood out at the time for their full implication in the new adventure. One was, as I mentioned, Fubini; the second was Yoichiro Nambu, at University of Chicago, who did important work on the DRM (co-inventor together with Fubini, D. Gordon and myself of the harmonic operators formalism), and was later responsible for the very start of string theory; then, last but not least, was indeed (and not surprisingly given what I mentioned earlier) Stanley. It was very important and very reassuring for the young generation to see that a few great theorists of the time were supportive of the new adventure. After all they were planning to spend on it some very crucial years of their career!

Since the early fall of 1968, Stanley and Korkut Bardakci started working on the DRM. As it turned out, they picked up, without knowing I guess, the same topic that Fubini and I had chosen to work on. Leaving that interesting phase to the next section, I will only mention an episode that came very soon after my arrival at MIT.

One day Fubini and/or I received a draft (or a preprint?) by Stanley where he was referring to my 1968 formula as the "Suzuki–Veneziano model." Fubini, quite surprised, took no time before calling up Stanley in order to find the reason for the double attribution. Stanley apparently told Fubini that M. Suzuki had shown him, on the blackboard, a formula very similar (or identical?) to mine. But he had not written (a fortiori published) anything on it. Fubini objected that this was hardly sufficient for Suzuki to get credit for the discovery and, as far as he reported to me, Stanley immediately agreed and never again made any attempt to share the credit. This shows well how fair and honest a person Stanley was. Another example will be given below.

4. A Fair Strong Competitor

Sergio and Stanley must have felt the same about the almost magic properties of my model. Nice, yes, but at what price? Sergio told me very explicitly: by his own experience it was almost impossible for the model to have all its nice properties without being affected by a very subtle, hidden problem: the spectrum of states implied by the model might contain negative norm states, dubbed ghosts. Ghosts would make the model inconsistent by predicting certain processes to proceed with negative probability. But how to find out? My amplitude, by itself, did not provide a sufficiently fine check of whether ghosts would be present. One needed formulae for a more complete set of scattering processes. Fortunately, by the late fall of 1968, several groups (Bardakci and Ruegg, Virasoro, Goebel and Sakita, Chan and Tsou, Koba and Nielsen) had produced the necessary extensions of my amplitude. Indeed the full set of those amplitudes defined the DRM.

At this point finding the exact state-content of DRMs was just matter of carrying out the right calculations. Fubini and I started doing that and, by end of 1968, we had a first glance at the outcome. One feature of the spectrum was unexpected and very interesting: the number of states was growing like an exponential of their mass, i.e. much faster than anyone (or at least Fubini and myself) could have guessed a priori. Amusingly, a similar spectrum had been postulated by Rolf Hagedorn about four years earlier using a "statistical bootstrap" model not so dissimilar, in spirit, from Chew's "Nuclear Democracy." That was an important discovery, we thought.

Unfortunately, however, our results also confirmed Fubini's original fear: the model did have ghosts. Before deciding to come out with this partly interesting, partly unwelcome conclusions we wanted to check whether a "ghost-decoupling" (or "ghost-killing") mechanism could be at work. We found that, indeed, a certain number of ghosts could be eliminated thanks to a mechanism similar to the one that saves QED from having ghosts, gauge invariance. However, eliminating all ghosts this way looked like an impossible dream. At this point we decided to publish what we had and to let the community draw its own conclusions.

While the paper was being written we received a letter (or a call) by Stanley saying that they had looked at the spectrum (had they heard about us working on the same problem?) and that they had found the number of states growing as a power of the mass. We replied that we were surprised since we had instead an exponentially growing spectrum. To which, a day or so after, Stanley in turn replied something like: "Sorry, you are absolutely right, I got mixed up. The spectrum is exponential, and actually Korkut knew this all the time." We had no doubt about his good faith and, since then, credit for the exponential (Hagedorn-like) spectrum of the DRM model (and of string theory) goes jointly to the two groups.

In the summer of 1969 I visited Berkeley and had the opportunity to discuss physics with many important theorists there, including Stanley, Chew, Bardakci, Halpern, and Stapp. With A. Schwimmer, then at SLAC, and L. Caneschi we found the so-called twist operator, but the road to eliminating all ghosts still looked somewhat blocked. Back at MIT, however, things started moving again with the discovery of an $SL(2,R)$ symmetry group responsible for the duality properties of the model, the partial elimination of ghosts, and the twisting of the duality diagrams. But the really good news came only months later when Virasoro made a crucial discovery: if the ρ Regge trajectory passed through $J=1$ at $t=0$ (implying unfortunately a massless ρ meson), the ghost-killing mechanism found by Fubini and myself could be enormously extended, making it now at least conceivable that all ghosts would be eliminated.

I will not go on to the next chapter of the story since Stanley was only marginally involved in it. But, for the curious outsider, the ghost-killing story went on for a number of years and culminated in 1972 with the proof (by P. Goddard and C. Thorn and, independently, by R. Brower) of a "No-Ghost" theorem. All ghosts could be eliminated albeit at two prices: the Regge trajectory had to satisfy, of course, Virasoro's condition and, on top, the dimensionality of space–time D had to be less than (or equal to) 26 (later, other constraints actually forced $D = 26$ for the bosonic string, and, even later, $D = 10$ for the superstring). This is in itself a beautiful story of physics-driven mathematics (Virasoro algebra, Vertex operators, DDF states, etc.), but lies outside the scope of this short article.

Incidentally, faithful to my original motivations, I became also quite interested in studying the high-energy behavior of DRM amplitudes and in comparing them with the predictions of a generalization of Regge–Chew–Mandelstam theory to multi-particle processes (later extended by A. Mueller to inclusive cross-sections). I still consider a pity that not so much effort has been going in this direction by the bulk of the string community which, instead, has concentrated its attention mainly on the low-energy (and essentially massless) spectra of different string theories. With the reinterpretation of string theory as a theory of quantum gravity, however, the high-energy behavior issue (and in particular the one associated with Regge–Chew–Mandelstam) has made an appreciable comeback.

5. A Superb Theorist, a Very Honest Man

I have only touched upon work by Stanley Mandelstam that somehow affected directly my research. But, as other authors in this collection of articles will recall, his contributions to theoretical physics are much more numerous and cover a much wider spectrum of subjects. Stanley gave fundamental contributions to both S-matrix and Quantum Field Theory.

I will only mention, in the first category, his light-cone approach to proving the finiteness of string theory loops. I believe that, to this date, this is the most powerful approach to proving the absence of ultraviolet infinities in string theory (as opposed to the infrared infinities which are basically those of QFT). Of course, it should be complemented by a proof that there is no Lorentz anomaly (this is how, historically, one found that string quantization, at tree level, requires special values for the intercept of the Regge trajectory and for the dimensionality of space–time).

Concerning Mandelstam's contributions to QFT it may suffice to recall his work about the equivalence of the Thirring and sine-Gordon two-dimensional models and his proof that the maximally supersymmetric gauge theory in four dimensions is scale invariant to all orders.

I have met many great physicists in my life, each one with his/her own scientific and human profiles. On the scientific side Stanley belonged to the category of deep, mathematically rigorous thinkers, but with physical relevance (rather mathematical beauty) representing the main motivation. On the human side, Stanley was very reserved and soft-spoken, certainly belonging more to the class of Paul Dirac than to the one of Dick Feynman.

Stanley was appreciated by his colleagues for his modest, kind, unassuming personality but, most of all, for his intellectual power and honesty. He will be long remembered.

Stanley Mandelstam and Me
and Life on the Light-cone

Lars Brink

Department of Physics, Chalmers University of Technology,
S-41296 Göteborg, Sweden
lars.brink@chalmers.se

Stanley Mandelstam has always been one of my heroes in physics. Here I will describe my meetings with Stanley over the years and how our respective work sometimes coincided.

1. Early Encounters with Stanley

When I started as a graduate student at Chalmers in the late 1960s we were still following the old tradition where your advisor told you to read the relevant books in the subject. Since we were only two hand-picked students that started there were no graduate courses, only books to read and then to have some oral exams on those books. I read whatever I could find in particle physics and it was clear to me that two names stuck out, Murray Gell-Mann and Stanley Mandelstam. There were, of course, many other famous names at the time that I learnt about, but somehow Murray and Stanley became my heroes in very different ways. It is my great privilege in life to later become very good friends with both. I have only collaborated with Murray. Stanley was a loner while Murray was an empire builder. However some of the work I have done which I am most proud of was also done by Stanley in his special way.

I came to CERN in the summer of 1971 as a fellow. I was extremely fortunate to meet John Schwarz and David Olive the first few weeks and they became my lifelong friends and collaborators over the years. I had a car and one evening we were going downtown to go to a cinema when we passed the bus that should go in to town. Stanley was just going to enter it when John saw him. He told me who was closest to call on him. I had never met him before and he represented a different generation and was a person I admired. I asked what I should say. Just say Stanley, John said. It was very difficult with my Swedish upbringing but I did it with very rosy cheeks. He then joined us and was just as nice and friendly as I had heard he should be. This was my first encounter with him. He then stayed for some days at CERN and joined us for lunches at the CERN restaurant not saying much, just sitting there.

2. Loops in Dual Models and Fermion Amplitudes

The first half of 1971 had seen tremendous progress in dual models as string theory was still called. Pierre Ramond had introduced supersymmetry and understood how to introduce fermions, André Neveu and John Schwarz had introduced a new dual model which was more realistic than the Veneziano Model and Charles Thorn had combined the ideas of Ramond, Neveu and Schwarz to formulate a dual model with N bosons and 2 fermions which became the Ramond Model. Claude Lovelace had also argued by studying loop amplitudes that the Veneziano model was only consistent in 26 dimensions of spacetime with an intercept of 1, i.e. when the lowest vector particle was massless, a result that not even he took too seriously at the time. Dual Models were supposed to describe pion physics with the vector particle being the massive ρ-particle.

However, in the beginning of 1972 Richard Brewer and Peter Goddard and Charles Thorn showed that the Veneziano Model was indeed ghost-free when $d = 26$ and the intercept is 1 and similarly the Neveu–Schwarz Model when $d = 10$. This was a very wanted result but even so of utmost importance. It did however have the side effect that the models could not really describe pion physics. Could one lower the intercept to 1/2 and the critical dimension to 4? This is still a good question. Another great insight was given by Jeffrey Goldstone, Peter Goddard, Claudio Rebbi and Charles Thorn, who quantised the Nambu–Gotō action for a relativistic string using the light-cone gauge and found that it could only be done again in 26 dimensions. They could then show that the spectrum agreed exactly with the spectrum of physical states in the Veneziano Model. For a description of all these results and references to all those work see Ref. 1.

During my first year I worked very hard with Holger Bech Nielsen on the question of a "correct" dual pion model. We devised various methods[2] which could indicate if such a model exists but we failed to find strong evidence for such a model and gave up the attempts, quite discouraged. In retrospect we should have interpreted our results such that the existing models are essentially unique. Some year later Werner Nahm[3] classified possible string theories and found no model which directly could be used for pion physics.

Instead I started to work with David Olive. David was one of the leaders in the field and extremely respected. He had a background in S-matrix theory. He was always interested in deeper problems not worrying too much about publishing quickly and often. We agreed that we should look at a seemingly very difficult problem. Could we construct correct one-loop amplitudes for the dual models? The ones constructed in the past were not unitary but more sketches what loop amplitudes could look like. Even so it had led to the discovery of critical dimensions. Let me sketch how we solved the problem in the Veneziano Model. Let me describe it with the notation of that time.

The model is described in the operator formalism by an infinite set of harmonic oscillators with commutation rules

$$[a_n{}^\mu, a^\dagger{}_m{}^\nu] = -\eta^{\mu\nu}\delta_{mn}. \tag{1}$$

The fundamental constructs to use are the Fubini–Veneziano fields which are extensions of the coordinate q^μ and momentum p^μ of a point particle

$$Q^\mu(z) = q^\mu - ip^\mu \ln z - \frac{i}{\sqrt{n}} \sum_{n=1}^{\infty} \left(a^\dagger{}_m{}^\nu z^n - a_n{}^\mu z^{-n}\right), \tag{2}$$

$$P\mu(z) = iz\frac{dQ^\mu(z)}{dz}, \tag{3}$$

where z is a complex variable that for an open string can be taken to run along the real axis. We can now introduce a vertex operator

$$V(k, z) =\ :e^{-ik\cdot Q(z)}:\ , \tag{4}$$

where : : means normal ordering.

We also need to introduce the Virasoro operators

$$L_n = -\frac{1}{2}\oint \frac{dz}{2\pi i z} z^n :P^2(z):\ , \tag{5}$$

where the contour encircles the origin. They satisfy the algebra

$$[L_n, L_m] = (n-m)L_{m+n} + \frac{d}{12}(n^3 - n)\delta_{m+n\cdot 0}. \tag{6}$$

The commutator of L_n with the vertex operator is found to be

$$[L_n, V(k, z)] = z^n\left(z\frac{d}{dz} - n\right)V(k, z), \tag{7}$$

which allows us to solve the z-dependence of the vertex operator as

$$V(k, z) = z^{L_0} V(k, 1) z^{-L_0}. \tag{8}$$

A typical scattering amplitudes for N scalars with momenta $k_i{}^\mu$ can now be written as a vacuum expectation value of a string of vertices and propagators like ($V(k) \equiv V(k, 1)$),

$$A_N = \langle 0 | e^{-ik_1\cdot q} V(k_2) \frac{1}{L_0 - 1} V(k_2) \frac{1}{L_0 - 1} \cdots \frac{1}{L_0 - 1} V(k_{N-1}) e^{-ik_N\cdot q} | 0 \rangle. \tag{9}$$

A physical state is defined by

$$L_n|\text{phys}\rangle = 0, \tag{10}$$

$$(L_0 - 1)|\text{phys}\rangle = 0. \tag{11}$$

It is straightforward to factorise the amplitude (9) and see that only physical states couple at the poles even though the whole spectrum of states contains an infinite set of negative norm states (see (1)).

The idea of Lovelace and others to construct one-loop amplitudes was then simply to sew together the amplitudes above taking a trace instead of a vacuum expectation value. This would clearly allow also the ghosts to travel around the loop and contribute. It is obvious that that the loop is not unitary. Even so Lovelace got his spectacular result by a clever guess. We then set out to seek a unitary result. This was at a time when gauge field theory was rediscovered and it was obvious that similar problems occur when constructing loop amplitudes for such theories. The problem was solved by Richard Feynman in 1963[4] and a few years later put on a precise mathematical ground by Ludwig Faddeev and Victor Popov.[5] Their result was that in order to find a unitary loop amplitude one has to introduce scalar ghosts, most often called Faddeev–Popov ghosts even though also Feynman mentioned them.

Our idea was first to try to find corresponding ghost fields for the dual amplitudes and we struggled for months to find them. In the end we gave up and went back to Feynman's work. This was a talk in a conference and he only mentioned it in some words as a reply to a question by Bryce DeWitt. We first reconstructed his argument and then applied it to the dual amplitudes.

One key point in Feynman's approach was to use the relation among propagators,

$$\frac{1}{k^2 - m^2 - i\epsilon k^0} = \frac{1}{k^2 + m^2 - i\epsilon} + 2\pi i \theta(k^0)\delta(k^2 - m^2), \quad (12)$$

i.e. a retarded propagator = a Feynman propagator + a cut amplitude put on shell. Hence if we construct a one-loop amplitude with only retarded propagators which must be zero it will equal a sum of loops where one is a correct one and the others tree amplitudes. Hence we can sum tree diagrams to get a loop amplitude and we are guaranteed to get the correct analytic behaviour. This is Feynman's tree theorem. The remaining point is to secure that only physical states flow through the loop when we cut a line. This can be achieved by introducing a physical state operator.

A complete set of physical state operators had been found by Emilio Del Guidice. Paolo Di Vecchio and Sergio Fubini,[6]

$$A_n^{i\dagger} = A_n^i = \oint \frac{dz}{2\pi i z} \epsilon^i \cdot P(z) e^{-ink \cdot Q(z)} \quad (13)$$

with ϵ^i a polarisation vector and k^μ a vector orthogonal to the vector p^μ which is the momentum of the state that the operator acts upon. It is clear that this is not a covariant description. Even so the operators satisfy

$$[A_n^i, A_m^{i\dagger}] = n\delta^{ij}\delta_{nm}, \quad (14)$$

which shows that these operators create a set of orthogonal physical states.

We can now construct a projection operator from the full Fock space to the space of transverse states. This was done in our first paper,[7]

$$\mathcal{T}(k) = \oint \frac{dy}{2\pi i y} y^{\mathcal{L}_0 - H}, \quad (15)$$

where the integration contour encircles the origin and

$$\mathcal{L}_0 = \sum_{n=1}^{\infty} \sum_{i=1}^{d-2} A_n^{i\dagger}(k) A_n^{i}(k) \tag{16}$$

and

$$H = \sum_{n=1}^{\infty} \sum_{\mu=1}^{d} a_n^{\mu\dagger} a_n^{\mu}. \tag{17}$$

An explicit calculation in $d = 26$ gives

$$\mathcal{L}_0 - H = (D_0 - 1)(L_0 - 1) + \sum_{n=1}^{\infty} (D_{-n} L_n + L_{-n} D_n) \tag{18}$$

with D_n a new set of operators.

It is immediately clear that this projection operator inserted into a tree graph gives just 1 proving that only physical states couple in a tree amplitude.

We used so the projection operator in a loop amplitude constructed using Feynman's tree theorem and could compute the effect of it to be that the original integral measure $d^{26}p$ becomes indeed $d^{24}p$.[8]

It was quite straightforward to perform the same analysis for the Neveu–Schwarz model and also for the closed string sectors. In the latter case it allowed us to show that the interaction vertex between open and a closed string also respected the gauge invariance.[9]

In order to also discuss the gauge properties of the Ramond model we also needed the projection operator onto the fermonic states which was constructed by Edward Corrigan and Peter Goddard.[10] With this operator we could then check that the Ramond model and the Neveu–Schwarz model were unitarily connected[11] and finally construct the four-fermion amplitude[12] although in a form which left an infinite determinant to be computed. It came about since we had to use operator product expansions which led to some infinite sums.

3. Stanley Mandelstam's Computations of Loops in Dual Models and Fermion Amplitudes

After our outburst of papers we organised a workshop at CERN in September of 1973. Only John Schwarz could come from the USA. He brought with him a computer calculation he had performed with a student of the infinite determinant in the four-fermion amplitude. However, that result was overshadowed by the results of three papers that Mandelstam had sent us to the workshop.[13–15] Stanley had very much wanted to come but for some reason he could not. After the groundbreaking paper by Goddard, Goldstone, Rebbi and Thorn it was clear that dual models were scattering amplitudes between states of a relativistic string. They had emphasised the use of the light-cone frame formulation and Mandelstam took that ad notam. He set up a functional integral over the transverse string coordinates $x^i(\sigma, \tau)$ defined on a rectangular domain $0 < \sigma < p^+ = (p^0 + p^z)/\sqrt{2}$, $0 < \tau = i(t+z)/\sqrt{2} < T$, with (imaginary time) action given by

$$S_{\text{l.c.}} = \frac{1}{2} \int_0^T d\tau \int_0^{P^+} d\sigma \left[\left(\frac{\partial x^i}{\partial \tau}\right)^2 + T_0^2 \left(\frac{\partial x^i}{\partial \sigma}\right)^2 \right]$$

with T_0, the string tension. By scaling σ as above he could map any diagram to a strip where the joining and splitting of strings are represented by horizontal lines in the strip. For a thorough description of Mandelstam's way of solving this problem see the contribution by Charles Thorn in this volume, Sec. 6.

He used some ingenious methods to compute the three-string vertex which had to enter into the functional integration and the calculations could only be done up to and including the one-loop graphs. (We faced the same problem.) The one-loop scattering amplitudes matched of course exactly the results we had obtained.

He could then use the same kind of setup to construct the corresponding amplitudes in the Ramond–Neveu–Schwarz Model. In particular he could deal quite simply with fermion-emission vertices since in the light-cone frame there is no problem with the gauge properties. In the covariant operator formulation it was the implementation of the physical state projection operator in the meson propagator that led to an infinite determinant since the gauge conditions bounced back and forth in the propagator of the four-fermion amplitude. For Mandelstam it was a straightforward calculation albeit with the use of some ingenious tricks to find the result that we had learnt just at the workshop with the computer calculation by John Schwarz and his student.

We never managed to compute higher loop amplitudes within the operator formalism. Feynman's tree theorem is very much streamlined for one-loop calculations. Mandelstam could eventually continue his scheme to arbitrary loop order which is described in the contributions by Charles Thorn and Nathan Berkovits. In this way he could investigate whether the string perturbation theories are finite to all order.

In 1976 we finally managed to find the action for the RNS-string.[16,17] We did it by using a σ-model formulation where the string is coupled to two-dimensional supergravity to implement reparametrization and local supersymmetry invariance. In 1981 Alexander Polyakov[18] used this action to formulate the functional integral for the Superstring. That became an alternative to compute higher loop order. However, as Mandelstam often pointed out it was only when it was shown to be equivalent to the obviously unitary formulation of his that one could fully trust the results using the covariant action.

4. Maximally Supersymmetric Yang–Mills Theories

In the fall of 1976 I came to Caltech to work with John Schwarz. Knowing that the zero-slope limit of the open superstring theory in $d = 10$ is the maximally supersymmetric Yang–Mills theory, we realised that this theory should be the mother of all supersymmetric Yang–Mills theories either by dimensional reduction or by that together with a truncation. Hence we set out to construct the ten-dimensional theory.[19] It is easy to write the action for it since it is really unique,

$$S = \int d^{10}x \left\{ -\frac{1}{4} F^a_{\mu\nu} F^{a\mu\nu} + i\bar{\lambda}^a \gamma \cdot D\lambda^a \right\} \tag{19}$$

with the usual definitions for the field strength and the covariant derivative.

In order for this to be supersymmetric the spinor must be both Majorana and Weyl which is possible in $d = 10$. In fact this form works in three, four, six and ten dimensions with the spinors properly chosen for the $N = 1$ theories. The supersymmetry transformations are easy to construct and also looks the same in all these dimensions.

Now the road was open to construct all possible supersymmetric Yang–Mills theories which we did. Here I only describe the maximally supersymmetric theory in $d = 4$ which is obtained by a straight dimensional reduction. We put x^4, $x^5, \ldots, x^9 = 0$. The 16-dimensional spinor is divided up into four Weyl spinors and some γ-gymnastics has to be performed leading to the following action (I write it as in the original paper.) The final action is then

$$S = \int d^4x \left\{ -\frac{1}{4} F^a_{\mu\nu} F^{a\mu\nu} + i\bar{\chi}^a \gamma \cdot DL\chi^a + \frac{1}{2} D_\mu \Phi^a_{ij} D\mu \Phi^a_{ij} \right.$$
$$\left. - \frac{i}{2} g f^{abc} \left(\bar{\chi}^{ai} L \chi^{jb} \Phi^c_{ij} - \bar{\chi}^a_j R \tilde{\chi}^b_j \Phi^{ijc} \right) - \frac{1}{4} f^2 f^{abc} g^{ade} \Phi^b_{ij} \Phi^c_{kl} \Phi^{ijd} \Phi^{kle} \right\}. \tag{20}$$

This is the "$N = 4$ theory" with an $SU(4)$ symmetry. If we had used Majorana spinors it would have been an $SO(4)$ symmetry. Note that it is quite hard to find the four-dimensional action from scratch. We were very much helped by our knowledge of higher dimensions. However, in most circles at this time it was not appropriate to talk about higher dimensions, so we were very careful to say that this was just a means to find the four-dimensional action.

Similar ideas were pursued at the same time by Gliozzi, Scherk and Olive[20] who concentrated on the ten-dimensional string to make it consistent and as a by-product they also got the supersymmetric Yang–Mills theories of above. When comparing notes with Joël Scherk we realised that we had the same models and hence we wrote our paper together with him.

When supersymmetric theories came around it was noticed that the quantum properties got improved. The issue rose if such a theory could be perturbatively finite but it looked implausible.

After having constructed the $N = 4$ theory we turned to other problems but in the summer of 1977 Murray Gell-Mann heard that the β-function for this theory is zero. He then prophetically declared that it is probably zero to all orders. He did not commit himself to write it in any report keeping up his promise to himself never to print anything that could be wrong. However, his comments were taken ad notam and a few months later Poggio and Pendleton[21] could report that the two-loop contribution to the β-function is indeed zero. Now there was a race to compute the three-loop contribution and three years later Caswell and Zanon[22] and Grisaru, Rocek and Siegel[23] could indeed confirm that it is zero. These were formidable calculations in Feynman diagrams. In the first case they had to consider some 600 diagrams while in the second case they worked with super-Feynman diagrams and had to consider 53 such diagrams. It became then clear that other techniques were needed for a general proof.

4.1. *The light-cone gauge formulation of $N = 4$ Yang–Mills theory*

Around 1980 and the years after I was busy with Michael Green and John Schwarz to set up the Superstring Theory and to check its physical properties. We did it mostly in a light-cone formulation, i.e. we were only using the dynamical degrees of freedom in a non-covariant way. In 1981 we asked what happens when we take the zero-slope limit of the one-loop graphs that we had constructed.[24] Both for the closed string that leads to maximal supergravity and for the open string where we could isolate the maximally supersymmetric Yang–Mills Theory we could check the one-loop graph in various dimensions and see how finiteness could come about. The results were remarkably simple. Both in the supergravity case and the Yang–Mills case the complete one-loop graph for a four-particle scattering is just the box diagram of a ϕ^3 theory with kinematical factors taking care of the spin of the particles appearing in the loop graph.

This gave me the idea to check the light-cone gauge formulation of the $N = 4$ theory more explicitly and I did it with my collaborators Olof Lindgren and Bengt Nilsson.[25] Since the formalism we used is still not too well known I will describe the paper rather carefully. We did it by starting with the action (19). We then chose the gauge $A^+ = \frac{1}{\sqrt{2}}(A^0 + A^3) = 0$ and solved for the kinematical field $A^- = \frac{1}{\sqrt{2}}(A^0 - A^3)$ leaving us with only the transverse degrees of freedom. (We used x^+, see below, as the evolution coordinate.) Similarly for the spinor field we used the decomposition

$$\lambda = \frac{1}{2}(\gamma_+\gamma_- + \gamma_-\gamma_+)\lambda = \lambda_+ + \lambda_-, \qquad (21)$$

where again we could solve for λ_- since again it satisfies a kinematical equation of motion. (We did it in a path integral formulation and integrated out the non-dynamical fields). We could then rewrite the action and dimensionally reduce it to $d = 4$. Finally we could introduce a superspace and rewrite the action in that space. An alternative way which we have used in many cases[26] later is to simply look for a representation of the super-Poincaré algebra. The final result is as follows which I now describe.

With the space–time metric $(-, +, +, \ldots, +)$, the light-cone coordinates and their derivatives are

$$x^{\pm} = \frac{1}{\sqrt{2}}(x^0 \pm x^3); \quad \partial^{\pm} = \frac{1}{\sqrt{2}}(-\partial_0 \pm \partial_3); \tag{22}$$

$$x = \frac{1}{\sqrt{2}}(x_1 + ix_2); \quad \bar{\partial} = \frac{1}{\sqrt{2}}(\partial_1 - i\partial_2); \tag{23}$$

$$\bar{x} = \frac{1}{\sqrt{2}}(x_1 - ix_2); \quad \partial = \frac{1}{\sqrt{2}}(\partial_1 + i\partial_2), \tag{24}$$

so that

$$\partial^+ x^- = \partial^- x^+ = -1; \quad \bar{\partial}x = \partial\bar{x} = +1. \tag{25}$$

In four dimensions, any massless particle can be described by a complex field, and its complex conjugate of opposite helicity, the $SO(2)$ coming from the little group decomposition

$$SO(8) \supset SO(2) \times SO(6). \tag{26}$$

Particles with no helicity are described by real fields. The eight vector fields in ten dimension reduce to

$$\mathbf{8}_v = \mathbf{6}_0 + \mathbf{1}_1 + \mathbf{1}_{-1}, \tag{27}$$

and the eight spinors to

$$\mathbf{8}_s = \mathbf{4}_{1/2} + \bar{\mathbf{4}}_{-1/2}. \tag{28}$$

The representations on the right-hand side belong to $SO(6) \sim SU(4)$, with subscripts denoting the helicity: there are six scalar fields, two vector fields, four spinor fields and their conjugates. To describe them in a compact notation, we introduce anticommuting Grassmann variables θ^m and $\bar{\theta}_m$,

$$\{\theta^m, \theta^n\} = \{\bar{\theta}_m, \bar{\theta}_n\} = \{\bar{\theta}_m, \theta^n\} = 0, \tag{29}$$

which transform as the spinor representations of $SO(6) \sim SU(4)$,

$$\theta^m \sim \mathbf{4}_{1/2}; \quad \bar{\theta}^m \sim \bar{\mathbf{4}}_{-1/2}, \tag{30}$$

where $m, n, p, q, \ldots = 1, 2, 3, 4$, denote $SU(4)$ spinor indices. Their derivatives are written as

All the physical degrees of freedom can be captured in one complex superfield

$$\phi(y) = \frac{1}{\partial^+}A(y) + \frac{i}{\sqrt{2}}\theta^m\theta^n \bar{C}_{mn}(y) + \frac{1}{12}\theta^m\theta^n\theta^p\theta^q \epsilon_{mnpq}\partial^+ \bar{A}(y)$$

$$+ \frac{i}{\partial^+}\theta^m \bar{\chi}_m(y) + \frac{\sqrt{2}}{6}\theta^m\theta^n\theta^p \epsilon_{mnpq}\chi^q(y). \tag{31}$$

In this notation, the eight original gauge fields A_i, $i = 1, \ldots, 8$ appear as

$$A = \frac{1}{\sqrt{2}}(A_1 + iA_2), \quad \bar{A} = \frac{1}{\sqrt{2}}(A_1 - iA_2), \tag{32}$$

while the six scalar fields are written as antisymmetric $SU(4)$ bi-spinors

$$C^{m4} = \frac{1}{\sqrt{2}}(A_{m+3} + iA_{m+6}), \quad \bar{C}^{m4} = \frac{1}{\sqrt{2}}(A_{m+3} - iA_{m+6}), \tag{33}$$

for $m \neq 4$; complex conjugation is akin to duality,

$$\bar{C}_{mn} = \frac{1}{2}\epsilon_{mnpq}C^{pq}. \tag{34}$$

The fermion fields are denoted by χ^m and $\bar{\chi}_m$. All have adjoint indices (not shown here), and are local fields in the modified light-cone coordinates

$$y = \left(x, \bar{x}, x^+, y^- \equiv x^- - \frac{i}{\sqrt{2}}\theta^m \bar{\theta}_m\right). \tag{35}$$

This particular light-cone formulation we call LC_2 since all the unphysical degrees of freedom have been integrated out, leaving only the physical ones.

Introduce the chiral derivatives,

$$d^m = -\partial^m - \frac{i}{\sqrt{2}}\bar{\theta}^m \partial^+; \quad \bar{d}_n = \bar{\partial}_n + \frac{i}{\sqrt{2}}\theta_n \partial^+, \tag{36}$$

which satisfy the anticommutation relations

$$\{d^m, \bar{d}_n\} = -i\sqrt{2}\delta^m{}_n \partial^+. \tag{37}$$

One verifies that ϕ and its complex conjugate $\bar{\phi}$ satisfy the chiral constraints

$$d^m \phi = 0; \quad \bar{d}_m \bar{\phi} = 0, \tag{38}$$

as well as the "inside-out" constraints

$$\bar{d}_m \bar{d}_n \phi = \frac{1}{2}\epsilon_{mnpq} d^p d^q \bar{\phi}, \tag{39}$$

$$d^m d^n \bar{\phi} = \frac{1}{2}\epsilon^{mnpq} \bar{d}_p \bar{d}_q \phi. \tag{40}$$

The Yang–Mills action is then simply

$$\int d^4x \int d^4\theta \, d^4\bar{\theta} \, \mathcal{L}, \tag{41}$$

where

$$\mathcal{L} = -\bar{\phi}\frac{\Box}{\partial^{+2}}\phi + \frac{4g}{3}f^{abc}\left(\frac{1}{\partial^+}\bar{\phi}^a \phi^b \bar{\partial}\phi^c + \text{complex conjugate}\right)$$

$$- g^2 f^{abc} f^{ade}\left(\frac{1}{\partial^+}(\phi^b \partial^+ \phi^c)\frac{1}{\partial^+}(\bar{\phi}^d \partial^+ \bar{\phi}^e) + \frac{1}{2}\phi^b \bar{\phi}^c \phi^d \bar{\phi}^e\right). \tag{42}$$

Grassmann integration is normalised so that $\int d^4\theta \, \theta^1 \theta^2 \theta^3 \theta^4 = 1$, and f^{abc} are the structure functions of the Lie algebra.

4.2. The perturbative finiteness $N = 4$ Yang–Mills theory

After having obtained the light-cone formulation of this theory we set out to check its perturbation expansion to see if it could be UV finite.[27] There are well-defined techniques to find super-Feynman rules and to construct supergraphs. One direct difficulty though is that the superfield satisfies different constraints (29), (30) and (31). This means that the functional derivatives used when computing the Feynman rules become a bit intricate.

$$\frac{\delta \phi^a(y,\theta)}{\delta \phi^b(y',\theta')} = \frac{1}{4!2} d^4 \delta^4(x-x') \delta^4(\theta-\theta') \delta^4(\bar{\theta}-\bar{\theta}') \delta_b^a, \qquad (43)$$

$$\frac{\delta \bar{\phi}^a(y,\theta)}{\delta \phi^b(y',\theta')} = 12 \frac{1}{4!^4} \frac{\bar{d}^4 d^4}{\partial^{+4}} \delta^4(x-x') \delta^4(\theta-\theta') \delta^4(\bar{\theta}-\bar{\theta}') \delta_b^a. \qquad (44)$$

With this knowledge one can derive the expressions for the propagator, the three-point and four-point vertices and build up super-Feynman diagrams. The explicit form can be seen in the paper. To estimate the naive dimension of a diagram one is helped by the fact that the δ-functions and the θ-integrals appearing in the propagator and in the vertex functions are not to be taken into account in computing the dimensionality. This fact is due to the property of supergraphs that they can always be reduced to a local expression in θ. One can now check that the naive dimension of any diagram is zero. To prove finiteness one has to show that for any diagram one can integrate out some momenta. In the paper we consider a general diagram and extract either a three-point vertex or a four-point one and show that for all contributions from them one can perform this extraction. Again I refer to the paper for the details.

The final key point is to show that one can make a Wick rotation to implement Weinberg's theorem.[28] The obstacles here are the poles in ∂^+. When we derived the formalism above we integrated out a determinant in ∂^+. This means the the remaining freedom we have is in the choice of the exact form of the poles in ∂^+. By making the choice that we interpret the pole as $(p^+ + i\epsilon p^-)^{-1}$ we indeed can make the Wick rotation and use Weinberg's theorem to complete the proof that the perturbation expansion is finite.

At the same time as we were doing this analysis Mandelstam[29] gave a similar proof using a slightly different light-cone formulation. I will compare the two formalisms in the next section.

There were also other arguments put forward for finiteness around this time. Sohnius and West[30] considered anomaly multiplets, and concluded that the $N = 4$ Yang–Mills theory is conformally invariant, if one can assume that the theory is supersymmetric and $O(4)$-invariant and that the structure of anomalies is given by the breakdown of conformal invariance in its coupling to supergravity. These have later been confirmed to be correct assumptions.

In the Soviet Union the group at ITEP[31] attacked the problem by studying instanton calculus and could also argue that the β-function should be zero.

The proofs by us and by Mandelstam depended on a non-linear realisation of supersymmetry. It was also very important to find a proof within the covariant formulation of supersymmetry. This was found by Howe, Stelle and Townsend.[32]

It became now an established fact that the $N = 4$ theory is indeed a perturbatively finite quantum field theory. For quite some time there was a discussion if this meant that the theory is trivial. For us with a superstring background this was obviously not true since it is the zero-slope limit of the open string theory and as such is an integral part of a theory construction that was getting more and more established as the correct framework for a unified theory of all the interactions. It was amply shown by Sen[33] how beautiful and intricate the structure of the $N = 4$ theory is when he showed the full structure of the dyons and monopoles in the theory.

When we had proved the finiteness I went into Murray Gell-Mann's office and told him that we now had proven his conjecture. He then replied that you cannot have a field theory without a scale. This was indeed an objection to be taken seriously. In the string theory there is a scale, the slope α', and the Yang–Mills particles couple to other massive particles. However, the crucial observation was made by Maldacena[34] when he suggested that the $N = 4$ Yang–Mills theory is dual to the superstring theory in the sense that a strong-coupling limit of one of them corresponds to the weak-coupling limit of the other. This has been very carefully studied since then and verified in all attempts. We should not think about the $N = 4$ Yang–Mills theory as just an ordinary quantum field theory with no dimensionful coupling but as a string theory in disguise and where conformal dimensions instead play an important role. Finally the $N = 4$ Yang–Mills theory had found its place in fundamental physics.

5. Stanley Mandelstam's Proof of the Perturbative Finiteness of the $N = 4$ Yang–Mills Theory

When we worked on the $N = 4$ theory we were quite nervous that we would overlook some complications in the proof. We then first published a paper with the light-cone action and deferred the study of the perturbation expansion to a second paper. We thought we had all the time in the world to do a very precise job, and we were very surprised when Stanley's paper reached us. He did in the paper essentially what we had also done and with his experience he could publish it immediately once he had the result while we sat on it for some time to be sure of the result.

In his paper he goes straight to four dimensions and he uses a slightly different superfield from us. This was a superfield we had used when we discussed the Superstring Theory. Consider the the superfield (31). If we Fourier transform the whole of y^- to p^+ we get a superfield completely in terms of θ with no $\bar{\theta}$. We need then only integrate over θ and the chirality constraint need not be implemented. We followed the more conventional road used for covariant chiral superfields. We found that to be more useful in many subsequent calculations but Stanley was only interested in the general structure of the perturbation expansion and for that the two formalisms were more or less identical. With his vast knowledge he could convince himself quicker than we and his proof is hence more sketchy than ours. However, in the end they amount to the same result.

6. More Personal Encounters with Stanley Mandelstam

In 1973 I had got a research position back home in Sweden. That was challenged by someone claiming that I was overqualified for the job. In a stroke that could only happen in Sweden the Government went on his line and I was stripped of the job which then went to this person who was three years older than me and definitely not overqualified. My situation was then rather serious and my professor, my former advisor, asked me if some famous person could write a supporting letter for me when he wrote to the research council about my situation. I said that perhaps Stanley Mandelstam could do it and I wrote to ask him. Very quickly I got a long letter back where he described essentially all my papers in great details (perhaps understanding them better than me) and I am sure that it was the letter that helped me get a temporary job so that I could survive.

In 1977 I gave a talk at Berkeley about our work on supergravity. At some stage I said that $N = 8$ supergravity might be a finite quantum theory since it is the ultimate supergravity theory. Quickly Stanley raised his hand and said: "Do you really believe that?" We still do not know the answer to that question but it was clear what Stanley thought.

After that we met quite often and Stanley was always nice, friendly and a bit reserved. Some time in the mid 1980s Abdus Salam invited the two of us for some three days to ICTP to have full days of brainstorming sessions. We spent probably eight hours every day in Salam's office and we told him everything we knew about Superstring Theory and supersymmetric field theories. Salam collected the relevant papers and went to his house in the evening to read them and next day he came back with new questions. Stanley and I spent three evenings in Trieste a bit exhausted. It was a bit difficult to keep up a conversation then.

In 1986 we arranged a Nobel symposium and in the aftermath of the Superstring revolution it was a very heavy interest to be invited to the meeting. Stanley was a key invitee and he talked about the possible finiteness of the perturbation expansion. Having so many famous people in town I used it to have a thesis defence the week after and asked Stanley to act as the opponent at the dissertation. He accepted and again did a marvellous job. He gave an overview of the field as an introduction and that was superb. From Gothenburg he went on to South Africa to visit his sister and he was a bit nervous about the trip and the situation in the country where he had grown up. Perhaps this might have been his last trip there since his sister with family later settled in the USA.

The last time I met Stanley was when we celebrated his 80th birthday in Santa Barbara in 2009. He looked the same and was the same friendly person. He had aged of course but it was still a delight to meet him.

Stanley was one of the real strongmen and at the same time gentlemen of particle physics of the last century. I feel very honoured to have had him as a close friend.

References

1. J. H. Schwarz, *Superstrings. The First 15-Years of Superstring Theory*, Vol. 1 (World Scientific, Singapore, 1985).
2. L. Brink and H. B. Nielsen, Two mass relations for mesons from string–quark duality, *Nucl. Phys. B* **89**, 118 (1975).
3. W. Nahm, Mass spectra of dual strings, *Nucl. Phys. B* **114**, 174 (1976).
4. R. P. Feynman, Quantum theory of gravitation, *Acta Phys. Polon.* **24**, 697 (1963).
5. L. D. Faddeev and V. N. Popov, Feynman diagrams for the Yang–Mills field, *Phys. Lett. B* **25**, 29 (1967).
6. E. Del Giudice, P. Di Vecchia and S. Fubini, General properties of the dual resonance model, *Ann. Phys.* **70**, 378 (1972).
7. L. Brink and D. I. Olive, The physical state projection operator in dual resonance models for the critical dimension of space-time, *Nucl. Phys. B* **56**, 253 (1973).
8. L. Brink and D. I. Olive, Recalculation of the the unitary single planar dual loop in the critical dimension of space time, *Nucl. Phys. B* **58**, 237 (1973).
9. L. Brink, D. I. Olive and J. Scherk, The gauge properties of the dual model pomeron–reggeon vertex — their derivation and their consequences, *Nucl. Phys. B* **61**, 173 (1973).
10. E. Corrigan and P. Goddard, The absence of ghosts in the dual fermion model, *Nucl. Phys. B* **68**, 189 (1974).
11. L. Brink, D. I. Olive, C. Rebbi and J. Scherk, The missing gauge conditions for the dual fermion emission vertex and their consequences, *Phys. Lett. B* **45**, 379 (1973).
12. D. I. Olive and J. Scherk, Towards satisfactory scattering amplitudes for dual fermions, *Nucl. Phys. B* **64**, 334 (1973).
13. S. Mandelstam, Interacting string picture of dual resonance models, *Nucl. Phys. B* **64**, 205 (1973).
14. S. Mandelstam, Interacting string picture of the Neveu–Schwarz–Ramond model, *Nucl. Phys. B* **69**, 77 (1974).
15. S. Mandelstam, Manifestly dual formulation of the Ramond model, *Phys. Lett. B* **46**, 447 (1973).
16. L. Brink, P. Di Vecchia and P. S. Howe, A locally supersymmetric and reparametrization invariant action for the spinning string, *Phys. Lett. B* **65**, 471 (1976).
17. S. Deser and B. Zumino, A complete action for the spinning string, *Phys. Lett. B* **65**, 369 (1976).
18. A. M. Polyakov, Quantum geometry of fermionic strings, *Phys. Lett. B* **103**, 211 (1981).
19. L. Brink, J. H. Schwarz and J. Scherk, Supersymmetric Yang–Mills theories, *Nucl. Phys. B* **121**, 77 (1977).
20. F. Gliozzi, J. Scherk and D. I. Olive, Supersymmetry, supergravity theories and the dual spinor model, *Nucl. Phys. B* **122**, 253 (1977).
21. E. C. Poggio and H. N. Pendleton, Vanishing of charge renormalization and anomalies in a supersymmetric gauge theory, *Phys. Lett. B* **72**, 200 (1977).
22. W. E. Caswell and D. Zanon, Zero three loop beta function in the $N = 4$ supersymmetric Yang–Mills theory, *Nucl. Phys. B* **182**, 125 (1981).

23. M. T. Grisaru, M. Rocek and W. Siegel, Zero three loop beta function in $N = 4$ super-Yang–Mills theory, *Phys. Rev. Lett.* **45**, 1063 (1980).
24. M. B. Green, J. H. Schwarz and L. Brink, $N = 4$ Yang–Mills and $N = 8$ supergravity as limits of string theories, *Nucl. Phys. B* **198**, 474 (1982).
25. L. Brink, O. Lindgren and B. E. W. Nilsson, $N = 4$ Yang–Mills theory on the light cone, *Nucl. Phys. B* **212**, 401 (1983).
26. A. K. H. Bengtsson, I. Bengtsson and L. Brink, Cubic interaction terms for arbitrarily extended supermultiplets, *Nucl. Phys. B* **227**, 41 (1983).
27. L. Brink, O. Lindgren and B. E. W. Nilsson, The ultraviolet finiteness of the $N = 4$ Yang–Mills theory, *Phys. Lett. B* **123**, 323 (1983).
28. S. Weinberg, High-energy behavior in quantum field theory, *Phys. Rev.* **118**, 838 (1960).
29. S. Mandelstam, Light cone superspace and the ultraviolet finiteness of the $N = 4$ model, *Nucl. Phys. B* **213**, 149 (1983).
30. M. F. Sohnius and P. C. West, Conformal invariance in $N = 4$ supersymmetric Yang–Mills theory, *Phys. Lett. B* **100**, 245 (1981).
31. V. A. Novikov, M. A. Shifman, A. I. Vainshtein and V. I. Zakharov, Exact Gell-Mann–Low function of supersymmetric Yang–Mills theories from instanton calculus, *Nucl. Phys. B* **229**, 381 (1983).
32. P. S. Howe, K. S. Stelle and P. K. Townsend, Miraculous ultraviolet cancellations in supersymmetry made manifest, *Nucl. Phys. B* **236**, 125 (1984).
33. A. Sen, Dyon-monopole bound states, selfdual harmonic forms on the multi-monopole moduli space, and $SL(2, Z)$ invariance in string theory, *Phys. Lett. B* **329**, 217 (1994), arXiv:hep-th/9402032.
34. J. M. Maldacena, The large N limit of superconformal field theories and supergravity, *Int. J. Theor. Phys.* **38**, 1113 (1999) [*Adv. Theor. Math. Phys.* **2**, 231 (1998)].

Reminiscences of Stanley Mandelstam

John H. Schwarz

Walter Burke Institute for Theoretical Physics,
California Institute of Technology, 452-48 Pasadena, CA 91125, USA
jhs@theory.caltech.edu

> I reminisce about my interactions with Stanley Mandelstam during my years as a graduate student at UC Berkeley (1962–66) and afterwards. His contributions to S-matrix theory, quantum field theory, and string theory are also discussed.

During his UC Berkeley postdoctoral period in 1959–61 Stanley Mandelstam collaborated with Geoffrey Chew on his "nuclear democracy" program. Following a brief interlude in 1962 at Birmingham University, where he had received his PhD in 1956, Stanley rejoined Berkeley as a professor in 1963. His discovery of the "Mandelstam representation"[1] was one of his key achievements that led to his 1962 election as a Fellow of the Royal Society as well as his appointment as a tenured professor of physics in Berkeley the following year.

This was an exciting time to be in Berkeley, both scientifically and politically. On the scientific side, new hadrons were being discovered almost every day at the Berkeley Bevatron, as well at CERN and Brookhaven, and there was an obvious need to make sense of this emerging zoo of new particles. Berkeley's junior faculty in theoretical physics at that time included Shelly Glashow and Steven Weinberg. On the political side, the "Free Speech Movement" emerged in Berkeley in 1965. It caused considerable turmoil, with far reaching implications. I'm pretty sure that Stanley was in sympathy with the goals of this movement, but he chose to focus on science. This was understandable, since he was quite shy and nonassertive. Stanley was deeply committed to his research and let his scientific achievements speak for themselves.

The goal of the nuclear democracy program was to derive the hadronic S matrix from very general considerations such as analyticity and unitarity. Another ingredient was the bootstrap conjecture, which was closely related to what was called "duality" in those days. I was a graduate student in Berkeley in the heyday of this activity (1962–66). My thesis advisor was Geoffrey Chew, but Stanley Mandelstam also played an important role in my education. Geoff and Stanley both influenced my scientific development profoundly.

Geoff was very skeptical about the possibility of quantum field theory describing the strong interactions. The underlying reason for this opinion was the realization that since the force is strong any coupling constants are of order unity, and therefore perturbation expansions could not be helpful. Of course, today we know that the strong nuclear force is described by QCD, which is a quantum field theory. In retrospect, it is clear that the dismissal of quantum field theory was mistaken. Nonetheless, a lot of good came out of this philosophy. Most importantly, it led to the discovery of string theory and supersymmetry. So I have no regrets about being a disciple of the Chew–Mandelstam S-matrix program. In fact, I feel that I was in the right place at the right time. String theory is mostly used these days to describe gravity and unification. However, it still seems likely that there is a string theory dual of QCD, which could be useful for understanding the properties of hadrons. Finding a precise formulation of this string theory remains an unsolved problem.

Geoff's views had a considerable influence on me and certainly played a role in my later scientific choices. Stanley helped to provide a more balanced picture. You might say that he had a foot in both camps — S-matrix theory and quantum field theory. David Gross was a Chew student at exactly the same time as me. Like Stanley, he was more independent in his view of the role of quantum field theory. This helps to account for his ability to play an important role in the 1973 discovery of asymptotic freedom, which convinced the world that QCD is the correct theory of the strong nuclear force. It should be noted that Stanley made important contributions to the study of nonabelian gauge theories, and quantum field theories more generally, in the mid 1970s. One dramatic result was his proof of the equivalence of the Sine-Gordon model and the massive Thirring model.[2] His most highly cited paper is a review article (published in 1976) entitled "Vortices and Quark Confinement in Nonabelian Gauge Theories".[3]

In 1963–64 I was enrolled in an excellent and challenging course in Quantum Field Theory that was taught by Stanley. One of my lasting memories of that course concerns the final exam at the end of the first semester. It was much tougher than I had anticipated, and I left the exam feeling dejected, fearful that I had done very poorly. In fact I was so upset that in order to help me calm down I climbed Mount Diablo, which is East of Berkeley, that same afternoon. As it turned out, I received the second highest score on Stanley's exam. Only my officemate David Gross did better. I guess this was a harbinger of things to come.

Stanley's name is associated with the Lorentz-invariant Mandelstam variables, denoted s, t, u, which he introduced to describe two-particle to two-particle scattering amplitudes. Much less trivial was his derivation of the "Mandelstam representation", which is a double dispersion relation in the stu plane. (It is a plane because $s+t+u$ is a constant.) The Mandelstam representation consists of three pieces: an st piece, an su piece, and a tu piece. Although the Mandelstam representation is not discussed much nowadays, it made a big splash at the time, as Stanley's honors and Berkeley appointment demonstrate.

Stanley also made important contributions to Regge-pole theory. In addition to helping to explain the implications of Regge poles for scattering amplitudes, he also explored the possibility of "Regge cuts," i.e., the existence of branch points in partial wave amplitudes $a_l(s)$ that have been analytically continued in angular momentum to give functions $a(l, s)$. It is plausible, by analogy with what happens in the s variable, that pairs of Regge poles should give rise to Regge cuts. Stanley showed that this is indeed what happens, but the details are quite subtle. In particular, it is the third double spectral function (the tu term in the Mandelstam representation) that is responsible for the existence of branch points in the angular momentum variable.[4]

In the 1960s, when the highest energy available experimentally was 20 GeV in the laboratory frame, it appeared that total cross sections approach constant values at high energies. This was neatly accounted for by postulating the existence of a Regge pole $\alpha_P(s)$, with vacuum quantum numbers, called the Pomeron or the Pomeranchukon, whose intercept is $\alpha_P(0) = 1$. However, it was unclear how robust this result is, because the multi-Pomeron Regge cuts also have the same intercept and reduced slopes. In fact, modern high-energy studies of total cross sections demonstrate that they increase asymptotically as $\log s$ or maybe even $(\log s)^2$. The latter is the maximum growth compatible with the Froissart bound, which follows from general considerations of unitarity and analyticity. It ought to be possible to use Mandelstam's work to decide theoretically what the right answer should be. However, as far as I am aware, this has not been done.

Stanley was an important early contributor to string theory. So it may be appropriate to recall here a bit of the early history of the subject, and the role that Stanley played. The original notion of the bootstrap was that the exchange of resonances in the t and u channels provide the attractive forces that are responsible for causing resonances to form in the s channel. However, when all amplitudes are considered, the spectrum should be the same in all channels. Therefore the hadron spectrum is the solution of a self-consistency requirement. In this sense, they pull themselves up by their own bootstraps. Eventually it was realized that this situation could be idealized to the situation in which the resonance widths are very small compared to their masses, in accord with what the experiments were finding. Thus, as an approximation, one could seek amplitudes $A(s,t)$ that are meromorphic and can be expanded either in terms of s-channel poles or t-channel poles.

In 1968 Veneziano found an explicit realization of duality and Regge behavior in the narrow-resonance approximation.[5] He introduced an amplitude of the form

$$T = A(s,t) + A(s,u) + A(t,u),$$

where

$$A(s,t) = g^2 \frac{\Gamma(-\alpha(s))\Gamma(-\alpha(t))}{\Gamma(-\alpha(s) - \alpha(t))},$$

$$\alpha(s) = \alpha(0) + \alpha' s.$$

The motivation was phenomenological. In fact, he discussed the reaction $\pi + \omega \to \pi + \pi$ or the related decay $\omega \to 3\pi$. Incredibly, the Veneziano amplitude turned out to be a tree approximation amplitude in a string theory! Soon thereafter Virasoro proposed, as an alternative,

$$T = \frac{g^2\, \Gamma(-\frac{\alpha(s)}{2})\Gamma(-\frac{\alpha(t)}{2})\Gamma(-\frac{\alpha(u)}{2})}{\Gamma(-\frac{\alpha(t)+\alpha(u)}{2})\Gamma(-\frac{\alpha(s)+\alpha(u)}{2})\Gamma(-\frac{\alpha(s)+\alpha(t)}{2})},$$

which has similar virtues.[6]

The N-particle generalization of the Veneziano formula (due to multiple contributors) is

$$A_N = g^{N-2} \int \mu(y) \prod_i dy_i \prod_{i<j} (y_i - y_j)^{\alpha' k_i \cdot k_j}.$$

For a suitable choice of the measure $\mu(y)$ this has cyclic symmetry in the N external lines. The complete amplitude can then be written as a sum over permutations $T = \sum C_N A_N$. The coefficients C_N are Chan–Paton factors which build in gauge group quantum numbers. Shapiro's generalization of the Virasoro formula[7]

$$T_N = g^{N-2} \int \mu(z) \prod_i d^2 z_i \prod_{i<j} |z_i - z_j|^{\alpha' k_i \cdot k_j}$$

has total symmetry in the N external lines. Quite quickly, it was realized that the Veneziano model describes tree-approximation amplitudes for scattering of open-string ground states and the Shapiro–Virasoro model does the same for closed strings.

Fubini and Veneziano[8] showed that these formulas have a consistent factorization on a spectrum of single-particle states described by an infinite number of oscillators

$$\{a_m^\mu\} \quad \mu = 0, 1, \ldots, d-1 \quad m = 1, 2, \ldots$$

where d is the dimension of spacetime. There is one set of such oscillators in the Veneziano case and two sets in the Shapiro–Virasoro case. These formulas describe the spectrum and amplitudes for the theory of a relativistic string: open strings in the first case and closed strings in the second case. Amazingly, the formulas for the amplitudes preceded the interpretation! Having found the spectrum and the tree amplitudes, it became possible to study radiative corrections (loop amplitudes).

In 1970 Gross, Neveu, Scherk, and I,[9] as well as Frye and Susskind,[10] studied one-loop amplitudes. Unexpected branch points that violate unitarity appeared in the "nonplanar" open-string loop amplitude. In 1971 Claud Lovelace observed that these singularities become poles if $\alpha(0) = 1$ and $d = 26$.[11] For these choices the theory is perturbatively consistent (though there are tachyons). The new poles are interpreted as closed-string modes. This was the discovery of open-string/closed-string duality.

In January 1971 Pierre Ramond introduced a string theory analog of the Dirac equation.[12] His proposal was that just as the string's momentum p^μ is the zero mode of a string density $P^\mu(\sigma)$, the Dirac matrices γ^μ should be the zero modes of densities $\Gamma^\mu(\sigma)$. Then he defined

$$F_n = \frac{1}{2\pi} \int_0^{2\pi} e^{-in\sigma} \Gamma \cdot P d\sigma \qquad n \in \mathbb{Z}.$$

In particular, $F_0 = \gamma \cdot p +$ oscillator terms. Ramond then proposed the wave equation

$$(F_0 + M)|\psi\rangle = 0,$$

which deserves to be called the Dirac–Ramond Equation. He also observed that the Virasoro algebra of the bosonic string theory generalizes to a super-Virasoro algebra with odd elements F_n and even elements L_n. This was one of the first superalgebras in the literature.

A couple of months later, André Neveu and I introduced a second interacting bosonic string theory, which we called the dual pion model.[13] It has a similar structure to Ramond's theory, but the periodic density $\Gamma^\mu(\sigma)$ is replaced by an antiperiodic one $H^\mu(\sigma)$. Then the modes

$$G_r = \frac{1}{2\pi} \int_0^{2\pi} e^{-ir\sigma} H \cdot P d\sigma \qquad r \in \mathbb{Z} + 1/2$$

are the odd elements of a similar super-Virasoro algebra. These bosons and Ramond's fermions combine into a unified interacting theory. This combined system, suitably treated, constitutes the RNS formulation of superstring theory.

Among his many contributions to string theory, Stanley derived the fermion–fermion amplitude in the RNS model,[14] and he contributed to the construction of light-cone gauge string field theory.[15] In 1974 he wrote a very nice review paper for Physics Reports entitled "Dual-Resonance Models," which is what string theory was called at the time.[16] A common theme throughout most of Stanley's string theory research was the use of light-cone gauge. As he correctly understood, it is a very effective tool for many purposes. Later he obtained evidence for the UV finiteness of multiloop string theory amplitudes.[17]

The maximally supersymmetric nongravitational theory in four dimensions is $\mathcal{N} = 4$ super-Yang–Mills theory, which was formulated by Brink, Scherk and myself.[18] It gradually became apparent that this theory might be ultraviolet finite to all orders, due to delicate cancellations between bosonic and fermionic contributions to loop diagrams. This would mean that the superconformal symmetry of the classical theory, described by the supergroup $PSU(2,2|4)$, remains an exact unbroken symmetry at the quantum level. In 1982 Mandelstam gave one of the first proofs of this result using methods based on light-cone superspace.[19]

In the mid 1990s, during a visit to the Aspen Center for Physics, Stanley, Peter van Nieuwenhuizen, and I hiked to Electric Pass, which is not far from Aspen. This hike is quite strenuous as the trail ascends from about 9500 feet to about 13,500 feet over a distance of about five miles. Peter and I were impressed that Stanley was able to keep up, even though he was more than a decade older than us. Near the top of the trail, where it is quite narrow with a steep dropoff on the side, Stanley seemed to become a bit unstable, which caused us some concern. Fortunately, Stanley kept his footing and the hike was a success.

I feel privileged to have known Stanley. He will be missed.

References

1. S. Mandelstam, Determination of the pion–nucleon scattering amplitude from dispersion relations and unitarity. General theory, *Phys. Rev.* **112**, 1344 (1958).
2. S. Mandelstam, Soliton operators for the quantized sine-Gordon equation, *Phys. Rev. D* **11**, 3026 (1975).
3. S. Mandelstam, Vortices and quark confinement in nonabelian gauge theories, *Phys. Rept.* **23**, 245 (1976).
4. S. Mandelstam, Cuts in the angular momentum plane. 1, *Nuovo Cim.* **30**, 1127 (1963); Cuts in the angular momentum plane. 2, *Nuovo Cim.* **30**, 1148 (1963).
5. G. Veneziano, Construction of a crossing-symmetric, Regge behaved amplitude for linearly rising trajectories, *Nuovo Cim. A* **57**, 190 (1968).
6. M. A. Virasoro, Alternative constructions of crossing-symmetric amplitudes with Regge behavior, *Phys. Rev.* **177**, 2309 (1969).
7. J. A. Shapiro, Electrostatic analog for the Virasoro model, *Phys. Lett. B* **33**, 361 (1970).
8. S. Fubini and G. Veneziano, Level structure of dual-resonance models, *Nuovo Cim. A* **64**, 811 (1969); S. Fubini, D. Gordon and G. Veneziano, A general treatment of factorization in dual resonance models, *Phys. Lett. B* **29**, 679 (1969).
9. D. J. Gross, A. Neveu, J. Scherk and J. H. Schwarz, Renormalization and unitary in the dual-resonance model, *Phys. Rev. D* **2**, 697 (1970).
10. G. Frye and L. Susskind, Non-planar dual symmetric loop graphs and the pomeron, *Phys. Lett. B* **31**, 589 (1970).
11. C. Lovelace, Pomeron form-factors and dual Regge cuts, *Phys. Lett. B* **34**, 500 (1971).
12. P. Ramond, Dual theory for free fermions, *Phys. Rev. D* **3**, 2415 (1971).
13. A. Neveu and J. H. Schwarz, Factorizable dual model of pions, *Nucl. Phys. B* **31**, 86 (1971).
14. S. Mandelstam, Manifestly dual formulation of the Ramond model, *Phys. Lett. B* **46**, 447 (1973).
15. S. Mandelstam, Interacting string picture of dual resonance models, *Nucl. Phys. B* **64**, 205 (1973); S. Mandelstam, Interacting string picture of the Neveu–Schwarz–Ramond model, *Nucl. Phys. B* **69**, 77 (1974).
16. S. Mandelstam, Dual-resonance models, *Phys. Rept.* **13**, 259 (1974).
17. S. Mandelstam, The n loop string amplitude: Explicit formulas, finiteness and absence of ambiguities, *Phys. Lett. B* **277**, 82 (1992).
18. L. Brink, J. H. Schwarz and J. Scherk, Supersymmetric Yang–Mills theories, *Nucl. Phys. B* **121**, 77 (1977).
19. S. Mandelstam, Light cone superspace and the ultraviolet finiteness of the $N = 4$ model, *Nucl. Phys. B* **213**, 149 (1983).

Stanley Mandelstam and My Postdoctoral Years at Berkeley

Steven Frautschi

Walter Burke Institute for Theoretical Physics,
California Institute of Technology, Pasadena, CA 91125, USA
scf@theory.caltech.edu

From 1959 to 1961 I was a postdoc at U.C. Berkeley. Personally, I regard my postdoctoral period as "Golden Years" — finished with course taking and thesis writing, not yet burdened with classes to teach, students to mentor, grants to apply for. Lots of freedom, lots of things to explore in the beautiful Bay Area.

Those were also Golden Years for research. As a member of Professor Geoffrey Chew's research group, I was placed in a four-desk office with Stanley Mandelstam, Marcel Froissart, and Dan Zwanziger. What an opportunity! Chew was a great group leader — he provided strong program guidance, weekly research talks by members of the group, frequent lunchtime opportunities to talk physics at the woodsy faculty club, chances for personal conversation with visiting research luminaries ranging from Francis Low to Bogoliubov, even the annual excursion to a baseball game at Candlestick Park. When he gave a talk he always took care to highlight how the individual contributions of each of his postdocs and grad students advanced the overall program. In my subsequent career I tried, but could never fully succeed in matching Chew's success at juggling all those balls.

During my years at Berkeley, Chew's program of S-Matrix studies was centered on understanding and utilizing Stanley Mandelstam's contributions. The first year the focus was on Stanley's double dispersion relations and s, t, u variables, where any one of s, t, u could serve as the energy variable and the other two as forward and backward exchange variables. This suggested all sorts of approaches, models, and approximations. Personally, I cut my teeth on a study with my college classmate Dirk Walecka (then at Stanford) of pion–nucleon scattering, where we used meson and baryon exchanges to set up effective direct and exchange potentials. This approach was extended to other situations in papers I wrote with Geof Chew and, in one case, with Stanley.

A focus of my second year at Berkeley was the struggle to make scattering amplitudes at asymptotically high energies behave when hadrons with spin greater than one were exchanged. Already around 1940 Werner Heisenberg had noted that such exchanges would cause the scattering cross-section to grow as a power of the energy at energies much greater than mc^2, approaching infinity in the high energy limit. That contradicted what we knew of scattering cross-sections for cosmic rays, it contradicted physical expectations for a short range force, and in 1961 Marcel Froissart proved a landmark theorem that such growth was impossible. So either no hadron existed with spin greater than one, or it seemed that the rapid growth of the one-particle high spin exchange cross-section would have to be cancelled by messy, hard-to-compute higher order corrections.

By 1961 experimental developments were making it imperative that this theoretical crisis be resolved. New facilities capable of accelerating protons to extreme high energies of 30 mc^2 were coming into operation at the CERN Lab in Geneva, Switzerland and Brookhaven Lab on Long Island, New York. And strongly interacting particles with higher spin were beginning to be discovered, presaging a flood of such discoveries during the 1960s.

In a seemingly unrelated 1959 paper, Tulio Regge had introduced new ideas into the theory of low energy (kinetic energy much less than mc^2) scattering of a particle by a potential. Regge considered angular momentum J as a complex variable. He showed that Schrödinger's equation allowed for a continuous set of solutions $J(E)$ as energy E increased, even though these mathematical solutions could represent physical bound states or resonances of the particle in the potential only at energies where J reached the quantum-allowed angular momentum values $J = 0, 1, 2, \ldots$. If the potential was attractive enough, one would get a sequence of physical states of progressively higher angular momentum and energy, which we dubbed a "Regge trajectory."

Remarkably, Stanley noticed the low energy Regge paper in the journal Nuovo Cimento and suggested that it might solve our high energy problems! Following up on Stanley's suggestion, Chew and I proposed that all hadrons lie on Regge trajectories. This meant that the appropriate mathematical object to exchange was not the individual J spin particle (with its dangerous high energy behavior if J exceeded one) but the whole moving Regge trajectory which, even if it represented one or more high spin particles, could have the property of falling below one in the kinematic region where the exchange scattering took place. It followed that the high energy behavior was automatically well-behaved without the necessity of making high energy corrections.

Our proposal had the following consequences:

(1) The conceptual problem of reconciling the existence of high spin particles with finite high energy cross-sections was resolved.
(2) We related the energy dependence of various high energy reactions such as $\pi^- + \text{p} \rightarrow \pi^0 + \text{n}$ to exchanged Regge trajectories, and predicted that high

energy forward scattering should shrink as the energy increases. These features were quickly verified experimentally as the CERN and Brookhaven accelerators came into action.

(3) Since we had at most two experimental points per Regge trajectory when we wrote our paper, we drew straight lines through them to represent the interpolating Regge trajectories. Over the next couple of years I attended several meetings where a speaker, unlike us, had taken the trouble to actually calculate the Regge trajectory for scattering in a Yukawa (or other hopefully realistic) potential. The displayed trajectory would rise with energy, then fall back along a curve; the speaker would say with a flourish "See how straight the trajectory is" and the audience would laugh. But our straight line choice proved serendipitous — as more data points became available it turned out that the Regge trajectories really were approximately straight lines, with a quasi-universal slope. Thus there was equal spacing in the energy variable s between successive states such as $J = 0, 2$, and 4. Since energy is proportional to frequency in wave mechanics, that meant we were observing equal spacing in frequency. And a familiar feature of waves such as those on a violin string is equal spacing between the frequencies of successive overtones. So, though we were not explicitly aware of it at the time, we had stumbled upon evidence that hadrons are not points but strings (mesons, for example, are now thought of as gluon strings with a quark at one end and an antiquark at the other end, and the universal slope of the Regge trajectories is related to the string tension). In this way our 1961–62 papers became part of the twisting road that led to string theory.

I later had the opportunity to check with other groups (such as Murph Goldberger and Dick Blankenbecler at Princeton, and a Russian group) that had been simultaneously investigating the role that Regge trajectories might play in high energy scattering. It seems that in every case their attention had first been drawn to Regge's work by Stanley. Yet while the subsequent maelstrom of activity was taking place, Stanley never tried to take credit for the idea. He remained his usual self — rather shy, rather lonely, yet consistently pleasant, always ready to go out to dinner with us and discuss the topics of the day — ranging from physics to the politics of his native South Africa to his little jokes such as the observation that (soprano Elizabeth) Schwarzkopf had white hair whereas (physicist Vicki) Weisskopf had black hair.

At some point during this period Chew, Mandelstam, and I attended a meeting at CERN. One morning before the sessions began we took a walk in one of the beautiful parks on the shore of Lake Geneva. Along the way, we encountered none other than Werner Heisenberg. After Professors Chew and Heisenberg had greeted each other, Heisenberg turned to Stanley and me and said "And these must be your students." That was an essentially accurate statement of my relationship to Chew, but it badly missed the mark with Stanley. Not only was Stanley several years older, and the previous author of a book with Wolfgang Yourgrau, but while Chew was the inspiring group leader, Stanley was the creator of most of the technical advances we were using. Probably Heisenberg simply did not know who Stanley was. But perhaps unfairly, I have always felt that Heisenberg's remark displayed a "top-down", hierarchical understanding of how research groups work.

In late summer 1961 I was about to leave for my new position as Assistant Professor at Cornell. I believe Stanley had already left for his new position at Birmingham, and Chew happened to be out of town. One evening in my apartment I received a phone call from Virendra Singh and B. M. Udgaonkar, members of Chew's research group, begging me to come to the Radiation Lab the next day because "Gell-Mann is coming back." Murray had visited the Lab that day asking lots of Regge pole questions and had intimidated them with his intensity. So I went up the hill to the Lab the next day, and presently Murray Gell-Mann appeared. One of this first questions was "Are there separate Regge trajectories for even and odd angular momentum l? It seems he had heard Murph Goldberger talk about this, and was unsure of the details.

By good fortune I had asked myself the same question ten days earlier and had thought it through carefully. I told Murray that a "direct potential" (exchange of particles in the t channel) normally gives a peak at small angles in the s channel and thus contributes the same sign to both even and odd l partial waves, but an "exchange potential" (exchange of particles in the u channel) normally gives a peak at angles near $180°$ and thus contributes with opposite sign to even and odd partial waves. So when exchange potentials are present (not considered by Regge), the even and odd l trajectories split. Gell-Mann got excited by my splitting argument and other answers I gave, saying "We have to work out all the implications of this and write it up." He invited me to visit Caltech which I was able to do Thanksgiving week that fall, leading to an invitation to transfer to Caltech where I have remained ever since.

Gell-Mann would never have sought my advice if Stanley had not pointed us to Regge poles. And I could not have answered Gell-Mann's question with the physically convincing exchange potential argument if I had not been immersed in Stanley's double dispersion stu approach through collaborations with Walecka, Chew, and Stanley himself. Clearly my contact with Stanley and the whole Chew group was a key turning point in my life.

Reminiscences on Stanley Mandelstam

Korkut Bardakci

Department of Physics,
University of California, Berkeley, CA 94720-7300, USA
kbardakci@lbl.gov

I first met Stanley during a summer visit to Berkeley around the year 1963. Of course, I knew of him before; he was already famous for the representation named after him. In my early career, I was also much influenced by his pioneering work on analyticity, Regge theory and S matrix theory in general. When I got a job offer from Berkeley the next year, Stanley's presence there, in addition to Berkeley's other assets, made the offer very attractive, and I accepted it.

From then on, Stanley continued to be a guiding light for my own work. His interests gradually changed from S matrix theory to field theory, string theory and supersymmetric models, and in each of these fields, he made fundamental contributions. Especially impressive is his pioneering work in the application of the light cone picture to string theory and supersymmetric models. Since these and his other work are ably reviewed elsewhere in this volume, I will turn to some personal recollections of him as a friend.

When I first met him, I found him a rather reserved and quiet person. Occasionally, he made brief comments on various topics in physics. These were always precise and right on target. As I got to know him better, I realized that behind the reserved exterior, he was actually a kind and friendly person. On social occasions, he clearly enjoyed himself, and when invited, he always brought an excellent bottle of wine. I also remember with pleasure the time when he invited me, my wife and Charles Thorn to a dinner in his flat. The meal was prepared by Stanley himself and it was excellent. Up to that time, I did not know that he was a good cook!

I would like to end by mentioning Stanley's kindness towards me. My first summer in Berkeley, I was without a car, and Stanley, who was away, let me use his car in his absence. Many years later, I had to take a sick leave, Stanley volunteered to take over the course I was teaching, adding to his teaching burden. With his death, the theory community lost a giant, and I lost a good friend and an inspirational figure.

Remembering a Gentle Giant of Physics

Charles Sommerfield
*Department of Physics, Yale University,
New Haven, CT 06511, USA*
charles.sommerfield@yale.edu

I first met Stanley Mandelstam in 1958 when he came to Berkeley to take up a faculty position at the University of California. I was a postdoc at the time, but since I was required to do a bit of teaching I was granted the impressive sounding title of acting assistant professor. Stanley's arrival had been excitedly announced the previous semester by Geoffrey Chew who had met him during a visit to Columbia University. Chew was so impressed by Stanley's work on double dispersion relations that he brought him to Berkeley to join in his group's efforts on the bootstrap approach to high-energy physics.

As I was an unreconstructed field theorist still under the influence of my mentor Julian Schwinger, our paths did not lead to ease of intellectual crossings; except that Stanley was interested in everything and that I was working with Bob Karplus and Eyvind Wichmann on dispersion relations in perturbation theory. One thing that we did have in common were cars. We both had Volvos. His was one step up the model line from mine.

Stanley and I would eat together fairly often. I think we must have explored every moderately priced restaurant in the East Bay. Marshall Baker, who was at Stanford at the time occasionally joined us.

During the spring vacation the three of us drove to Death Valley to see the desert in bloom. We intended to camp at Furnace Creek. There was a very strong wind that night buffeting us around. At some point Stanley's ultracivilized nature took over and he sensibly moved indoors to the Inn in order to get some sleep.

I moved back east after that year and our subsequent encounters were limited to the occasional cross-continental visit and professional meetings.

My only vaguely negative recollection about Stanley relates to an occasion when my wife Linda and I had him over for dinner in Cambridge during the time he was visiting Harvard. In the course of an after-dinner conversation I told him about some work I was doing about deriving off-shell 2- and 3-point functions associated with the Virasoro model. He was not at all encouraging and so, relying on my extreme respect for Stanley, I missed out getting involved with Joel Shapiro's work on the subject.

Stanley was a deep thinker. He listened attentively with what one would tell him, giving no hint of a reaction, except for the occasional grunt. And then he would proceed to give an accurate and often profound assessment of the situation based both on general principle and specific detail.

And as far as I know he was still driving a Volvo to the end.

Grad School with Stanley Mandelstam

Joseph Polchinski
Kavli Institute for Theoretical Physics,
University of California, Santa Barbara, CA 93106-4030, USA
joep@kitp.ucsb.edu

I was a graduate student at Berkeley from 1975–1980, and worked with Stanley for the last three years of my stay. I will call him Stanley here, but I am pretty sure that I always called him Professor Mandelstam at the time, as was more customary then.

Stanley actually had three grad students at that time: Susan Elma Moore, Omer Kaymakcalan, and me. It was more than he would usually take on; it was my impression that he had already started working with Susan and Omer, and I had to be a bit persuasive.

This was an interesting time to be doing fundamental physics. It was a few years after asymptotic freedom. The basics of strongly coupled hadronic physics and its phenomenology had taken form. Fascinating new ideas in quantum field theory — instantons, monopoles, quark confinement — were appearing, and Stanley was deeply interested in these. I believe that quark confinement was the central problem for Stanley at the time, but it tied many of his ideas together, as I will recall.

Each of Stanley's students had a different project, but all were connected in various ways through quantum field theory and QCD. Susan's project was to find variational states of heavy quarks, based on a Wilson loop model of the states.[1] In his own work, Stanley was never following the herd, and he guided his students the same way. This could be challenging, as I will recall more later. In Susan's case, after writing her dissertation she changed fields — first spending some time trying acting, and then ended up as a doctor.

Omer's graduate project was to understand the non-Abelian properties of the Higgs phase.[2] After his thesis, Omer went to Syracuse and worked on various projects with the members of the Syracuse group. Several of his papers, dealing with chiral Lagrangians, proton decay, and strongly coupled Higgs dynamics, were well-cited. Sadly, Omer was diagnosed with cancer at around this time, and passed away shortly after.

My central project was to understand the 't Hooft loop operator, the magnetic analog of the Wilson operator. The motivation, of course, was quark confinement. Mandelstam and 't Hooft had both recognized that confinement could be understood as an electric–magnetic dual of the superconducting phase. So one should develop the 't Hooft dual of the Wilson operator as completely as possible.

To be honest, I have never thought very highly of my dissertation. I think one could say that it is a hodge-podge of results with no main point.[3] Perhaps this was inevitable. Rather strikingly, my central problem was not solved until 25 years later, by Anton Kapustin. This required several new ideas, such as conformal invariance, that had not yet been applied. Remarkably, the first paper that one studies in the Langlands program today is the first paper that Stanley gave me to read 40 years ago.[4] This shows how far ahead he was in much of his thinking.

Most of my graduate time was spent learning the ins and outs of quantum field theory. Somehow out of these I cobbled together a dissertation, based on questions raised by Stanley and on my own curiosities — especially about renormalization, which of course played a big role in my later work.

Stanley was always generous with his time when one went in with questions. This did not mean that one would understand the answers, though. As we have seen from Kapustin's example, the problem may have been years ahead of its time.

Another striking example was 11-dimensional supergravity. I have this distinct memory that on several occasions we would be discussing quantum field theory and quark confinement, and Stanley would make some remark about 11-dimensional supergravity or string theory. This was a bit mind-blowing for someone struggling with four-dimensional quantum field theory, and for whom string theory was assumed to be an artifact from the past. I did not know what to make of this, so I just waited for Stanley to come back to earth. But a few years later string theory was back in the center of things, so Stanley was as usual well ahead of the game.

Unfortunately I did not continue to work with Stanley after graduation. It was not his style to collaborate (or mine, at the time: I still had to learn to do this effectively). But it is interesting for me to look over his history. While I was his grad student, his main focus was the rich new ideas in quantum field theory, especially confinement and duality. But before and after, this was embedded in a wider web, including in particular string theory and supergravity.

References

1. S. E. Moore, Gauge invariant description of heavy quark bound states in Quantum Chromodynamics, August 1980, 111pp., LBL-11365, UMI-81-13134.
2. O. S. Kaymakcalan, Higgs phase in non-Abelian gauge theories, June 1981, LBL-12840.
3. J. G. Polchinski, Vortex operators in gauge field theories, July 1980, LBL-11295, UMI-81-13165.
4. P. Goddard, J. Nuyts and D. I. Olive, Gauge theories and magnetic charge, *Nucl. Phys.* B **125**, 1–28 (1977).

Remembering a Gentle Giant of Physics

Mary K. Gaillard

Department of Physics, University of California,
Berkeley, CA 94720, USA
Theory Group, Lawrence Berkeley National Laboratory,
Berkeley, CA 94720, USA
mkgaillard@lbl.gov

I don't remember when I first learned about Mandelstam variables, but it must have been very early on, because even though I didn't work on pre-QCD strong interaction physics except for the first three months or so of my time as a graduate student with a basement office at CERN, I certainly knew that Stanley Mandelstam was a very important figure in our field.

During the summer of 1981 I was trying to decide whether to move to Berkeley or to Fermilab. I ran into Stanley outside the CERN cafeteria and felt extremely honored when he told me, in his habitual self-effacing manner, how much he hoped I would come to Berkeley.

Stanley was a remarkable colleague — a constant presence at seminars and other scientific activities of our group, but who spoke only when he had something significant to say or a meaningful question to ask. I can best illustrate his modesty in the face of his remarkable contributions to physics with a brief excerpt from my book:[1]

> "...it was speculated that $N=4$ supersymmetry might actually be finite... This theory was in fact proven to be finite by our Berkeley colleague, Stanley Mandelstam. Stanley, who was already famous for his work on scattering theory, was so modest that nobody in our group knew that he had made this discovery until John Schwarz came to visit our department. I met John in the hallway, coming out of Stanley's office, and he told me that he had asked Stanley what he was working on. The answer was no less than this very important result."

Stanley went on to make significant contributions to superstring theory, including a proof that the perturbation expansion is finite. Perhaps I was more aware of this work as it was going on because I had learned that one had to ask.

References

1. M. K. Gaillard, *A Singularly Unfeminine Profession: One Woman's Journey in Physics* (World Scientific, Singapore, 2015).

Mandelstam & NAL

Pierre Ramond

*Institute for Fundamental Theory, Department of Physics,
University of Florida, Gainesville, FL 32611, USA*
ramond@phys.ufl.edu

A few years after I entered graduate school in 1965, I found myself knee-deep in the complex plane, trying to make sense of crossing relations. To that effect I spent some time between the s and t channels, leaving the s channel and enter the "Mandelstam Triangle," and then trying to emerge without cutting myself into the t channel. The name Mandelstam meant a lot to me, but not necessarily in a positive way.

Fast forward a few years later when I was a postdoc at the National Accelerator Laboratory, today's FermiLab. Our small group of theorists was used by its director Bob Wilson as ushers for the many distinguished visitors to the Laboratory. There was not much to see, and our task was to take these physics VIPs to climb a silo with stairs to take a look at the main ring in the early stages of construction. For that we were rewarded by the management to a free lunch at an off-site restaurant. Most visitors did not express any interest in our research, but not Mandelstam, who visited NAL in early 1971.

At that time I had finished my papers on what became the superstring, by suitably generalizing the Dirac equation to the Veneziano model. I was super-excited, although the preprints had been sent out for several months and I had not heard any feedback from the community. Mandelstam came to the theory house to ask each of us about our research. I was very happy to tell him about my work, since he was also a leading light in the field of dual resonance models. He listened and after some time we broke for lunch. I was quite put off by his lack of response since I thought I had stumbled on something very deep.

Later that afternoon, I could not contain myself, and asked him what he thought. His reply surprised me. He said that I claim to have solved the problem of adding spin to the Veneziano model, adding that the same had been tried by his colleagues. He added that he needed time to study what I had proposed, and until then he was not ready to say anything. All uttered in a somewhat monotone "matter of fact" voice; needless to say I was left very deflated. The next morning he left the laboratory.

But I had underestimated Stanley: he meant every word he said. In the following months he did read my papers, understood the approach and never shied from giving me credit when due. The first of the big physicists to do so.

Physicists knew Mandelstam the physicist, how could they not; this episode which has stayed with me all these years told me about Stanley the human being.

The Influence of Stanley Mandelstam

Michael B. Green

Department of Applied Mathematics and Theoretical Physics,
Wilberforce Road, Cambridge CB3 0WA, UK
School of Physics and Astronomy, Queen Mary University of London,
Mile End Road, London E1 4NS, UK
m.b.green@damtp.cam.ac.uk

I hardly knew Stanley Mandelstam — our paths rarely crossed, and when they did our discussions were restricted to technical, rather than personal, issues. It is, however, an honour to be asked to write this short memoir since his work was hugely innovative and he was one of the pioneers who laid the foundations for much of the subject of my research.

My interests as a PhD student in Cambridge in the late 1960s were strongly influenced by the ideas of S-matrix theory that had emerged from Berkeley, largely following the work of Chew and Mandelstam. Stanley made use of his exceptional understanding of the analytic properties of perturbative quantum field theory to motivate a more rigorous approach to S-matrix theory. This led him to the covariant formulation of scattering amplitudes in terms of "Mandelstam variables" and to the "Mandelstam representation," which provided an elegant framework for discussing dispersion relations. He was one of the early contributors to Regge pole theory and its relation to sums of Feynman diagrams. These were the basic ingredients for much of the S-matrix programme that attempted to explain the strong interactions in the absence of a quantum field theory description.

Stanley was prominent in the series of developments relating to the strong interactions that grew out of the S-matrix programme and were taking place during my period as a graduate student. He realised that the dual relation between resonances and Regge poles necessitates the presence of an infinite number of narrow resonance poles lying on linear Regge trajectories. This was subsequently explicitly realised in Veneziano's 1968 paper that introduced the dual resonance model that was supposed to describe meson scattering and later developed into string theory. Stanley immediately began making seminal contributions to the development of dual models and then to their formulation as string theories. He and his Berkeley group formulated dual model n-particle tree amplitudes that included spinning external particles, as well as internal symmetries, which are closely related to the fermionic string theory of Ramond, Neveu and Schwarz.

I visited Berkeley in the Summers of 1971 and 1972 but, to my regret, I did not interact with Stanley. In the Summer of 1973 I participated in a small and very interesting conference on dual models at CERN, in which the highlight was the arrival of preprints of two remarkable papers by Stanley — although he himself did not attend the meeting. These papers demonstrated how to calculate scattering amplitudes in bosonic string theory and in Ramond–Neveu–Schwarz string theory by making use of the light-cone gauge. The light-cone gauge for free string theory had been introduced in a beautiful paper by Goldstone, Goddard, Rebbi and Thorn a year earlier and had led to a great simplification in the classification of the spectrum of physical states in dual models. Stanley's two papers provided a meticulous derivation of string scattering amplitudes that involved a detailed discussion of two-dimensional conformal field theory in the presence of suitable world-sheet boundary conditions. The second paper determined the four-fermion amplitude, which had been the subject of much discussion at the conference since the covariant rules for constructing such amplitudes were causing confusion (which was sorted out more than a decade later when the role of world-sheet BRST ghosts was developed by Friedan, Martinec and Shenker). Mandelstam's light-cone formalism bypassed this confusion — in retrospect this is achieved by virtue of the fact that BRST ghosts do not enter in the light-cone gauge treatment.

Stanley's implementation of light-cone gauge methods was hugely influential. They provided a practical method for tackling a variety of otherwise obscure problems. This was particularly relevant to my own research in several contexts. Firstly for studying a possible route to a realistic string theory of hadrons in the late 1970s, and then for understanding the structure of superstring tree and one-loop amplitudes constructed by John Schwarz and me in the early 1980s.

With the advent of QCD and asymptotic freedom in 1973, Stanley turned his attention to properties of Yang–Mills theory and quark confinement, which was a most challenging subject that also had clear connections to a possible string theory description of hadrons. He (as well as 't Hooft) made the crucial observation that the QCD vacuum exhibits a dual Meissner effect, in which colour electric flux is confined into string-like flux tubes by a condensate of magnetic monopoles and vortices.

In 1983 Stanley again made use of the simple properties of light-cone gauge superspace in his proof of the finiteness of $\mathcal{N} = 4$ supersymmetric Yang–Mills theory. This (together with a related proof by Lars Brink and collaborators) provided further inspiration for the corresponding superspace formulation of light-cone gauge superstring field theory by Lars Brink, John Schwarz and myself.

Stanley returned to work on string theory in the late 1980s, a period in which I met him more frequently. As ever, he was a mine of information, which was often difficult to understand but always merited careful attention. In 1992 he produced a notable proof of the ultraviolet finiteness of superstring amplitudes to all orders in perturbation theory, which was based on a particularly subtle combination of arguments. I did not feel confident that I understood Stanley's arguments, although I felt confident that Stanley did (as did his ex-students Charles Thorn and Nathan Berkovits with whom I had many interactions). In recent times his proof has been refined and generalised to take into account infrared issues.

I last met Stanley at the conference to celebrate Bruno Zumino's 90th birthday in 2013. It was delightful to find him basically unchanged — quiet as a mouse but as enthusiastic as ever about physics, despite having retired many years ago. He was a remarkably original thinker, who was also exceptionally modest and clearly enjoyed the process of discovery, down to every nitty-gritty detail.

My Interaction with Stanley

Paolo Di Vecchia

The Niels Bohr Institute, University of Copenhagen,
Blegdamsvej 17, DK-2100 Copenhagen Ø, Denmark
Nordita, KTH Royal Institute of Technology and Stockholm University,
Roslagstullsbacken 23, SE-10691 Stockholm, Sweden
divecchi@nbi.dk

I saw Stanley for the first time at Caltech where he came during fall 1968, if I remember correctly, to give a seminar on the newly discovered Veneziano model, but I do not remember the details. I remember, however, that Murray Gell-Mann was quite excited during the seminar and kept asking one question after the other, while Stanley was calmly answering them. I got the impression of a very deep physicist and I was also impressed by his modesty. I was a young postdoc and I did not dare to talk to him. After I came back from Caltech at the end of 1968, I applied to a NATO fellowship and I thought to use it at Berkeley with him, but then, when I got it, I went instead to work with Sergio Fubini at MIT.

I met Stanley again in 1980 because I invited him to give a seminar at the Freie Universität in Berlin. He brought to me the sad news of the death of Joël Scherk. Then, he wanted to visit Potsdam and I investigated how to go from West Berlin to Potsdam, but it turned out to be too complicated and we had to give up this idea. Then, I moved to Wuppertal and when he came to Europe, in the beginning of the eighties, I invited him again to give a seminar. I met him once more in the summer of 1985 in Santa Barbara for the program organized by Mike Green and David Gross and in June 1986 in connection with the Nobel symposium organized by Lars Brink, together with other people from Göteborg in Marstrand. In both occasions he talked on multiloop calculations, mainly in the bosonic string, using the light-cone formalism.

I would like to remind the reader that Stanley, together with Daniele Amati, Sergio Fubini and Yoichiro Nambu, were the only somewhat older and well established people who understood the depth of the newly discovered dual resonance model, actively and enthusiastically worked on it and encouraged students and postdocs to work on it too. Other people of the same generation were either neutral or against these new developments, with the exception of Murray Gell-Mann who never actively worked on string theory, but he was advocating its importance in several occasions.

In the following, I want to describe the research of Stanley that overlapped with my own research. At the beginning of 1973, with the work of Goddard, Goldstone, Rebbi and Thorn,[1] it was clear that the Lagrangian of the bosonic string reproduced the spectrum obtained by factorizing the previously constructed dual resonance model for intercept of the Regge trajectory $\alpha_0 = 1$ and critical dimension $D = 26$. In two subsequent papers, one by Ademollo *et al.*[2] and the other by Stanley,[3] it was shown how to reproduce the N-point amplitude of the dual resonance model from string theory. In particular, Stanley[3] and Cremmer and Gervais[4] computed the three-point amplitude involving three arbitrary strings and showed that, on shell, they obtained the same amplitude, computed by Ademollo, Del Giudice, Di Vecchia and Fubini,[5] by using three vertex operators corresponding to arbitrary DDF states. At this point, it was clear that the old dual model was completely equivalent to the bosonic string!

Multiloop amplitudes were already computed in 1969–1970 by many people as a way to unitarize the string tree diagrams, but, here, I would like to concentrate on the effort of Lovelace who constructed the so-called N-Reggeon vertex[6] and, by using the sewing procedure, constructed,[7] together with others including Alessandrini and Amati,[8] an expression for the multiloop N-point amplitude that contained quantities belonging to Riemann surfaces, as the prime form, the Abelian differential and the period matrix, well before its connection with string theory was clear. A summary of this approach can be found in Ref. 9.

This approach was, however, not complete because it was not clear how to compute the measure of integration over the moduli. At one loop it was guessed that one should cancel two powers of the bosonic partition function, but it was not known what to do in the case of higher loops. We must remember that, at that time, the formalism of the ghosts and the BRST invariance were not yet known and this prevented the use of a manifestly Lorentz covariant formalism for the computation of the measure of integration over the moduli. On the other hand, it would have been possible in the light-cone formalism that Mandelstam introduced and that does not require the introduction of ghosts, but, for some reason not clear to me, he did not manage to get the complete result in Ref. 10.

The complete result was published in Refs. 11 and 12 after the introduction of the ghosts and the construction of a BRST invariant formalism for the bosonic string. It is based on the sewing procedure of Lovelace *et al.* where the Riemann surface appears in the Schottky parametrization and therefore in a form that is not manifestly modular invariant. However, the numerical calculations performed by Petersen, Roland and Sidenius[13] left no doubt that the expression was modular invariant. Mandelstam takes up again this calculation in a later paper[14] where he presents the multiloop N-point amplitude in different forms. One of them is the one discussed in Refs. 11 and 12. Then, he writes "By extending the methods used in Ref. 10 we have obtained a proof of the formula given in Refs. 11 and 12. Details of the proof of this and other formulas will be given in a further paper." I was unable to find this paper and probably he never published it. I remember, however, that he had a preprint or something like that where he finally derived the expression in the Schottky parametrization of the Riemann surface[11,12] with the light-cone formalism, but I could not find any mention of this in the literature. He used his light-cone formalism to also compute multiloops in the RNS string, but I leave to Nathan Berkovits to describe this in detail.

The last time I was in contact with Stanley was about ten years ago. in connection with the book "The Birth of String Theory" that I was editing together with Andrea Cappelli, Elena Castellani and Filippo Colomo. When we were preparing for the book we contacted him and asked him to write a contribution to the book but he answered that he did not want to. Then we sent our scheme for the book to the Cambridge University Press and one of the referees criticized that we had not included Mandelstam. We contacted Stanley again mentioning the comment of the referee and then he accepted to write his contribution that he sent to the high-energy archive in 2008. We were very pleased that he wrote something for our book. This is his last contribution appearing in the high-energy archive.

References

1. P. Goddard, J. Goldstone, C. Rebbi and C. Thorn, *Nucl. Phys. B* **56**, 109 (1973).
2. M. Ademollo, A. D'Adda, A. D'Auria, P. di Vecchia, F. Gliozzi, R. Musto, E. Napolitano, F. Nicodemi and S. Sciuto, *Nucl. Phys. A* **21**, 77 (1974).
3. S. Mandelstam, *Nucl. Phys. B* **64**, 205 (1973).
4. E. Cremmer and J. L. Gervais, *Nucl. Phys. B* **76**, 209 (1974); *ibid.* **90**, 410 (1975).
5. M. Ademollo, E. Del Giudice, P. Di Vecchia and S. Fubini, *Nuovo Cimento* **19**, 181 (1974).
6. C. Lovelace, *Phys. Lett. B* **32**, 490 (1970).
7. C. Lovelace, *Phys. Lett. B* **32**, 703 (1970).
8. V. Alessandrini, *Nuovo Cimento A* **2**, 321 (1971); V. Alessandrini and D. Amati, *Nuovo Cimento A* **4**, 793 (1971).
9. V. Alessandrini, D. Amati, D. Le Bellac and D. Olive, *Phys. Rep.* **1**, 269 (1971).
10. S. Mandelstam, *Unified String Theories*, eds. M. B. Green and D. Gross (World Scientific, 1986), pp. 46–102.
11. P. Di Vecchia, M. Frau, A. Lerda and S. Sciuto, *Phys. Lett. B* **199**, 49 (1987).
12. J. L. Petersen and J. Sidenius, *Nucl. Phys. B* **301**, 247 (1988).
13. J. L. Petersen, K. Roland and J. Sidenius, *Phys. Lett. B* **205**, 262 (1988).
14. S. Mandelstam, *Phys. Lett. B* **277**, 82 (1992).
15. S. Mandelstam, *The Birth of String Theory*, eds. A. Cappelli *et al.* (Cambridge University Press, 2012), pp. 294–311, arXiv:0811.1247 [hep-th].

My Advisor Stanley

Sang-Jin Sin

Hanyang University, Seoul, 133-791, Republic of Korea
sjsin@hanyang.ac.kr

1. Working With Him

In the fall of 1983, I was a first year graduate student in Berkeley studying theoretical high energy physics. I sat in at the Quantum Mechanics course taught by Stanley. I did not take his course formally because I was one of the students who had passed the preliminary tests and was exempted from course work. At the end of the year, I asked Stanley whether I can work with him. But I learned that he planned to leave for Paris to spend his sabbatical year. So I could start with him only in the summer of the 1985. In the beginning, I was given a problem by him to calculate the n-loop closed bosonic string scattering amplitude in Fuchsian representation of the Riemann surfaces. At that time, he has just finished the calculation in the Schottky representation.

Several weeks later, I understood what he had done and I solved the problem easily and submitted him the result. After a look, he did not say anything whether I could publish the result or not. Instead, he gave me another problem, which he had actually raised to the world. The problem was to prove the equivalence of the Polyakov approach which is manifestly relativistic and the lightcone approach which is manifestly unitary. Proving the equivalence would deliver the nontrivial fact of each side to the other without pain. After spending a year, I realised that there is one lightcone diagram for any Riemann surfaces. Since I had already calculated the n-loop amplitudes before, the rest of the problem is just how to express the zero-mode determinants of n-loop amplitudes of the covariant approach in terms of the lightcone diagram variables. I reported the status to Stanley and he encouraged me and said: "That is promising!" I had been working alone for a year and it was clear the answer was not far away. So I took a month's break in Seoul and came back to attend the Strings Conference at San Diego. There, I attended a talk by

Steven Giddings who actually had finished all the essential calculation of what I was trying to finish. He had collaborated with a mathematician for the measure and with other physicists for the zero-mode determinants. Stanley said what should be done is essentially done. So I missed the chance to graduate early as well as the ticket for the postdoc job I planned for. You see that you should not take a break before you actually finish what you think you could do!

But I did not have too long a time to worry because Stanley suggested another problem. This time it was to prove the Lorentz invariance for the lightcone string theory at the off-shell level. I did it without much difficulty and I also extended it to the superstring case. That was my Ph.D. thesis. It was an interesting result but was regarded as less important by the community, because everyone else were interested in the covariant approach or conformal field theory itself. It was not a popular subject at that time. Thirty years later, when I look back, I do not see many important results of the string theory other than my thesis which was suggested by Stanley. Important things are not always popular.

Throughout the entire period as a student, I never had a chance to publish with Stanley. I was reluctant to propose to publish together with him partly because I was not sure whether the work was of good quality or not although the problem was originally suggested by him. My reluctance to propose to him was also due to the fact that there was no other student who had published with him before. He always smiled like a child and was kind to me every time I knocked on his door. I could feel that he was happy whenever we discussed physics. However, I do not remember whether we had any chance to chat on anything else outside physics. To me, he was a saint with a simple mind in a lonely castle sitting at the desk with endless calculations.

2. The Minimalist

I still remember two of his styles: his lecture and clothing fashion. In 1986, I was a teaching assistant for Stanley's Quantum Field Theory course. Actually I had taken the QFT course of Prof. Bruno Zumino, who passed away a few years ago. So I had the chance to listen to the QFT lectures from two masters, who are actually at the end of two extremes. Bruno's lectures were elegant. He covered only those parts that he wanted, stressing the algebraic structure and beauty of the formality. On the other hand, Stanley eliminated all the unnecessary structure and used only minimum degrees of freedom. For example, for the quantization of QED, the simplest gauge theory, Bruno used BRST approach while Stanley used Coulomb gauge fixing. I may say without much exaggeration that he was a minimalist in physics style. If one needs to solve a new type of theory for the first time, it is obvious that one would attempt to get the result with a shortest cut using the minimal degree of freedom.

During my stay in Berkeley, I could see him dressed in check-style jackets which were essentially the same. He wanted only minimum amount of variations in everyday life, too. He simplified everything to the end. Actually it became my definition of physics: Physics is the simplification of an object to the limit where the mathematics can be applied. I miss Stanley's child-like smile which I saw whenever I opened his office door.

Stanley Mandelstam My Graduate Supervisor

Arjun Berera

*School of Physics and Astronomy, University of Edinburgh,
Edinburgh, EH9 3FD, UK*
ab@ph.ed.ac.uk

I was in the Physics Graduate School at U.C. Berkeley from 1986 to 1992. I was a PhD student of Stanley Mandelstam for the last four years of this. I had been awarded a National Science Foundation Graduate Fellowship and then another Fellowship from the Department of Education and was enjoying my time at Berkeley and studying Physics, so was in no rush to complete my PhD. During these years, aside from my PhD work with Stanley, I had also done some research work in statistical mechanics and material science.

When I started my work with Stanley, he was studying string perturbation theory, with the ultimate goal to make string amplitudes explicitly calculable. I was not really that interested in the subject of string theory, but I had decided I wanted Stanley as my supervisor if he would take me. I told him this much, and fortunately he agreed to supervise my research. The primary problem Stanley was addressing was to show that the divergence associated with the dilaton tadpole cancels upon summing over spin structures in Type-II Superstring theory. Alongside this problem, there were some smaller issues that needed to be resolved to demonstrate that string theory was finite and allows for explicit calculations. Some of these smaller problems I was tasked to look into. For one thing, string amplitudes as calculated from the standard formalism are real, since they are calculated in imaginary time. They contain no imaginary parts and are not unitary. My task was basically to devise an analytic continuation of the amplitudes to real time, which I did at the one-loop level and demonstrated some general features for higher loop orders.[1] There is also a divergence in string amplitudes when single particle intermediate states have the same energy as an external state, requiring a sort of mass renormalization. I solved the problem using the light-cone formalism.[2] This result complemented the treatment of the same problem through the covariant formalism.[3–5]

I used to have meetings with Stanley at his office in Birge Hall about once a week, at least in the beginning and then it became less frequent in later years. He was generous with his time and always very pleasant to talk with. The majority of my discussions with Stanley had to do with anything aside from my thesis topic. I was interested in many areas of theoretical physics. Stanley would be willing to discuss whatever topic of interest I brought up in a given week and he rarely ever told me to focus more time on my main thesis project. He was very good at leaving convention aside and would readily be willing to think differently. At times our discussions could get quite lively. Stanley had simple ways of seeing the most complex problems and he was good at reducing complicated analytics to their basic steps.

At the time when I was doing my PhD, string theory had entered a new era. The theory started in the early 70s, with Stanley as one of the pioneers, and had as its main goal to explain strong interaction physics. Regge trajectories, the linear relation between the mass squared of strong interaction resonances and their angular momentum, found a dynamical model in string theory, which was a key motivation for its initial development. However by the mid-80s this direction seemed to have been exhausted and abandoned and the emerging view was string theory may offer a renormalizable theory of gravity. My interest in physics has always been in areas that were at least partially in reach of empirical testing, and so working on theories of quantum gravity seemed a step too far. This was a key reason I initially had some reservations about working on string theory. Before returning to string theory in the 80s, Stanley had made significant contributions to the study of QCD and confinement and in the 60s of course was a central figure in what back then was called strong interaction physics. These were area I was particularly interested in and many of my sessions with Stanley were taken by discussions along these lines. Sometimes after a very fruitful discussion on this subject with Stanley, getting into the thick of calculation and then relating that to hadron spectrum, Wilson loops or other measurable quantities, it became a running joke between us where I would kid with him that perhaps someday he may consider doing testable physics again, to which he would reply with some amusement, "Ok, right".

Stanley was always friendly, patient, and at times could be humorous. However there was always a formal divide of supervisor and student between him and me. Rarely did he ever talk about anything at a personal level with me. When I had completed my PhD and was about to leave Berkeley, my parents came and invited Stanley to dinner, which he graciously accepted. They had never met each other before, but I have never seen Stanley talk this much and so openly about his past and his life, as during that evening with my parents. Perhaps since all three were from about the same generation it made a difference, but the focal point seemed to be that all of them had lived in the UK during the 50s. Stanley talked with fondness of his time from back then, as did my parents.

Once I completed my PhD, I started working on perturbative QCD and then later on cosmology. Aside for finishing the papers from my PhD work and getting them published, I did not work on string theory again. Thus from a research perspective my overlap with Stanley had ended. Nevertheless, in the years since I left Berkeley, I continued my contact with Stanley, mostly through occasional emails and less occasional visits back to Berkeley. There was always a consistency about Stanley. His emails were brief, to the point, yet had a pleasantness about them. I met him on a handful of occasions in the intervening years since leaving Berkeley. Even though for much of that time I have now been in the UK, as I still had other links in California, it provided opportunities to go up to Berkeley. The same consistency would persist in Stanley when I would meet him in person, and each time I would feel like we were back to my graduate days. The gradual ageing was of course there in him (and me). But his sharpness of mind, his friendly disposition, and that tinge of humor were never changing.

In writing this essay, I was paging through my thesis, which basically was just combining together the drafts of the two papers I was writing at the time and a third paper for the future. In the acknowledgement when I thanked the several colleagues and professors from Berkeley with whom I had interacted, I expressed only the briefest gratitude to Stanley by saying, my thanks to him will come from every paper and result in Physics I produce, each of which he can consider as a product of his training. In hindsight I think that was a very accurate statement and I would hope Stanley felt the time he gave me was worthwhile. I am sure he has influenced so many others through his thoughts, ideas and his gentle but determined demeanor. He is amongst that rare category of powerful thinkers who morph into spirits that continue to live on in many others. I always found him humble about his scientific contributions but his insights have been influential to Physics. Stanley Mandelstam, Man of Science, a gentleman and a scholar, rest in peace.

References

1. A. Berera, Unitary string amplitudes, *Nucl. Phys.* **411**, 157 (1994).
2. A. Berera, All-order mass renormalization of string theory, *Phys. Rev. D* **49**, 6674 (1994).
3. S. Weinberg, The Oregon meeting, in *Proceedings of the Annual Meeting of the Division of Particles and Fields of the APS*, Eugene, Oregon, 1985, ed. R. C. Hwa (World Scientific, 1986).
4. N. Seiberg, Anomalous dimensions and mass renormalization in string theory, *Phys. Lett. B* **187**, 56 (1987).
5. A. Sen, Mass renormalization and (BRST) anomaly in string theories, *Nucl. Phys. B* **304**, 403 (1988).

Reprints and Abstracts of Selected Publications

Determination of the Pion-Nucleon Scattering Amplitude from Dispersion Relations and Unitarity. General Theory

S. MANDELSTAM*
Department of Physics, Columbia University, New York, New York
(Received June 27, 1958)

A method is proposed for using relativistic dispersion relations, together with unitarity, to determine the pion-nucleon scattering amplitude. The usual dispersion relations by themselves are not sufficient, and we have to assume a representation which exhibits the analytic properties of the scattering amplitude as a function of the energy and the momentum transfer. Unitarity conditions for the two reactions $\pi+N \to \pi+N$ and $N+\bar{N} \to 2\pi$ will be required, and they will be approximated by neglecting states with more than two particles. The method makes use of an iteration procedure analogous to that used by Chew and Low for the corresponding problem in the static theory. One has to introduce two coupling constants; the pion-pion coupling constant can be found by fitting the sum of the threshold scattering lengths with experiment. It is hoped that this method avoids some of the formal difficulties of the Tamm-Dancoff and Bethe-Salpeter methods and, in particular, the existence of ghost states. The assumptions introduced are justified in perturbation theory.

As an incidental result, we find the precise limits of the region for which the absorptive part of the scattering amplitude is an analytic function of the momentum transfer, and hence the boundaries of the region in which the partial-wave expansion is valid.

1. INTRODUCTION

IN recent years dispersion relations have been used to an increasing extent in pion physics for phenomenological and semiphenomenological analyses of experimental data,[1] and even for the calculation of certain quantities in terms of the pion-nucleon scattering amplitude.[2] It is therefore tempting to ask the question whether or not the dispersion relations can actually replace the more usual equations of field theory and be used to calculate all observable quantities in terms of a finite number of coupling constants—a suggestion first made by Gell-Mann.[3] At first sight, this would appear to be unreasonable, since, although it is necessary to use all the general principles of quantum field theory to derive the dispersion relations, one does not make any assumption about the form of the Hamiltonian other than that it be local and Lorentz-invariant. However, in a perturbation expansion these requirements are sufficient to specify the Hamiltonian to within a small number of coupling constants if one demands that the theory be renormalizable and therefore self-consistent. It is thus very possible that, even without a perturbation expansion, these requirements are sufficient to determine the theory. In fact, if the "absorptive part" of the scattering amplitude, which appears under the integral sign of the dispersion relations, is expressed in terms of the scattering amplitude by means of the unitarity condition, one obtains equations which are very similar to the Chew-Low[4] equations in static theory. These equations have been used by Salzman and Salzman[5] to obtain the pion-nucleon scattering phase shifts.

It is the object of this paper to find a relativistic analog of the Chew-Low-Salzman method, which could be used to calculate the pion-nucleon scattering amplitude in terms of two coupling constants only. As in the static theory, the unitarity equation will involve the transition amplitude for the production of an arbitrary number of mesons, and, in this case, of nucleon pairs as well. In order to make the equations manageable, it is necessary to neglect all but a finite number of processes; as a first approximation, the "one-meson" approximation, we shall neglect all processes except elastic scattering.

The equations obtained from the dispersion relations and the one-meson approximation differ from the static Chew-Low equations in two important respects. Whereas, in the static theory, there was only P-wave scattering, we now have an infinite number of angular momentum states, and the crossing relation, if expressed in terms of angular momentum states, would not converge. Further, in the relativistic theory, the dispersion relations involve the scattering amplitude in the "unphysical" region, i.e., through angles whose cosine is less than -1. For these reasons, the method of procedure will be more involved than in the static theory. We shall require, not only the analytic properties of the scattering amplitude as a function of energy for fixed momentum transfer, which are expressed by the dispersion relations, but its analytic properties as a function of both variables. The required analytic properties have not yet been proved to be consequences of microscopic causality. In order to carry out the proof,

---

* Now at the Department of Physics, University of California, Berkeley, California.
[1] Chew, Goldberger, Low, and Nambu, Phys. Rev. **106**, 1345 (1957). This paper contains further references.
[2] Chew, Karplus, Gasiorowicz, and Zachariasen, Phys. Rev. **110**, 265 (1958).
[3] M. Gell-Mann, *Proceedings of the Sixth Annual Rochester Conference High-Energy Physics, 1956* (Interscience Publishers, Inc., New York, 1956), Sec. III, p. 30.

[4] G. F. Chew and F. E. Low, Phys. Rev. **101**, 1570 (1956).
[5] G. Salzman and F. Salzman, Phys. Rev. **108**, 1619 (1957).

one would almost certainly have to consider simultaneously several Green's functions together with the equations connecting them which follow from unitarity. It is unlikely that such a program will be carried through in the immediate future. However, if the solution obtained by the use of these analytic properties were to be expanded in a perturbation series, we would obtain precisely those terms of the usual perturbation series included in the one-meson approximation. The assumed analytic properties are, therefore, probably correct, at any rate in the one-meson approximation.

As we have to resort to perturbation theory in order to justify our assumptions, we do not yet have a theory in which the general principles of quantum theory are supplemented only by the assumption of microscopic causality. Nevertheless, the approximation scheme used has several advantages over the approximations previously applied to this problem, such as the Tamm-Dancoff or Bethe-Salpeter approximations. It refers throughout only to renormalized masses and coupling constants. The Tamm-Dancoff equations, by contrast, are unrenormalizable in higher approximations and the Bethe-Salpeter equations, while they are covariant and therefore renormalizable in all approximations, present difficulties of principle when one attempts to solve them. Further, we may hope that the one-meson approximation is more accurate than the Tamm-Dancoff approximations. The latter assumes that those components of the state vector containing more than a certain number of bare mesons are negligibly small—an approximation that is known to be completely false for the experimental value of the coupling constant. The one-meson approximation, on the other hand, assumes that the cross section for the production of one or more *real* mesons is small except at high energies. While this approximation is certainly not quantitatively correct, it is nevertheless probably a good deal more accurate than the Tamm-Dancoff approximation. Finally, the one-meson approximation, unlike the Tamm-Dancoff or Bethe-Salpeter approximations, possesses crossing symmetry. Now it is very probable that the "ghost states" which have been plaguing previous solutions of the field equations are due to the neglect of crossing symmetry. As evidence of this, we may cite the case of charged scalar theory without recoil, for which the one-meson approximation has been solved completely.[6,7] The solution obtained with neglect of the crossing term possesses the usual ghost state if the source radius is sufficiently small. The Lee model,[8] which has no crossing symmetry, shows a similar behavior. If the crossing term in the charged scalar model is included, however, there is no ghost state.

It has been pointed out by Castillejo, Dalitz, and Dyson[7] that the dispersion relations, at any rate in the charged scalar model, do not possess a unique solution.

This might have been expected, since it is possible to alter the Hamiltonian without changing the dispersion relations. One simply has to introduce into the theory a baryon whose mass is greater than the sum of the masses of the meson and nucleon. Such a baryon would be unstable, and would therefore not appear as a separate particle or contribute a term to the dispersion relations. In perturbation theory, the simplest of the solutions found by Castillejo, Dalitz, and Dyson, i.e., the solution without any zero in the scattering amplitude, agrees with the solution obtained from a Hamiltonian in which there are no unstable particles, and the more complicated solutions correspond to the existence of unstable baryons. We shall assume that this is so independently of perturbation theory, and shall concern ourselves with the simplest solution. There is no physical reason why one of the other solutions may not be the correct one, but it seems worthwhile to try to compare with experiment the consequences of a theory without unstable particles. It should in any case be emphasized that the ambiguity is not a specific feature of this method of solution, but is inherent in the theory itself. The difference is that, in other methods, it occurs in writing down the equations, whereas in this method it occurs in solving them.

In Sec. 2 we shall discuss the analytic properties of the scattering amplitude, and, in Sec. 3, we shall show how these properties can be used together with the unitarity condition to solve the problem. We shall in this section ignore the "subtraction terms" in the dispersion relations. As in the corresponding static problem, we have to use an iteration procedure in which the crossing term is taken from the result of the previous iteration. The details of this solution will be entirely different from the static problem, the reason being that the part of the amplitude corresponding to the lowest angular momentum states, which is a polynomial in the momentum transfer, actually appears as a subtraction term in the dispersion relation with respect to this variable and has thus not yet been taken into account. In this and the next section we shall also be able to specify details of the analytic representation that were left undetermined in Sec. 2, in particular, we shall be able to give precise limits to the values of the momentum transfer within which the partial-wave expansion converges. In Sec. 4 we shall investigate the subtraction terms in the dispersion relations. We shall find that, in order to determine them, we shall require the unitarity condition for the lowest angular momentum states, not only in pion-nucleon scattering, but also in the pair-annihilation reaction $N+\bar{N} \rightarrow 2\pi$, which is represented by the same Green's function. The coupling constant for meson-meson scattering is thus introduced into the theory; as its value is not known experimentally it will have to be determined by fitting one of the results of the calculation, such as the sum of the S-wave scattering lengths at threshold, with experiment. The calculations

---

[6] T. D. Lee and R. Serber (unpublished).
[7] Castillejo, Dalitz, and Dyson, Phys. Rev. **101**, 453 (1956).
[8] K. W. Ford, Phys. Rev. **105**, 320 (1957).

of these low angular momentum states would be done in the same spirit as the Chew-Low calculations, and the details will not be given in this paper. We thus have a procedure in which the first few angular momentum states are calculated by methods similar to those used in the static theory, while the remaining part of the scattering amplitude, which will be called the "residual part," is calculated by a different procedure which does not make use of a partial-wave expansion. Needless to say, the two parts of the calculation become intermingled by the iteration procedure.

It is only in the calculation of the subtraction terms that use has to be made of the unitarity condition for the pair-annihilation reaction. For the residual part, it is only necessary to use the unitarity condition for pion-nucleon scattering. Had it been possible to use the unitarity condition exactly instead of in the one-meson approximation, the result would also satisfy the unitarity condition for the annihilation reaction in a consistent theory. As it is, we find that the residual part consists of a number of terms which correspond to various intermediate states in the annihilation reaction. In Sec. 5 it is pointed out that the calculation is greatly simplified if we keep only those terms of the residual part corresponding to pair annihilation through states with fewer than a certain number of particles. Such an approximation has already been made in calculating the subtraction terms. The unitarity condition for pion-nucleon scattering is no longer satisfied except for the low angular momentum states. However, the terms neglected are of the order of magnitude of, and probably less than, terms already neglected. The two reactions of pion-nucleon scattering and pair annihilation are now treated on an equivalent footing.

It will be found that the unitarity condition, in the one-meson approximation, cannot be satisfied at all energies if crossing symmetry and the analytic properties are to be maintained. The reason is that the unitarity condition for the scattering reaction is not completely independent of the unitarity condition for the "crossed" reaction with the two pions interchanged, and they contradict one another if an approximation is made. There is, of course, no difficulty in the region where the one-meson approximation is exact. For sufficiently small values of the coupling constant, we shall still be able to obtain a unique procedure. For values of the coupling constant actually encountered, one part of the crossing term may have to be cut off at the threshold for pair production in pion-nucleon scattering. It is unlikely that the result will be sensitive to the form and the precise value of the cutoff.

2. DISPERSION RELATIONS AND ANALYTICITY PROPERTIES OF THE TRANSITION AMPLITUDE

The kinematical notation to be used in writing down the dispersion relations will be similar to that of Chew et al.[1] The momenta of the incoming and outgoing pions will be denoted by q_1 and q_2, those of the incoming and outgoing nucleons by p_1 and p_2. We can then define two invariant scalars

$$\nu = -(p_1+p_2)(q_1+q_2)/4M, \quad (2.1)$$

$$t = -(q_1-q_2)^2. \quad (2.2)$$

The latter is minus the square of the invariant momentum transfer. The laboratory energy will be given by the equation

$$\omega = \nu - (t/4M). \quad (2.3a)$$

It is more convenient to use, instead of the laboratory energy, the square of the center-of-mass energy (including both rest-masses), which is linearly related to it by the equation

$$s = M^2 + \mu^2 + 2M\omega. \quad (2.3b)$$

The Green's function relevant to the process under consideration,

$$\pi_1 + N_1 \to \pi_2 + N_2, \quad \text{(I)}$$

also gives the processes

$$\pi_2 + N_1 \to \pi_1 + N_2 \quad \text{(II)}$$

and

$$N_1 + \bar{N}_2 \to \pi_1 + \pi_2. \quad \text{(III)}$$

The matrix elements for the process II can be obtained from those for the process I by crossing symmetry; the laboratory energy and the square of the center-of-mass energy will now be

$$\omega_c = -\nu - (t/4M) = -\omega - (t/2M), \quad (2.4a)$$

$$s_c = M^2 + \mu^2 + 2M\omega_c = 2M^2 + 2\mu^2 - s - t. \quad (2.4b)$$

The square of the momentum transfer will be $-t$ as before. For the process III, the square of the center-of-mass energy will be t. The square of the momentum transfer between the nucleon N_1 and the pion π_2 will be s_c and that between the nucleon N_1 and the pion π_1 will be s.

The kinematics for the three reactions are represented diagrammatically in Fig. 1 in which t has been plotted against ν. AB represents the line $s=(M+\mu)^2$, or $\omega = \mu$, and lines for which s is constant will be parallel to it. The region for which the process I is energetically possible is therefore that to the right of AB. However, only the shaded part of this area is the "physical region"; in the unshaded part, though the energy of the meson is greater than its rest-mass, the cosine of the scattering angle is not between -1 and $+1$. The physical region is bounded above by the line $t=0$, i.e., the line of forward scattering, and below by the line of backward scattering. Similarly CD is the line $s_c = (M+\mu)^2$; the region for which the process II is energetically possible is that to the left of CD, and the shaded area represents the physical region for this reaction. Lines of constant energy for the reaction III are horizontal lines. The reaction will be energetically possible above the line EF, at which $t=4M^2$, and again the shaded area represents the physical region.

We now examine the analytic properties of the scattering amplitude. To simplify the writing, we shall first neglect spin and isotopic spin; the transition amplitude will then be a scalar function $A(\nu,t)$ of the two invariants ν and t. Its analytic properties as a function of ν, with t constant, are exhibited by the usual dispersion relations

$$A(\nu,t) = \frac{g^2}{2M}\left(\frac{1}{\nu_B-\nu}+\frac{1}{\nu_B+\nu}\right) + \frac{1}{\pi}\int_{\mu+(t/4M)}^{\infty} d\nu' \frac{A_1(\nu',t)}{\nu'-\nu}$$
$$-\frac{1}{\pi}\int_{-\infty}^{-\mu-(t/4M)} d\nu' \frac{A_2(\nu',t)}{\nu'-\nu}, \quad (2.5)$$

where $\nu_B = -(\mu^2/2M)+(t/4M)$. In this and all subsequent such equations, the energy denominators are taken to have a small imaginary part. A_1 and A_2 are the "absorptive parts" associated with the reactions I and II, respectively, and are given by the equations

$$(2\pi)^4 A_1(\nu,t)\delta(p_1+q_1-p_2-q_2) = (2\pi)^6 \left(\frac{4p_{01}p_{02}q_{01}q_{02}}{M^2}\right)^{\frac{1}{2}}$$
$$\times \sum_n \langle N(p_1)\pi(q_1)|n\rangle\langle n|N(p_2)\pi(q_2)\rangle, \quad (2.6)$$

$$(2\pi)^4 A_2(\nu,t)\delta(p_1+q_1-p_2-q_2) = (2\pi)^6 \left(\frac{4p_{01}p_{02}q_{01}q_{02}}{M^2}\right)^{\frac{1}{2}}$$
$$\times \sum_n \langle N(p_1)\pi(-q_2)|n\rangle\langle n|N(p_2)\pi(-q_1)\rangle. \quad (2.7)$$

The symbol $\langle N(p_1)\pi(q_1)|$ denotes a state with an ingoing nucleon of momentum p_1 and an ingoing pion of momentum q_1. The sum $\sum_n$ is to be taken over all intermediate states. A_1 and A_2 are nonzero to the right of AB, and to the left of CD, respectively.

Equation (2.5) indicates that A is an analytic function of ν in the complex plane, with poles at $\pm\nu_B$, and cuts along the real axis from $\mu+(t/4M)$ to ∞ and from $-\infty$ to $-\mu-(t/4M)$.

On Fig. 1, (2.5) will be represented by an integration along a horizontal line below the ν axis. The poles will occur where this line crosses the dashed lines; apart from them, the integrand will be zero between AB and CD. Except for forward scattering, the region where the integrand is nonzero will lie partly in the unphysical region, where the energy is above threshold but the angle imaginary.

Equation (2.5) is only true as it stands if the functions A, A_1, and A_2 tend to zero sufficiently rapidly as ν tends to infinity; otherwise it will be necessary to perform one or more subtractions in the usual way. Whenever such a dispersion relation is written down,

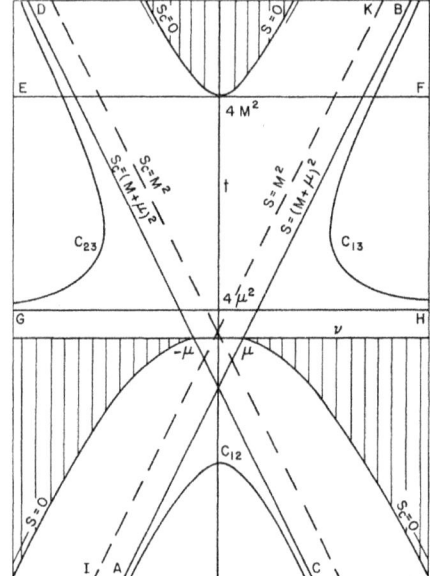

FIG. 1. Kinematics of the reactions I, II, and III.

the possibility of having to perform subtractions is implied.

We next wish to obtain analytic properties of A as a function of t. In order to do this we shall write the scattering amplitude, not as the expectation value of the time-ordered product of the two meson current operators between two one-nucleon states, as is done in the proof of the usual dispersion relations,[9,10] but as the expectation value of the product of a meson current operator and a nucleon current operator between a nucleon state and a meson state. Thus

$$(2\pi)^4 A\delta(p_1+p_2-q_1-q_2) = (2\pi)^3 \left(\frac{2p_{01}q_{02}}{M}\right)^{\frac{1}{2}} i\int dx\,dx'$$
$$\times e^{-iq_1 x+ip_2 x'}\langle N(p_1)|T\{j(x)\bar{a}(x')\}|\pi(q_2)\rangle, \quad (2.8)$$

where $a(x')$ is a nucleon current operator. From this expression, we can obtain dispersion relations in which the momentum transfer between the incoming nucleon and the outgoing pion, rather than between the two nucleons, is kept constant—the proof is exactly the same as the usual heuristic proof of the ordinary dispersion relations.[9,10] As this momentum transfer is just s_c, we obtain dispersion relations in which s_c is kept constant; if A is written as a function of s_c and t, they

---

[9] M. L. Goldberger, Phys. Rev. **99**, 979 (1955).
[10] R. H. Capps and G. Takeda, Phys. Rev. **106**, 1337 (1956).

take the form

$$A(s_c,t) = \frac{g^2}{s_c+t-M^2-2\mu^2} - \frac{1}{\pi}\int_{-\infty}^{(M-\mu)^2-s_c} dt' \frac{A_1(s_c,t')}{t'-t}$$
$$+ \frac{1}{\pi}\int_{4\mu^2}^{\infty} dt' \frac{A_3(s_c,t')}{t'-t}. \quad (2.9)$$

The absorptive parts in the integrand are as usual obtained by replacing the time-ordered product in (2.8) by half the commutator. The first term, in which the operators are in the order $j(x)\bar{a}(x')$, is exactly A_1, and will therefore be nonzero to the right of AB and have a δ function along IK. The second term, however, in which the operators are in the order $\bar{a}(x')j(x)$, will now be related to the process III. It will be given by the equation

$$(2\pi)^4 A_3(s_c,t)\delta(p_1+q_1-p_2-q_2) = (2\pi)^6 \left(\frac{4p_{01}p_{02}q_{01}q_{02}}{M^2}\right)^{\frac{1}{2}}$$
$$\times \sum_n \langle N(p_1)\bar{N}(-p_2)|n\rangle\langle n|\pi(-q_1)\pi(q_2)\rangle. \quad (2.10)$$

The state n of lowest energy will now be the two-meson state. A_3 will therefore be nonzero above the line $t=4\mu^2$, represented by GH in Fig. 1 (since t is square of the center-of-mass energy of the process III). The dispersion relation (2.10) is represented by an integration along a line parallel to CD and to the right of the line $s_c=0$. It implies that A is an analytic function of t for fixed s_c, with a pole at $t=M^2+2\mu^2-s_c$, and cuts along the real axis from $-\infty$ to $(M-\mu)^2-s_c$ and from $4\mu^2$ to ∞.

As in the usual dispersion relation, part of the range of integration in Eq. (2.9) will lie in the unphysical region. This region now includes, besides imaginary angles at permissible energies, the entire area between the lines $t=4\mu^2$ and $t=4M^2$, where there are contributions to A_3 from intermediate states with two or more pions. The rigorous proof of (2.9) is therefore much more difficult than that of (2.5), and probably cannot be carried out without introducing the unitarity equations.

By interchanging the two pions in the expression (2.8), we can obtain a third dispersion relation in which s is kept constant:

$$A(s,t) = \frac{g^2}{s+t-M^2-2\mu^2} - \frac{1}{\pi}\int_{-\infty}^{(M-\mu)^2-s} dt' \frac{A_2(s,t')}{t'-t}$$
$$+ \frac{1}{\pi}\int_{4\mu^2}^{\infty} dt' \frac{A_3(s,t')}{t'-t}. \quad (2.11)$$

On Fig. 1, this would be represented by an integration along a line parallel to AB, and to the left of the line $s=0$.

Let us now try to obtain the analytic properties of A considered as a function of two complex variables. The simplest assumption we could make is that it is analytic in the entire space of the two variables except for cuts along certain hyperplanes. We can then determine the location of the cuts from the requirement that A must satisfy the dispersion relations (2.5), (2.9), and (2.11); there will be a cut when s is real and greater than $(M+\mu)^2$, a cut when s_c is real and greater than $(M+\mu)^2$, and a cut when t is real and greater than $4\mu^2$. The discontinuities across these cuts will be, respectively, $2A_1$, $2A_2$, and $2A_3$. In addition, A will have poles when $s=M^2$ and when $s_c=M^2$. By a double application of Cauchy's theorem, it can be shown that a function with cuts and poles in these positions can be represented in the form

$$A = \frac{g^2}{M^2-s} + \frac{g^2}{M^2-s_c} + \frac{1}{\pi^2}\int_{(M+\mu)^2}^{\infty} ds' \int_{4\mu^2}^{\infty} dt' \frac{A_{13}(s',t')}{(s'-s)(t'-t)}$$
$$+ \frac{1}{\pi^2}\int_{(M+\mu)^2}^{\infty} ds_c' \int_{4\mu^2}^{\infty} dt' \frac{A_{23}(s_c',t')}{(s_c'-s_c)(t'-t)}$$
$$+ \frac{1}{\pi^2}\int_{(M+\mu)^2}^{\infty} ds' \int_{(M+\mu)^2}^{\infty} ds_c' \frac{A_{12}(s',s_c')}{(s'-s)(s_c'-s_c)}. \quad (2.12)$$

This is a generalization of a representation first suggested by Nambu.[11] While we have for convenience used the three variables s, s_c, and t, which are the energies of the three processes, they are connected by the relation

$$s+s_c+t = 2(M^2+\mu^2), \quad (2.13)$$

so that A is really a function of two variables only. A_{13}, A_{23} and A_{12}, which will be referred to as the "spectral functions," are nonzero in the regions indicated at the top right, top left and bottom of Fig. 1. The precise boundaries C_{13}, C_{23}, and C_{12} of the regions will be determined by unitarity in the following sections; from the reasoning given up till now, all that can be said is that the regions must lie within the respective triangles as indicated, and that the boundary must approach the sides of the triangles asymptotically (or it could touch them at some finite point). The spectral functions are always zero in the physical region.

As in the case of ordinary dispersion relations, the representation (2.12) will not be true as it stands, but will require subtractions. The subtractions will modify one or both of the energy denominators in the usual way and, in addition, they will require the addition of extra terms. These terms will not now be constants, but functions of one of the variables, e.g., if there is a subtraction in the s integration of the first term, the extra term will be a function of t. These functions must then have the necessary analytic properties in their

---

[11] Y. Nambu, Phys. Rev. **100**, 394 (1955).

variables, so that they will have the form

$$\frac{1}{\pi}\int_{(M+\mu)^2}^{\infty}ds'\frac{f_1(s')}{s'-s}+\frac{1}{\pi}\int_{(M+\mu)^2}^{\infty}ds_c'\frac{f_2(s_c')}{s_c'-s_c}$$
$$+\frac{1}{\pi}\int_{4\mu^2}^{\infty}dt'\frac{f_3(t')}{t'-t}. \quad (2.14)$$

If more than one subtraction is involved, we may have similar terms multiplied by polynomials. Even if the spectral functions in (2.12) tend to zero as one of the variables tends to infinity, so that no subtraction in that variable is necessary, it is still not precluded that the corresponding term in (2.14) does not appear, as the function still has the required analytic properties. For pion-nucleon scattering, however, there is no underdetermined over-all term, independent of both variables, to be added, as the requirement that the scattering amplitude for each angular momentum wave have the form $e^{i\delta}\sin\delta/k$, with $\mathrm{Im}\delta<0$, forces A to tend to zero in the physical region when both s and t become infinite.

The Nambu representations for the complete Green's functions are known to be invalid, even in the lowest nontrivial order of perturbation theory. The representation quoted here, however, restricts itself to the mass shells of the particles, and has not been shown to be invalid. In fact, in the case of Compton scattering, the fourth-order terms, which have been worked out by Brown and Feynman,[12] are found to have this representation, and, as we have stated in the introduction, all the perturbation terms included in the one-meson approximation can be similarly represented.

The dispersion relations are an immediate consequence of the representation (2.12). To obtain the usual dispersion relation (2.5), the third integral in (2.12) must be written as[13]

$$-\frac{1}{\pi^2}\int_{(M+\mu)^2}^{\infty}ds'\int_{-\infty}^{t_2(s)}dt'\frac{A_{12}(s',t')}{(s'-s)(t'-t)}$$
$$-\frac{1}{\pi^2}\int_{(M+\mu)^2}^{\infty}ds_c'\int_{-\infty}^{t_2(s_c)}dt'\frac{A_{12}(s_c',t')}{(s_c'-s_c)(t'-t)}.$$

It then follows that

$$A=\frac{g^2}{M^2-s}+\frac{g^2}{M^2-s_c}+\frac{1}{\pi^2}\int_{(M+\mu)^2}^{\infty}ds'\frac{A_1(s',t)}{s'-s}$$
$$+\frac{1}{\pi^2}\int_{(M+\mu)^2}^{\infty}ds_c'\frac{A_2(s_c',t)}{s_c'-s_c}, \quad (2.15)$$

---

[12] L. M. Brown and R. P. Feynman, Phys. Rev. **85**, 231 (1952).
[13] When we make a change of variables, we imply of course that the spectral functions still have the same value at the same point, and not that we must take the same function of the new variables.

where

$$A_1(s,t)=\frac{1}{\pi}\int_{t_1(s)}^{\infty}dt'\frac{A_{13}(s,t')}{t'-t}$$
$$-\frac{1}{\pi}\int_{-\infty}^{t_2(s)}dt'\frac{A_{12}(s,t')}{t'-t}, \quad (2.16)$$

$$A_2(s_c,t)=\frac{1}{\pi}\int_{t_1(s_c)}^{\infty}dt'\frac{A_{23}(s_c,t')}{t'-t}$$
$$-\frac{1}{\pi}\int_{-\infty}^{t_2(s_c)}dt'\frac{A_{12}(s_c,t')}{t'-t}. \quad (2.17)$$

Equation (2.15) is, however, just the dispersion relation (2.5), since s, s_c, and ν are connected by the relations (2.4) and t is being kept constant. We also see that the absorptive parts A_1 and A_2 themselves satisfy dispersion relations in t, with s (or s_c) constant; the imaginary parts which appear in the integrand are now simply the spectral functions. Equation (2.16) will be represented in Fig. 1, by an integration along a line parallel to AB and to the right of it. The limits t_1 and t_2 are the points at which this line crosses the curves C_{13} and C_{12}. They satisfy the inequalities

$$t_1>4\mu^2, \quad (2.18a)$$
$$t_2<(M-\mu)^2-s. \quad (2.18b)$$

A_1 will be nonzero for $s>(M+\mu)^2$, as it should, as long as the curves C_{13} and C_{12} approach the line AB at some point and do not cross it.

The dispersion relations (2.9) and (2.11) can be proved from (2.12) in a similar way; the absorptive part A_3 will then satisfy a dispersion relation in ν with s constant:

$$A_3=\frac{1}{\pi}\int_{\nu_3(t)}^{\infty}d\nu'\frac{A_{13}(\nu',t)}{\nu'-\nu}-\frac{1}{\pi}\int_{-\infty}^{-\nu_3(t)}d\nu'\frac{A_{23}(\nu',t)}{\nu'-\nu}. \quad (2.19)$$

This dispersion relation will be represented by an integration along a horizontal line above GH. ν_3 and $-\nu_3$ will be the points at which the line of integration crosses C_{13} and C_{23}.

Finally, then, the scattering amplitude A satisfies dispersion relations in which any of the quantities t, s_c, and s are kept constant. Further, it follows from (2.12), by the reasoning just given, that the values of the quantity which is being kept constant need no longer be restricted in sign. Thus, for example, we now know the analytic properties of A, as a function of momentum transfer, for fixed energy greater than (as well as less than) $(M+\mu)^2$. They are given by the dispersion relation (2.11), so that A is an analytic function of the square of the momentum transfer, with a pole at $t=M^2+2\mu^2-s$, and cuts along the real axis from $t=4\mu^2$ to ∞ and from $t=-\infty$ to $(M-\mu)^2-s$. For $s>(M+\mu)^2$, these cuts and poles are entirely in the nonphysical region. It has already been shown rigorously by Lehmann[14] that A is analytic in t in an area including the physical region. The absorptive parts A_1, A_2 and A_3 will themselves satisfy dispersion relations, provided that the correct variable be kept constant (s, s_c, and t for A_1, A_2, and A_3, respectively). The weight functions for these dispersion relations are entirely in the nonphysical region, and the boundaries of the areas in which they are nonzero are yet to be determined. In particular, we see that the absorptive part A_1 has the same analytic properties as a function of the momentum transfer [for s constant and greater than $(M+\mu)^2$] as the scattering amplitude, except that there is now no pole, and the cuts only extend from t_1 to ∞ and from $-\infty$ to t_2. According to the inequalities (2.14), these cuts do not reach as far inward as the cuts of A considered as a function of the momentum transfer. This agrees with another result of Lehmann[14] who showed that the region of analyticity of A_1 as a function of t was larger than the region of analyticity of A as a function of t.

The modifications introduced into the theory by spin and isotopic spin are trivial. The transition amplitude will now be given by the expression

$$-A+\tfrac{1}{2}i\gamma(q_1+q_2)B, \quad (2.20)$$

and both A and B will have representations of the form (2.12). There will, further, be two amplitudes corresponding to isotopic spins of $\tfrac{1}{2}$ and $\tfrac{3}{2}$. It is sometimes more convenient to use the combinations

$$A^{(+)} = \tfrac{1}{3}(A^{(\frac{1}{2})} + 2A^{(\frac{3}{2})}), \quad (2.21a)$$

$$A^{(-)} = \tfrac{1}{3}(A^{(\frac{1}{2})} - A^{(\frac{3}{2})}), \quad (2.21b)$$

and similar combinations $B^{(+)}$ and $B^{(-)}$. We then have the simple crossing relations

$$A^{(\pm)}(\nu,t) = \pm A^{(\pm)}(-\nu,t), \quad (2.22a)$$

$$B^{(\pm)}(\nu,t) = \mp B^{(\pm)}(-\nu,t), \quad (2.22b)$$

or, in terms of the spectral functions,

$$A_{13}^{(\pm)}(s,t) = \pm A_{23}^{(\pm)}(s_c,t), \quad (2.23a)$$

$$A_{12}^{(\pm)}(s,s_c) = \pm A_{12}^{(\pm)}(s_c,s), \quad (2.23b)$$

$$B_{13}^{(\pm)}(s,t) = \mp B_{23}^{(\pm)}(s_c,t), \quad (2.23c)$$

$$B_{12}^{(\pm)}(s,s_c) = \mp B_{12}^{(\pm)}(s_c,s). \quad (2.23d)$$

The poles in (2.12) and in the dispersion relations will only occur in the representation for $B^{(\pm)}$ (in pseudoscalar theory), and the second term will have a minus or plus sign in the equations for $B^{(+)}$ and $B^{(-)}$, respectively.

3. COMBINATION OF THE DISPERSION RELATIONS WITH THE UNITARITY CONDITION

The dispersion relations given in the previous section must now be combined with the unitarity equations in

[14] H. Lehmann (to be published).

order to determine the scattering amplitude. We shall again begin by neglecting spin and isotopic spin; the unitarity condition (2.7) then becomes, in the one-meson approximation,

$$A_1(s,\cos\theta_1) = \frac{1}{32\pi^2}\frac{q}{W}\int \sin\theta_2 d\theta_2 d\phi_2\, A^*(s,\cos\theta_2)$$
$$\times A(s,\cos(\theta_1,\theta_2)),$$

or

$$A_1(s,z_1) = \frac{1}{32\pi^2}\frac{q}{W}\int_{-1}^{1} dz_2 \int_0^{2\pi} d\phi\, A^*(s,z_2)$$
$$\times A\{s, z_1 z_2 + (1-z_1^2)^{\frac{1}{2}}(1-z_2^2)^{\frac{1}{2}}\cos\phi\}, \quad (3.1)$$

where $z=\cos\theta$ and $\theta_i (i=1,2)$ is a unit vector in the (θ_i,ϕ_i) direction. W is the center-of-mass energy (equal to $\sqrt{s}$), and q is the momentum in the center-of-mass system, given by the equation

$$q^2 = \{s-(M+\mu)^2\}\{s-(M-\mu)^2\}/4s. \quad (3.2)$$

z is related to the momentum transfer by the simple relation

$$z = 1 + (t/2q^2). \quad (3.3)$$

The unitarity requirements only prove that Eq. (3.2) is true in the physical region. A_1 must then be obtained in the unphysical region by analytic continuation. In order to do this, A can be expressed as an analytic function of t or, equivalently, of z, by means of Eq. (2.11), in which the energy is kept fixed. Equation (3.3) shows that we can simply replace t by z in (2.12), so that we may write

$$A^*(s,z_2) = \frac{1}{\pi}\int dz_2' \frac{A_2^*(s,z_2') + A_3^*(s,z_2')}{z_2' - z_2}, \quad (3.4a)$$

$$A\{s, z_1 z_2 + (1-z_1^2)^{\frac{1}{2}}(1-z_2^2)^{\frac{1}{2}}\cos\phi\}$$
$$= \frac{1}{\pi}\int dz_3' \frac{A_2(s,z_3') + A_3(s,z_3')}{z_3' - z_1 z_2 - (1-z_1^2)^{\frac{1}{2}}(1-z_2^2)^{\frac{1}{2}}\cos\phi}. \quad (3.4b)$$

For simplicity we have included the absorptive parts A_2 and A_3 under the same integral sign, but they will of course contribute in different regions of the variable of integration. $A_2(s,z)$ will be nonzero only if $z<1-\{s-(M-\mu)^2\}/2q^2$, apart from a δ function at $z=1-(s-M^2-2\mu^2)/2q^2$, and $A_3(s,z)$ will be nonzero only if $z>1+2\mu^2/q^2$. The dispersion relations have been written down on the (incorrect) assumption that there are no subtractions necessary; we shall see in the following section how the theory must be modified to take them into account.

On substituting (3.4) into (3.2) and performing the integrations over z_2 and ϕ, we are left with the equation

$$A_1(s,z_1) = \frac{1}{16\pi^3}\frac{q}{W}\int dz_2' \int dz_3' \frac{1}{\sqrt{k}}\ln\frac{z_1 - z_2'z_3' + \sqrt{k}}{z_1 - z_2'z_3' - \sqrt{k}}$$
$$\times \{A_2^*(s,z_2') + A_3^*(s,z_2')\}\{A_2(s,z_3') + A_3(s,z_3')\}, \quad (3.5)$$

where
$$k = z_1^2 + z_2'^2 + z_3'^2 - 1 - 2z_1 z_2' z_3'. \quad (3.6)$$

We must take that branch of the logarithm which is real in the physical region $-1 < z_1 < 1$. Equation (3.5) then gives the value of A_1 in the entire complex z_1 plane.

According to Eq. (2.16), $A(s,z_1)$ must be an analytic function of t, and therefore of z, with discontinuities of magnitude $2A_{13}$ and $2A_{12}$ as z_1 crosses the positive and negative real axes. It is easily seen that the expression for A_1 in (3.5) has this property, and, on identifying the discontinuities along the real axis with A_{13} and A_{12}, we arrive at the equations

$$A_{13}(s,z_1) = \frac{1}{8\pi^2} \frac{q}{W} \int dz_2 \int dz_3\, K_1(z_1,z_2,z_3)$$
$$\times \{A_3^*(s,z_2)A_3(s,z_3) + A_2^*(s,z_2)A_2(s,z_3)\}, \quad (3.7a)$$

$$A_{12}(s,z_1) = \frac{1}{8\pi^2} \frac{q}{W} \int dz_2 \int dz_3\, K_2(z_1,z_2,z_3)$$
$$\times \{A_2^*(s,z_2)A_3(s,z_3) + A_3^*(s,z_2)A_2(s,z_3)\}. \quad (3.7b)$$

The primes on z_2 and z_3 have been suppressed. K_1 and K_2 are defined by the equations

$$K_1(z_1,z_2,z_3)$$
$$= -1/[k(z_1,z_2,z_3)]^{\frac{1}{2}}, \quad z_1 > z_2 z_3 + (z_2^2-1)^{\frac{1}{2}}(z_3^2-1)^{\frac{1}{2}}$$
$$= 0 \qquad\qquad z_1 < z_2 z_3 + (z_2^2-1)^{\frac{1}{2}}(z_3^2-1)^{\frac{1}{2}} \quad (3.8a)$$

$$K_2(z_1,z_2,z_3)$$
$$= 1/[k(z_1,z_2,z_3)]^{\frac{1}{2}}, \quad z_1 < z_2 z_3 - (z_2^2-1)^{\frac{1}{2}}(z_3^2-1)^{\frac{1}{2}}$$
$$= 0, \qquad\qquad z_1 > z_2 z_3 - (z_2^2-1)^{\frac{1}{2}}(z_3^2-1)^{\frac{1}{2}}. \quad (3.8b)$$

The points $z_1 = z_2 z_3 \pm (z_2^2-1)^{\frac{1}{2}}(z_3^2-1)^{\frac{1}{2}}$ are the points at which k changes sign.

Let us now transform back from z to our original variables. As we shall use the dispersion relations (2.17) and (2.19), it is convenient to express A_2 and A_{12} as functions of s and s_c and A_3 and A_{13} as functions of s and t. Equations (3.7) then become

$$A_{13}(s,t) = \frac{1}{32\pi^2 q^3 W} \left[\int dt_2 \int dt_3 \right.$$
$$\times K_1(s; t_1,t_2,t_3) A_3^*(s,t_2) A_3(s,t_3) \quad (3.9a)$$
$$\left. + \int ds_{c2} \int ds_{c3}\, K_1(s; t_1,s_{c2},s_{c3}) A_2^*(s,s_{c2}) A_2(s,s_{c3}) \right],$$

$$A_{12}(s,s_{c1}) = \frac{1}{32\pi^2 q^3 W} \int dt_2 \int ds_{c3}\, K_2(s; s_{c1},t_2,s_{c3})$$
$$\times [A_3^*(s,t_2) A_2(s,s_{c3}) + A_2^*(s,s_{c3}) A_3(s,t_2)]. \quad (3.9b)$$

Note that s is fixed in these equations, while s_c and t vary. K must be re-expressed as a function of the new variables by (3.3) and (2.13).

The use of Eq. (3.9), together with the dispersion relations, in order to determine the spectral functions is greatly facilitated by the fact that K is zero unless the variables satisfy certain inequalities; for all s,

$$K_1(s; t_1,t_2,t_3) = 0 \text{ unless } t_1^{\frac{1}{2}} > t_2^{\frac{1}{2}} + t_3^{\frac{1}{2}}, \quad (3.10a)$$

$$K_1(s; t_1,s_{c2},s_{c3}) = 0 \text{ unless } t_1^{\frac{1}{2}} > s_{c2}^{\frac{1}{2}} + s_{c3}^{\frac{1}{2}}, \quad (3.10b)$$

$$K_2(s; s_{c1},t_2,s_{c3}) = 0 \text{ unless } s_{c1}^{\frac{1}{2}} > t_2^{\frac{1}{2}} + s_{c3}^{\frac{1}{2}}. \quad (3.10c)$$

(For any particular s, the restrictions on the variables could be strengthened.) Equations (3.10) are true as long as s_{c2}, s_{c3}, t_2, and t_3 are in the regions $s_c > M^2$, $t > 4\mu^2$, outside which A_2 and A_3 vanish. It follows from (3.9) that, *for any given value of t (or s_c), $A_{13}(s,t)$ [or $A_{12}(s,s_c)$] can be calculated in terms of $A_3(s,t')$ and $A_2(s,s_c')$, where the values of t' and s_c' involved are all less than t (or s_c).* On the other hand, by writing the dispersion relations (2.17) and (2.19) in the form

$$A_2(s,s_t) = \frac{1}{\pi} \int_{s_2(s_c)}^{\infty} ds' \frac{A_{12}(s',s_c)}{s'-s}$$
$$+ \frac{1}{\pi} \int_{t_1(s_c)}^{\infty} dt' \frac{A_{23}(s_c,t')}{t'-t}, \quad (3.11a)$$

$$A_3(s,t) = \frac{1}{\pi} \int_{s_3(t)}^{\infty} ds' \frac{A_{13}(s',t)}{s'-s}$$
$$+ \frac{1}{\pi} \int_{s_3(t)}^{\infty} ds' \frac{A_{23}(s',t)}{s_c'-s_c}, \quad (3.11b)$$

it is evident that $A_3(s,t)$ and $A_2(s,s_c)$ can be found in terms of $A_{12}(s',s_c)$ and $A_{13}(s',t)$, if for the moment we neglect the second term in these equations. We can therefore calculate A_{13}, A_{12}, A_3, and A_2 for all values of s and successively larger values of s_c and t. The lowest value of s_c or t for which either A_2 or A_3 is nonzero is $s_c = M^2$, at which there is a contribution of $g^2 \delta(s_c - M^2)$ to A_2 from the one-nucleon state. From (3.9) and (3.10) it follows that A_{13} and A_{12} are zero if t and s_c are less than $4M^2$; for a range of values of t above this, A_{13} is nonzero and can be calculated by inserting the δ-function contribution to A_2 into (3.9a). The rest of A_2 and A_3 will still not contribute owing to (3.10). Once we have the procedure thus started, we can proceed to larger and larger values of t and s_c by alternate application of (3.9) and (3.11).[15]

Before discussing how to take the second terms of (3.11) into account, let us study in more detail the form of the functions A_{13} and A_{12} calculated thus far. In order to do this, we require the precise values of t and s_c, at a given value of s, for which the kernels K vanish; we find that

---

[15] It will be noticed that, though we have brought the pole in the crossing term from the one-nucleon intermediate state into our calculations, we have not yet introduced the pole in the direct term. This pole is actually a subtraction term of Eq. (2.11) and will be treated in the following section.

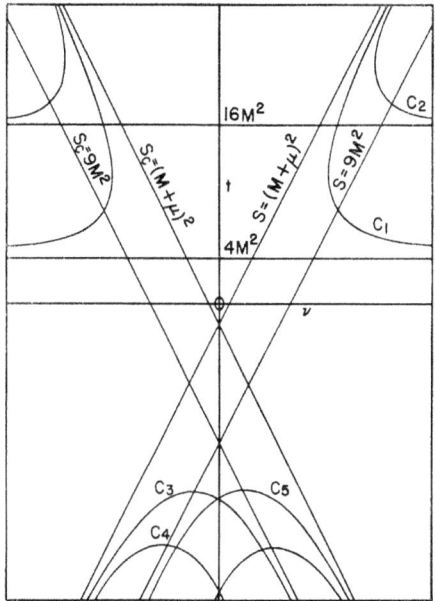

FIG. 2. Properties of the spectral functions.

$K_1(s; t_1, t_2, t_3) = 0$ unless

$$t_1^{\frac{1}{2}} > t_2^{\frac{1}{2}}(1 + t_3/4q^2)^{\frac{1}{2}} + t_3^{\frac{1}{2}}(1 + t_2/4q^2)^{\frac{1}{2}}, \quad (3.12a)$$

$K_1(s; t_1, s_{c2}, s_{c3}) = 0$ unless

$$t_1^{\frac{1}{2}} > (s_{c2} - u)^{\frac{1}{2}}\{1 + (s_{c3} - u)/4q^2\}^{\frac{1}{2}}$$
$$+ (s_{c3} - u)^{\frac{1}{2}}\{1 + (s_{c2} - u)/4q^2\}^{\frac{1}{2}}, \quad (3.12b)$$

$K_2(s; s_{c1}, t_2, s_{c3}) = 0$ unless

$$(s_1 - u)^{\frac{1}{2}} > t_2^{\frac{1}{2}}\{1 + (s_{c3} - u)/4q^2\}^{\frac{1}{2}}$$
$$+ (s_{c3} - u)^{\frac{1}{2}}(1 + t_2/4q^2)^{\frac{1}{2}}, \quad (3.12c)$$

where

$$u = (M^2 - \mu^2)^2/s. \quad (3.13)$$

As the smallest value of s_c or t which contributes to the integrand in Eq. (3.9a) is $s_c = M^2$, where A_2 has a δ-function singularity, it follows from (3.12b) that the smallest value of t for which $A_{13}(s,t)$ is nonzero (for any given value of s) is given by

$$t^{\frac{1}{2}} = 2(M^2 - u)^{\frac{1}{2}}\{1 + (M^2 - u)/4q^2\}^{\frac{1}{2}}. \quad (3.14)$$

For very large s, this value of t approaches $4M^2$, but, as s decreases, t becomes larger and larger until, at $s = (M+\mu)^2$, it becomes infinite. Equation (3.14) has been plotted as C_1 in Fig. 2. A_{13} will be nonzero above C_1, and, near it, it will behave like $(t-t_0)^{-\frac{1}{2}}$, where t_0 is the value of t given by (3.14). It follows from (3.11b) that $A_3(s,t)$ is nonzero if $t > 4M^2$, and behaves like $(t - 4M^2)^{\frac{1}{2}}$ just above this limit. The value $t = 4M^2$ is precisely the threshold for the process III, and we would have obtained the same results from our general reasoning in the previous section if we had neglected intermediate states containing two or more mesons but no nucleon pairs. This indicates that our assumptions are probably correct, as we have not considered the process III explicitly in this section. When we treat the subtraction terms in the dispersion relations, we shall see that A_{13} is also nonzero between $t = 4\mu^2$ and $t = 4M^2$, and that the region in which A_{13} is nonzero must be enlarged. The curve C_1 is therefore not yet the curve C_{13} of Fig. 1.

For a range of values of t above the curve C_1, the entire contribution to the integrand in (3.9a) comes from the δ function in A_2. At a certain point, however, the other terms in A_2 and A_3 begin to contribute. If for the moment we neglect the second term in (3.9a), the new contribution begins at the value of t obtained by putting $t_2 = t_3 = 4M^2$ in (3.12a), since this is (at the present stage of the calculation) the lowest value of t for which A_3 is nonzero. The result has been plotted against s in Fig. 2 to give the curve C_2. As this curve approaches the line $t = 16M^2$ asymptotically, there will be a corresponding new contribution to A_3 above this value, and, near it, the new contribution will behave like $(t - 16M^2)^{\frac{1}{2}}$. The value $t = 16M^2$ is just the threshold for the production of an additional nucleon pair in the process III, and A_3 would be expected to show such a behavior at this threshold.

We find similar discontinuities in the higher derivatives of A_{13} at series of curves (there will now be more than one for each threshold) approaching asymptotically the lines $t = 4n^2M^2$, so that A_3 will have the expected behavior at the thresholds for producing n nucleon pairs.

The functions A_{12} and A_2 will exhibit the same sort of characteristics. In Eq. (3.9b), the lowest values of t_2 and s_{c3} which contribute to the integrand are $t_2 = 4M^2$, $s_{c3} = M^2$, so that the boundary of the region in which A_{12} is nonzero is obtained by inserting these values into (3.12c). The result is represented by the curve C_3 in Fig. 2; it approaches the line $s_c = 9M^2$ as s tends to infinity. As with A_{13}, the region in which A_{12} is nonzero will be widened in the following section. From (3.19a), it follows that A_2 will (at present) be nonzero for $s_c > 9M^2$, which is the threshold for pair production in the reaction II. A_{12} will also have discontinuities in the higher derivatives at series of curves such as C_4 which approach asymptotically the lines $s_c = (2n+1)^2M^2$. Finally, it can be seen that the second term of (3.9a) will give rise to further curves at which the higher derivatives of A_{13} are discontinuous, but these curves will all approach asymptotically the lines $t = 4n^2M^2$.

We must now return to the second term in the Eq. (3.11), which we have so far neglected in the calculation. It can be taken into account by introducing the requirement of crossing symmetry, which has not yet been used. As in the static theory, one now has to use an iteration procedure. The function A_{23}, which only affects the crossing term in the dispersion relation (2.5), is first neglected, and the calculation done as described. A_{23} is then found from the calculated value of A_{13} and the crossing-symmetry relations (2.23), and inserted into Eq. (3.11) for the next iteration. However, the scattering amplitude calculated by this procedure would still not satisfy the equations of crossing symmetry since, while A_{13} and A_{23} are connected by (2.23a), A_{12} does not satisfy (2.23b). We have seen that the dispersion relations together with the equation of unitarity determine A_{12} uniquely, and the result is not a symmetric function of s and s_c; even the region in which it is nonzero is not symmetric. It therefore appears that we cannot satisfy simultaneously the requirements of analyticity, unitarity (in the one-meson approximation), and crossing symmetry.

The reason why this is so is easily seen in perturbation theory. Among the graphs included in the first iteration of the one-meson approximation is Fig. 3(a). The topologically similar graph Fig. 3(b) will also be included, since Fig. 3(a) by itself would have square roots in the energy denominators and would not have the necessary analytic properties. If, therefore, crossing symmetry is to be maintained, Fig. 3(c) must also be included. In this graph, however, there is an intermediate state of a nucleon and a pair, so that the unitarity condition in the one-meson approximation is not satisfied.

This example also indicates how we should modify our iteration procedure. In addition to inserting a term A_{23}, obtained by crossing symmetry from the previous iteration, into (3.11), we must insert a term $A_{12}'(s,s_c)$ equal to $A_{12}(s_c,s)$ as calculated in the previous iteration. The contribution from this term is to be added to the contribution from $A_{12}(s,s_c)$ calculated in the normal way. A_{12}' will be nonzero above the curve C_5 in Fig. 2, and, in particular, it will be zero for all values of s_c if s is less than $9M^2$. Complete crossing symmetry is now maintained, but the addition of A_{12} violates the unitarity condition (in the one-meson approximation) for values of s greater than $9M^2$, and a perturbation expansion would include graphs such as Fig. 3(c). As these graphs will appear in higher approximations, the fact that we are forced to include them here should not be considered a disadvantage of our method. In any case, the unitarity condition is only violated where the one-meson approximation is far from correct.

The iteration procedure is found to give rise to further curves, like C_2 and C_4 (Fig. 2), at which the higher derivatives of the spectral functions are discontinuous. These new discontinuities correspond to the production of mesons together with nucleon pairs. We still do not have discontinuities at all possible thresholds.

The inclusion of the spin does not change any of the essential features of the theory, though the details are

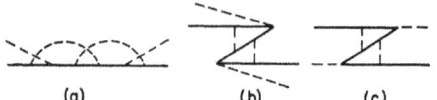

Fig. 3. Graphs which bring in intermediate states with pairs.

rather more complicated. Following Chew et al.,[1] we write the pion-nucleon T matrix in the form

$$T = -\frac{2\pi W}{Ew}(a + \boldsymbol{\sigma}\cdot\mathbf{q}_2\boldsymbol{\sigma}\cdot\mathbf{q}_1 b), \quad (3.15)$$

where E is the center-of-mass energy of the nucleon and w that of the pion. a and b are related to the quantities A and B in the expression (2.20) by the formulas

$$a = \frac{E+M}{2W}\left(\frac{A+(W-M)B}{4\pi}\right), \quad (3.16a)$$

$$b = \frac{E-M}{2W}\left(\frac{-A+(W+M)B}{4\pi}\right). \quad (3.16b)$$

The unitarity condition corresponding to (3.7) can now be worked out in terms of a and b; the equation obtained is

$$a_{13(12)}(s,z_1) = \sum_\alpha \frac{q}{\pi}\int dz_2 \int dz_3\, K_{1(2)}(z_1,z_2,z_3)$$

$$\times \left\{ a_\alpha^*(s,z_2)a_\alpha(s,z_3) + \frac{z_2-z_3z_1}{1-z_1^2}b_\alpha^*(s,z_2)a_\alpha(s,z_3)\right.$$

$$\left. + \frac{z_3-z_2z_1}{1-z_1^2}a_\alpha^*(s,z_2)b_\alpha(s,z_3)\right\}, \quad (3.17a)$$

$$b_{13(12)}(s,z_1) = \sum_\alpha \frac{q}{\pi}\int dz_2 \int dz_3\, K_{1(2)}(z_1,z_2,z_3)$$

$$\times \left\{\frac{z_3-z_2z_1}{1-z_1^2}b_\alpha^*(s,z_2)a_\alpha(s,z_3) + \frac{z_2-z_3z_1}{1-z_1^2}\right.$$

$$\left. \times a_\alpha^*(s,z_2)b_\alpha(s,z_3) + b_\alpha^*(s,z_2)b_\alpha(s,z_3)\right\}, \quad (3.17b)$$

where $\sum_\alpha$ indicates that terms of the form $a_\alpha^*a_\alpha$ are to be replaced by $a_2^*a_2 + a_3^*a_3$ in the calculation of a_{13} and b_{13} and by $a_2^*a_3 + a_3^*a_2$ in the calculation of a_{12} and b_{12}, exactly as in (3.7). a_2 and b_2, a_3 and b_3, a_{12} and b_{12}, and a_{13} and b_{13} are related respectively to A_2 and B_2, A_3 and B_3, A_{12} and B_{12}, and A_{13} and B_{13} by Eqs. (3.16). The unitarity condition (3.17) can be rewritten in terms of A and B; it then becomes

$$A_{13(12)}(s,z_1) = \sum_\alpha \frac{q}{4\pi^2 W} \int dz_2 \int dz_3 \, K_{1(2)}(z_1,z_2,z_3)$$

$$\times \left\{ \left(1 - \frac{w}{2W} \frac{1-z_2-z_3+z_1}{1+z_1}\right) A_\alpha^*(s,z_2) A_\alpha(s,z_3) \right.$$

$$+ \left(\frac{\omega}{2} \frac{1-z_2+z_3-z_1}{1-z_1} + \frac{Mw}{2W} \frac{1-z_2-z_3+z_1}{1+z_1}\right)$$

$$\times A_\alpha^*(s,z_2) B_\alpha(s,z_3) + \left(\frac{\omega}{2} \frac{1+z_2-z_3-z_1}{1-z_1}\right.$$

$$+ \frac{Mw}{2W} \frac{1-z_2-z_3+z_1}{1+z_1}\bigg) B_\alpha^*(s,z_2) A_\alpha(s,z_3)$$

$$+ \frac{W^2 - M^2}{2W} \frac{1-z_2-z_3+z_1}{1+z_1} B_\alpha^*(s,z_2) B_\alpha(s,z_3) \bigg\}, \quad (3.18a)$$

$$B_{13(12)}(s,z_1) = \sum_\alpha \frac{q}{4\pi^2 W} \int dz_2 \int dz_3 \, K_{1(2)}(z_1,z_2,z_3)$$

$$\times \left\{ \frac{E}{2MW} \frac{1-z_2-z_3+z_1}{1+z_1} A_\alpha^*(s,z_2) A_\alpha(s,z_3) \right.$$

$$+ \left(\frac{1+z_2-z_3-z_1}{2(1-z_1)} - \frac{E}{2W} \frac{1-z_2-z_3+z_1}{1+z_1}\right)$$

$$\times A_\alpha^*(s,z_2) B_\alpha(s,z_3) + \left(\frac{1-z_2+z_3-z_1}{1-z_1}\right.$$

$$- \frac{E}{2W} \frac{1-z_2-z_3+z_1}{1+z_1}\bigg) B_\alpha^*(s,z_2) A_\alpha(s,z_3)$$

$$+ \left(\omega - \frac{(w^2-M^2)E}{2MW} \frac{1-z_2-z_3+z_1}{1+z_1}\right)$$

$$\times B_\alpha^*(s,z_2) B_\alpha(s,z_3) \bigg\}. \quad (3.18b)$$

Equations (3.17) and (3.18) will hold separately for the amplitudes corresponding to isotopic spin $\frac{1}{2}$ and $\frac{3}{2}$.

It remains to justify the claim that the result calculated by our procedure, if expanded in a perturbation series, would give a subset of the usual perturbation series. The proof is somewhat awkward because we were unable to satisfy the unitarity condition in the one-meson approximation at all values of the energy. Let us first ignore this. The nth term in the perturbation series $A^{(n)}$ is then determined uniquely in the physical region by the following two requirements:

(i) For sufficiently small values of the momentum transfer $\{$less than $2\mu[\frac{2}{3}(2M+\mu)/(2M-\mu)]^{\frac{1}{2}}\}$, $A^{(n)}$ must satisfy the dispersion relation (2.5), a result

which has been proved rigorously.[14] The absorptive part A_1 (and hence, by crossing symmetry, A_2) is known, since it is determined by unitarity in terms of lower order perturbation terms in the physical region, and by analytic continuation (with s constant) outside it.[14]

(ii) For a fixed value of s, $A^{(n)}$ is an analytic function of the momentum transfer throughout the physical region.[14]

As the functions calculated by our method certainly fulfil these requirements, they must generate the correct perturbation series.

However, our result does not satisfy the unitarity condition in the one-meson approximation at all energies, and we must examine more closely how A_1 is to be determined. Let us assume that our method gives the correct perturbation series up to the $(n-1)$th order. The reasoning developed in this section then shows that the nth-order contribution to A_1 will be of the form

$$A_1^{(n)} = \frac{1}{\pi} \int dt' \frac{A_{13}^{(n)}(s,t')}{t'-t} - \frac{1}{\pi} \int dt' \frac{A_{12}^{(n)}(s,t')}{t'-t}, \quad (3.19)$$

where $A_{13}^{(n)}$ and $A_{12}^{(n)}$ are certainly zero below C_1 and above C_3, respectively, in Fig. 2. Inserting this expression into (2.5), we find that

$$A_d^{(n)} = \frac{1}{\pi^2} \int ds' \int dt' \frac{A_{13}^{(n)}(s',t')}{(s'-s)(t'-t)}$$

$$- \frac{1}{\pi^2} \int ds' \int dt' \frac{A_{12}^{(n)}(s',t')}{(s'-s)(t'-t)}. \quad (3.20)$$

The suffix d indicates that we are considering the direct and not the crossing term. The second term of (3.20) will not be an analytic function of t in the physical region, but it will have a branch point at the largest value of t for which A_{12} is nonzero. We can make it analytic by adding to A_2 the expression

$$-\frac{1}{\pi} \int dt' \frac{A_{12}^{(n)}(s_c,t')}{t'-t}, \quad (3.21)$$

which we would expect from (2.17), if our representation is correct. By inserting this into (2.5) and adding the result to the second term of (3.20), we obtain

$$\frac{1}{\pi^2} \int ds' \int ds_c' \frac{A_{12}^{(n)}(s',s_c')}{(s'-s)(s_c'-s_c)}, \quad (3.22)$$

which is analytic in the physical region. The contribution (3.21) to $A_2^{(n)}$ is uniquely determined from the requirement that $A^{(n)}$ be an analytic function of the momentum transfer in the physical region, and is nonzero only for $s_c > 9M^2$. It corresponds to adding a graph such as Fig. 3(b) to Fig. 3(a); as A_1 for Fig. 3(c) is nonzero for $s>9M^2$, A_2 for Fig. 3(b) will be nonzero for $s_c>9M^2$.

Finally, then, the nth-order perturbation term can be determined from the lower order perturbation terms without using any unproved properties of the scattering amplitude as follows:

(i) Calculate A_1 by unitarity, and extend it into the nonphysical region for momentum transfers less than $2\mu[\frac{2}{3}(2M+\mu)/(2M-\mu)]^{\frac{1}{2}}$ by analytic continuation.

(ii) Calculate a contribution $A_{2d}^{(n)}$ to $A_2^{(n)}$, for $s_c>9M^2$, from the requirement that if it, together with A_1, be inserted into (2.5), the resulting function $A_d^{(n)}$ must be an analytic function of the momentum transfer in the physical region. By doing this we partially include intermediate states with nucleon pairs, which is necessary if we are to maintain the required analytic properties and crossing symmetry.

(iii) Now calculate $A_2^{(n)}$ and the extra contribution to $A_1^{(n)}$ by crossing symmetry from $A_1^{(n)}$ and the extra contribution to $A_2^{(n)}$.

(iv) Find $A^{(n)}$ from (2.5) for values of the momentum transfer less than $2\mu[\frac{2}{3}(2M+\mu)/(2M-\mu)]^{\frac{1}{2}}$, and calculate it in the rest of the physical region by analytic continuation in t.

This procedure defines a one-meson approximation in perturbation theory. From what has been said, it is clear that our solution will give precisely this perturbation expansion, so that our assumptions are justified in perturbation theory.

4. SUBTRACTION TERMS IN THE DISPERSION RELATIONS

We have thus far assumed that the dispersion relations are true without any subtractions. As we have pointed out in the first section, by doing this we neglect what is physically the most important part of the scattering amplitude. In this section we shall investigate how many subtractions are necessary for each dispersion relation and shall outline how they can be calculated, leaving the details for a further paper.

Let us first consider Eqs. (2.11) and (2.16), which were used in obtaining the unitarity condition (3.9) [or (3.18) for nucleons with spin]. Even if these dispersion relations are written with subtraction terms, it is found that (3.9) is unchanged, so that the subtraction terms are only needed in the final evaluation of A from A_2 and A_3 by means of (2.11), or of A_1 from A_{12} and A_{13} by means of (2.16). The number of subtractions will depend on the behavior of A_{12}, A_{13}, A_2, and A_3, as calculated by our procedure, as s_c and t tend to infinity—we shall have to perform at least enough subtractions for (2.11) and (2.16) to converge.

It is difficult to make an estimate of the behavior of these functions at infinite values of s_c and t from the equations determining them, and we shall use indirect arguments which, though not rigorous, are very plausible. We shall find that, if the coupling constant is small enough, the functions tend to zero at infinity, so that one can write the dispersion relations without any subtractions. For larger values of the coupling constant, more and more subtractions will be needed. The reader who is prepared to accept this may omit the following two paragraphs.

We consider only the first iteration, since subsequent iterations proceed in a similar way and the results are unlikely to be qualitatively different. The result can then be expanded in a perturbation series. If the solutions obtained for this problem by other methods, such as the Tamm-Dancoff or Bethe-Salpeter methods, are expanded in a perturbation series, it is found that the series for each angular momentum state converges as long as the coupling constant is within a certain radius of convergence, and that this radius of a convergence tends to infinity with the angular momentum.[16] Our perturbation series would be different from the perturbation series obtained by these methods, partly because the intermediate states with pairs which we include are not the same as those included by either of them, and partly because, in calculating the subtraction terms (other than those at present under discussion), we shall not take into account terms corresponding to all graphs included by these approximations. Such differences would not be expected to affect qualitatively the convergence properties of the angular momentum states, and we shall assume that the results quoted above are true for our perturbation series too.

The transition amplitude for the state of total angular momentum j and orbital angular momentum $j\pm\frac{1}{2}$ can be shown to be

$$f_{j\pm} = \int_{-1}^{1} dz\, a(s,z) P_{j\pm\frac{1}{2}}(z) + \int_{1}^{1} dz\, b(s,z) P_{j\mp\frac{1}{2}}(z), \quad (4.1)$$

where a and b are the functions defined in (3.15) and (3.16). Now it is easily seen that each term in the perturbation series for $a_2(s,z)$, $a_3(s,z)$, $b_2(s,z)$, and $b_3(s,z)$ tends to zero like $1/z$ as z tends to infinity, so that the dispersion relation (2.11) for each term can be written down without any subtractions. Hence

$$f_{j\pm}^{(n)} = \int_{-1}^{1} dz \int dz' \left\{ \frac{a_2^{(n)}(s,z')+a_3^{(n)}(s,z')}{z'-z} P_{j\pm\frac{1}{2}}(z) \right.$$
$$\left. + \frac{b_2^{(n)}(s,z')+b_3^{(n)}(s,z')}{z'-z} P_{j\mp\frac{1}{2}}(z) \right\} \quad (4.2)$$

$$= \int dz' \{[a_2^{(n)}(s,z')+a_3^{(n)}(s,z')]\phi_{j\pm\frac{1}{2}}(z')$$
$$+ [b_2^{(n)}(s,z')+b_3^{(n)}(s,z')]\phi_{j\mp\frac{1}{2}}(z')\}, \quad (4.3)$$

---

[16] Note that the "potential" in the Tamm-Dancoff or Bethe-Salpeter equation involved includes only the crossing term and not the direct term, which has still to be brought into the calculation.

where
$$\phi_n(z') = \int_{-1}^{1} dz \frac{P_n(z)}{z'-z} \quad (4.4)$$
$$\approx 1/z'^{n+1} \text{ as } z' \to \infty.$$

Let us suppose that the value of the coupling constant is such that the perturbation series for states of angular momentum j_1 converges. If each term in the perturbation series for this angular momentum state is expressed by (4.3), and if we assume that we can interchange the order of summation and integration, we arrive at the equation

$$f_{j1\pm} = \int dz' \{\sum_n [a_2^{(n)}(s,z') + a_3^{(n)}(s,z')] \phi_{j1\pm\frac{1}{2}}(z')$$
$$+ \sum_n [b_2^{(n)}(s,z') + b_3^{(n)}(s,z')] \phi_{j1\mp\frac{1}{2}}(z')\}. \quad (4.5)$$

In order for the integrand to exist, we see from (4.4) that a and b must be smaller than $z^{j_1-\frac{1}{2}}$ at infinite z. The dispersion relations can therefore be written down with not more than $j-\frac{1}{2}$ subtractions. In particular, if the coupling constant is small enough the dispersion relations can be written down without any subtractions.[17]

If the coupling constant is such that n subtractions are required, the unitarity condition for the states of angular momentum $\frac{1}{2}$ to $n-\frac{1}{2}$ will have to be applied separately. The wave functions for these states are polynomials of degree not greater than $n-1$ in the variable z (or s_c and t), and are not determined from the absorptive parts in the dispersion relations (2.11) and (2.16).

The calculation must be done after each iteration, as the result will be needed for the next iteration. The details of the calculation will not be discussed here, but they will in principle be similar to those of Chew and Low[4] and Dalitz, Castillejo, and Dyson,[5] and will involve considering the reciprocal of the scattering amplitude. The analytic properties of the individual angular momentum states are not as simple as in the static theory, but they can be determined from the assumed analytic properties of the transition amplitude, and, as in the static theory, the singularities not on the positive real axis can be found from the previous iteration.

The precise number of subtractions required cannot be determined without calculating the result, but it is almost certainly not less than two. It is difficult to see how the observed resonant behavior of the $P_{\frac{3}{2}}$ state could be reproduced by means of the calculations described in the last section, whereas it follows quite naturally from a Chew-Low-type calculation. If the coupling constant were large enough to bind the (3,3) resonance state, and for a certain range of values of the coupling constant below this, we would definitely have to perform two subtractions. The precise range involved is difficult to determine, but it would be expected to include those values of the coupling constant for which the (3,3) state still has the appearance of an unstable isobar. Until we state otherwise, however, we shall suppose that the coupling constant is sufficiently small for the functions $A(s,z)$ and $B(s,z)$ to tend to zero at infinite z, as the situation with regard to the other subtractions is much simpler in this case. Even then, we would have to perform one subtraction for each of A and B, since the calculations of the previous section did not include the pole of the scattering amplitude from the one-nucleon intermediate state; only the pole in the crossing term was included. The pole affects the states with $j=\frac{1}{2}$ alone, so that, if we apply the unitarity condition for these states separately by the Chew-Low method, we can include it correctly. We thereby change A and B by a quantity independent of z.

When we calculate the scattering amplitudes for the states with $j=\frac{1}{2}$, we find a ghost state in the first iteration, just as in all other models. In subsequent iterations, however, where the crossing terms contribute, it does not follow from the form of the equations that we shall necessarily find a ghost state, and, judging from the charged scalar model, we may hope that the ghost state does not in fact occur.

We now turn to consider the subtraction terms in the other dispersion relations used in the calculations, Eq. (3.11). By putting the δ-function contribution to A_2 into (3.18), it can be seen that the lowest order term in $A_{13}(s,t)$ tends to a constant as s tends to infinity, whereas the lowest order term in $B_{13}(s,t)$ behaves like $1/s$. For a certain range of values of t, only the lowest order term contributes to A_{13} and B_{13}, so that there will certainly be one subtraction in Eq. (3.11b) for A_3, while the equation for B_3 could be written down without any subtractions. We find similarly that both $A_{12}(s,s_c)$ and $B_{12}(s,s_c)$ tend to zero like $1/s$ as s tends to infinity. It would therefore appear that the dispersion relations (3.11a) did not require any subtractions. However, we have seen that $A_1(s,s_c)$ and $B_1(s,s_c)$ behave like a constant for large s_c with s constant, even for small values of the coupling constant, so that, by crossing symmetry, $A_2(s,s_c)$ and $B_2(s,s_c)$ will behave like a constant for large s. There will therefore be one subtraction term in Eqs. (3.11a) for both A_2 and B_2.

The determination of the subtraction terms in Eq. (3.11a) is not difficult, since the contributions to A_2 and B_2 from the states with $j=\frac{1}{2}$ (with the energy s_c of the reaction II kept constant) can be found by crossing symmetry from the corresponding contributions to A_1 and B_1 in the previous iteration. However, for the subtraction terms in Eq. (3.11b), we require

---

[17] We should emphasize that it is only in the first iteration that we relate the number of subtractions needed for the convergence of the angular momentum states. We say nothing at all about the convergence of the perturbation series in subsequent iterations, but assume simply that the behavior of the spectral functions at infinite values of z is not likely to be qualitatively different from their behavior in the first iteration.

the unitarity condition for A_3, which involves the reaction III. As there is one subtraction, only the S waves will be involved. Again we have to limit the intermediate states considered; in this first approximation we would consider the two-meson states ("two-meson approximation") and perhaps the nucleon-antinucleon intermediate states ("two-meson plus pair approximation") as well. We shall then require the meson-meson scattering amplitude (and the nucleon-antinucleon scattering amplitude if nucleon-antinucleon intermediate states are being considered). The determination of these scattering amplitudes would be as extensive a calculation as the determination of the pion-nucleon scattering amplitude, but neglect of the crossing term would probably not give rise to too great an error in our final result, in which case the S-wave amplitudes could be written down immediately in the two-meson or two-meson plus pair approximations. The meson-meson coupling constant is thereby introduced into the calculation, as has been mentioned in the introduction. Once the meson-meson and nucleon-antinucleon scattering amplitudes are known, the transition amplitude for the reaction III can be calculated. Since the integral equation is now linear, the details will be different from those of the Chew-Low calculations, but, as in their case, the solution could be written down exactly if there were no other singularities of the transition amplitude, and we can use an iteration procedure for the actual problem. The iterations will again be interspersed between the iterations of the main calculation. The S-wave portion of A_3, as calculated by this procedure, will be nonzero for $t > 4\mu^2$, so that the scattering amplitude now has the expected spectral properties. The boundaries of the regions in which the spectral functions are nonzero will thereby also be changed; this will be discussed in more detail at the end of the section.

We have seen that, as long as the coupling constant is sufficiently small, we require one subtraction for each of the dispersion relations except the dispersion relation (3.11b) for B_3, for which we do not require any subtractions. It is also easily seen that this behavior is consistent—the functions as calculated in the last section, with the calculations modified by the subtraction terms, will not at any stage become too large at infinity. If, however, one were to make any additional subtractions, one would find that, on performing the calculations, one would need more and more subtractions as the work progressed, and one could not obtain any final result. The number of subtractions to be performed is therefore determined uniquely. There is one exception to this statement: we could perform one subtraction in Eq. (3.11b) for B_3. Such a subtraction is, however, excluded by the requirement that the theory remain consistent when the interaction with the electromagnetic field is introduced. If one were to make this subtraction, the scattering amplitude would behave like $f(t)\gamma(q_1+q_2)$ for large values of s. It then follows from gauge invariance that the matrix element for the processes

$$\pi^\pm + n \to \pi^\pm + n + \nu \quad \text{or} \quad \pi^0 + p \to \pi^0 + p + \nu$$

will contain a term which behaves like $f(t)\gamma$ for large s, where t is now minus the square of the momentum transfer of the neutral particle.[18] The contribution to B_1 and B_{13} from the $\pi - N - \gamma$ intermediate state therefore tends to infinity at least as fast as s for infinite s, so that one would require two subtractions for the dispersion relation in question and the theory would not be consistent.

Since the unitarity conditions for the two $j = \frac{1}{2}$ states of the pion-nucleon system, and for the S state of the pion-pion system, have to be applied separately by the Chew-Low method, there will be Castillejo-Dalitz-Dyson ambiguities associated with these states. The ambiguities will of course affect all states in subsequent iterations. They correspond to the existence of unstable baryons of spin $\frac{1}{2}$ and either parity, or of heavy unstable mesons of spin zero. There are no ambiguities associated with states of higher angular momentum; this is in agreement with perturbation theory, according to which it is impossible to renormalize systems containing particles of spin 1 or more. Had there been no interaction with the electromagnetic field, we could have introduced a further subtraction term which would have necessitated a separate application of the unitarity condition for the P state of the pion-pion system. The resulting Castillejo-Dalitz-Dyson ambiguity would have been associated with a heavy unstable meson of spin 1. This corresponds to the Bethe-Beard mixture of vector and scalar mesons, which can be renormalized in perturbation theory as long as there is no interaction with the electromagnetic field.

Now let us consider the situation that occurs in practice, when the coupling constant is sufficiently large for the scattering amplitude and its absorptive parts to tend to infinity with z (or s_c and t) when s remains constant. The function A_{12}' which, according to our procedure, must be added to A_{12} in iterations other than the first, will now tend to infinity with s, so that A_2, as calculated from (3.11a), would show a similar behavior. In practice, when the unitarity condition for states with $j = \frac{3}{2}$ as well as with $j = \frac{1}{2}$ must be applied separately, $A_{12}'(s,s_c)$ and $A_{23}(s_c,t)$ will tend to infinity faster than s or t, and the dispersion relation (3.11a) will require two subtractions. The subtraction terms can be determined by crossing symmetry as before. However, we have seen that, if A_2 tends to infinity with s, we cannot consistently perform the calculation, so that we shall have to introduce some further modifications.

The reason for the difficulty is probably the in-

---

[18] This can be shown by using a generalization of the Ward identity due to H. S. Green, Proc. Phys. Soc. (London) **66**, 873 (1953), and T. D. Lee, Phys. Rev. **95**, 1329 (1954), and proved by Y. Takahashi, Nuovo cimento **6**, 372 (1957).

adequacy of the one-meson approximation. The breakdown occurs just at the value of the coupling constant for which the contribution to the scattering amplitude from A_{12}' is comparable to the remainder of the scattering amplitude when s is large. Since that part of A_1 calculated from A_{12}' represents a partial effect of states with one or more pairs, the contribution of these intermediate states is now important at high energies and it seems reasonable that, if one could take them into account properly, one could still perform the calculations for large values of the coupling constant. In the one-meson approximation, one would have to make some sort of a cutoff to the contribution to A_2 from the crossing term above $s=9M^2$. As this entails modifying the unitarity condition in the region where it is in any case inaccurate, it is consistent with our approximations, and it may be hoped that the theory is not very sensitive to the precise location and form of the cutoff. If one were to go to further approximations in which intermediate states with pairs were included, the cutoff would always be applied only at or above the threshold for processes which were neglected.

Once we are prepared to introduce cutoffs into our approximations, we might legitimately ask whether or not we should perform more than one subtraction in Eq. (3.11b). This could only be determined by examining the behavior of the scattering amplitude and its absorptive parts at large values of s when we go beyond the one-meson approximation. However, if A and B have the behavior assumed thus far (A remains constant and B behaves like $1/s$), the cross section would tend to zero like $1/s$ at large s, whereas the experimental results indicate that the cross section remains constant. It therefore may be necessary to perform an additional subtraction and to introduce the unitarity condition of the reaction III in P states.

At first sight it would seem as though there were Castillejo-Dalitz-Dyson ambiguities associated with all states for which the unitarity condition has to be applied separately, not only with the $j=\frac{1}{2}$ states. However, it is also possible that only the solution without any of the extra terms in the higher angular momentum waves would converge as we introduced more and more states into the unitarity equations. This

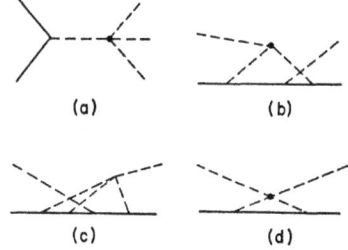

FIG. 4. Graphs involving the pion-pion interaction.

solution would be an analytic continuation of the solution obtained for small values of the coupling constant, whereas the other solutions could not be continued below a certain value of the coupling constant and would have no perturbation expansion. While we can by no means exclude such a behavior, it nevertheless gives us grounds to suppose that the ambiguity exists only for meson-nucleon states with $j=\frac{1}{2}$ and for S-wave meson-meson states, even when the coupling constant is large.

Before leaving this section, let us state the boundaries of the region in which the spectral functions A_{13}, A_{23}, and A_{12} are nonzero, i.e., the position of the curves C_{13}, C_{23}, and C_{12} in Fig. 1. Since A_3 is now nonzero for $t^2>4\mu^2$, C_{13} in the one-meson approximation is obtained by putting $t_2=t_3=4\mu^2$ in (3.12a), so that

$$t_{1a}^{\frac{1}{2}}=4\mu(1+\mu^2/q^2)^{\frac{1}{2}},$$

or

$$t_{1a}=\frac{16\mu^2(s-M^2+\mu^2)^2}{[s-(M+\mu)^2][s-(M-\mu)^2]}. \quad (4.6)$$

For any given value of s, A_{13} will be nonzero if $t>t_{1a}$. We notice that, as s tends to infinity, t_{1a} approaches the value $16\mu^2$. This is not the expected result—we have shown in Sec. 2 that it should approach the value $4\mu^2$. The reason for the discrepancy is that, in our approximation, the reaction III takes place purely through S waves for $4\mu^2<t<16\mu^2$, and A_3 will be a function only of t in this region. Had it been possible for the reaction III to go through an intermediate state of one pion, A_3 would have had a δ function at $t=\mu^2$, and, on putting this value into (3.12a), we would have obtained the expected result. As it is, however, we shall have to go beyond the one-meson approximation to get the correct boundary of A_{13}.

The reaction $N+\bar{N}\to 3\pi$ can go through a one-pion intermediate state by means of the process represented in Fig. 4(a). If, therefore, we treat the outgoing pions in the reaction $N+\pi\to N+2\pi$ as one particle with fixed energy and angular momentum, and represent the transition amplitude in the same way as we have represented the transition amplitude for pion-nucleon scattering, the absorptive part corresponding to A_3 will have a δ function at $t=\mu^2$. We can work out the resulting contribution to A_{13} (of the pion-nucleon scattering amplitude) by unitarity in the same way as we worked out the contributions from the one-meson approximation. z_2 and z_3 in Eqs. (3.4)–(3.8) will now refer to the center-of-mass deflection of the nucleon in the production reaction, and will be connected with the momentum transfer by the relation

$$z=\{q^2+q_1^2+t-[(M^2+q^2)^{\frac{1}{2}}-(M^2+q_1^2)^{\frac{1}{2}}]^2\}/2qq_1,$$

where q_1, is the center-of-mass momentum of the outgoing nucleon. The value of q_1 will depend on the relative energy of the two pions; we shall require the maximum value of q_1 (for a fixed s), which occurs when the pions are at rest with respect to one another and is given by

$$q_{1m}^2 = \{s - (M+2\mu)^2\}\{s - (M-2\mu)^2\}/4s. \quad (4.7)$$

We then find that the boundary of this contribution to A_{13} has the equation

$$t_{1b} = \frac{4\mu^2(s - M^2 - 2\mu)^2}{[s-(M+2\mu)^2][s-(M-2\mu)^2]}. \quad (4.8)$$

The curve represented by (4.8) approaches asymptotically the lines $t = 4\mu^2$ and $s = (M+2\mu)^2$. Thus, as would be expected, this contribution to A_{13} only occurs above the threshold for pion production.

A_{13} is therefore nonzero for $t > t_1$, where

$$\begin{aligned} t_1 &= t_{1a}, & (M+\mu)^2 < s < (M+2\mu)^2; \\ t_1 &= \min(t_{1a}, t_{1b}), & (M+2\mu)^2 < s < \infty; \end{aligned} \quad (4.9)$$

and $t = t_1$ is the curve C_{13} of Fig. 1. We cannot be sure that contributions from other intermediate states will not extend beyond this curve, but this is unlikely owing to the greater mass of these states.

The curve C_{23} is obtained from C_{13} simply by changing s to s_c. C_{12} can be calculated in a similar way; we find that

$$\begin{aligned} s_{c2} &= s_{c2a}, & (M+\mu)^2 < s < (M+2\mu)^2 \\ &= \min(s_{c2a}, s_{c2b}), & (M+2\mu)^2 < s < \infty, \end{aligned} \quad (4.10)$$

where

$$(s_{c2a} - u)^{\frac{1}{2}} = 2\mu \left\{ \frac{s^2 - s(3M^2 + 2\mu^2) + 2(M^2 - \mu^2)^2}{[s-(M+\mu)^2][s-(M-u)^2]} \right\}^{\frac{1}{2}}$$

$$+ \left\{ \frac{[M^2 s - (M^2 - \mu^2)^2][s^2 - 2s(M^2 + 3\mu^2) + (M^2 - \mu^2)^2]}{s[s-(M+\mu)^2][s-(M-\mu)^2]} \right\}^{\frac{1}{2}}, \quad (4.11)$$

$$s_{c2b}(s) = s(s_{c2a}). \quad (4.12)$$

The equation $s_c = s_{c2b}$ represents in fact the boundary of the region in which A_{12}' is nonzero. We observe that, once the pion-pion interaction has been included, this region approaches asymptotically the line $s = (M+2\mu)^2$ rather than the line $s = 9M^2$. The reason is that processes represented by graphs such as Fig. 4(b) are now included in our approximation, so that the crossing term will include the contribution from Fig. 4(c), the intermediate state of which involves a nucleon and two pions.

For a given real value of s, the absorptive part A of the scattering amplitude will be an analytic function of the momentum transfer as long as

$$t_2 < t < t_1, \quad (4.13a)$$

where t_1 is given by (4.9), and t_2 by (4.10) and (2.13). The expansion in partial waves will converge if

$$-t_1 - 4q^2 < t < t_1, \quad (4.13b)$$

as $-t_1 - 4q^2$ is always greater than t_2.

We may note finally one interesting point concerning the spectral properties of the scattering amplitude. The unitarity condition should, strictly, be used in the physical region only, and the results extended to the unphysical region by analytic continuation. This has actually been done for the reaction I, as well as for the reaction III with $t > 4M^2$. For the reaction III in the region $4\mu^2 < t < 4M^2$, we should apply the unitarity condition with the nucleon masses taken, not on the mass shell, but at some smaller value where all the momenta would be real. The result should then be continued analytically onto the mass shell. In our case this is found to make no difference, but if, in addition to the nucleon, we had a baryon whose mass M_B satisfied the inequality

$$M_B^2 < M^2 - \mu^2, \quad (4.14)$$

it would be necessary to do the calculation in this way. On making the continuation to the mass shell, it would be found that the absorptive part A_3 extended below the limit $t^2 = 4\mu^2$. It has been shown by several workers[19] that, if an inequality such as (4.14) is satisfied, the vertex function would show similar spectral properties. The simplest graph to exhibit them in our case would be Fig. 4(d), which will obviously have properties similar to those of a vertex graph. It is thus seen that these spectral abnormalities would not limit the applicability of our method, but, on the contrary, follow from it.

5. APPROXIMATION SCHEME FOR OBTAINING THE SCATTERING AMPLITUDE

In the methods developed in the previous sections, the unitarity condition for the reaction I is satisfied for all angular-momentum states in the one-meson approximation. The unitarity condition for the reaction III is satisfied only for S states in the two-meson or two-meson plus pair approximations. The unitarity condition for higher angular momentum states of the reaction III is not satisfied, but the scattering amplitude shows the expected behavior at the threshold for competing real processes.

These properties suggest immediately a further approximation which would be consistent with our other approximations. The major portion of the work, and certainly the major part of the computing time, would be employed in calculating the spectral functions, as this involves finding double integrals which are themselves functions of two variables. The calculations would therefore be simplified if we neglected those contributions to the spectral functions which begin at the threshold for processes involving more than two particles. The only contributions to A_{13} and A_{23} left would be those beginning at $t = 4M^2$, and they could be obtained by inserting the δ-function contribution to B_2

---

[19] Karplus, Sommerfield, and Wichman, Phys. Rev. **111**, 1187 (1958); Y. Nambu, Nuovo cimento **9**, 610 (1958); R. Oehme, Phys. Rev. **111**, 1430 (1958).

into (3.18). The spectral function A_{12} would be zero in this approximation.

The unitarity condition for the higher angular momentum states of the reaction I is no longer satisfied. However, the terms neglected appear by their form to arise from intermediate states of the reaction III with more than two particles, so that the approximation is in the spirit of the approximations already made. We have in fact made precisely this approximation in the unitarity condition for the S waves of the reaction III. The unitarity condition for the low angular momentum states of the reaction I, and in particular for the states with $j=\frac{1}{2}$ or $\frac{3}{2}$, is still satisfied, as it has been introduced separately. The present approximation treats the reactions I, II, and III on the same footing.

To summarize, then, our method of procedure will be the following: The first few angular momentum states of A_1 and A_3 are found on the assumption that each angular momentum state is an analytic function of the square of the center-of-mass energy except for the perturbation singularities and the cuts on the positive real axis. This calculation can be done exactly if the discontinuity across the cut along the positive real axis is determined by unitarity (complications arise, as the relations connecting a and b with A and B involve square roots of kinematical factors, but the methods can be modified accordingly). A_{13} and A_{12} are also found as just described. The analytic properties of the low angular momentum states are now determined from the analytic properties of the scattering amplitude given by (2.12). The singularities can be calculated in terms of A_1, A_2, A_3. These absorptive parts can in turn be found from A_{13} and A_{23} by means of the dispersion relations (2.16), (2.17), (2.19), with subtraction terms which can be obtained from the low-angular-momentum states. In the next iteration, all the singularities of the low angular momentum states except that along the positive real axis are found from the quantities calculated in the first iteration, and the singularity along the positive real axis is redetermined from the unitarity condition. The iteration procedure is repeated until it converges. As in the calculations of Sec. 4, it is found necessary to cut off the absorptive parts A_1, A_2 and A_3 at high energies, before calculating the singularities of the low angular momentum states in the next iteration. However, the cutoff is only applied above the threshold for processes neglected in the unitarity condition, and in particular, above the threshold for pair production in the reaction I.

This approximation could be regarded as the first of a series of approximations in which more and more of the contributions to the spectral functions are included, until we ultimately reach a solution in which the unitarity condition in the one-meson approximation is satisfied for every angular momentum state. In the higher approximations the spectral functions are no longer determined by perturbation theory, but, once the contribution from the crossing term enters, they will have to be recalculated after each iteration. However, it would be more worthwhile to go beyond the one-meson approximation at the same time as we took the higher contributions to the spectral functions into account. In other words, we continue to put the reactions I, II, and III on the same footing, bringing in the higher intermediate states of all three together. If the approximation scheme converged, the exact unitarity condition of the three reactions would finally be satisfied for all angular momentum states. Needless to say, one would not in practice be able to go beyond the first one or two approximations.

The number of angular momentum states for which the unitarity condition is applied separately will, as has been explained in the last section, depend on the behavior of A and B as t (or s_c) tends to infinity with s constant. However, in our first approximation, it should be sufficient to treat separately only states with $j=\frac{1}{2}$ and $j=\frac{3}{2}$, as the other angular momentum states will not be important below the threshold for pion production. If we went beyond the one-meson approximation we would probably have to treat some higher angular momentum states separately in any case, since, for instance, two pions both in a (3,3) resonance state with a nucleon could form a $D_{\frac{1}{2}}$ state. For reaction III, one would have to treat separately S states and possibly P states as well.

If one neglected the nucleon-antinucleon intermediate state in the reaction III and only took the two-pion intermediate state into account, all three spectral functions A_{13}, A_{23}, and A_{12} would be zero, since they all begin above the threshold for processes which are being neglected. The entire scattering amplitude would then consist of "subtraction terms" for one or other of the dispersion relations. This may be the best first approximation from the point of view of the amount of work required and the accuracy of the result, as the nucleon-antinucleon intermediate state is a good deal heavier than multipion states which are being neglected. Though the spectral functions are not now brought in at all, it will of course be realized that the only justification for the approximation is that it is the first of a series of approximations which do involve the spectral functions. In this approximation, if the crossing term is neglected in the calculation of the pion-pion scattering amplitude, only intermediate S states occur in reaction III, so that the unitarity condition for the P states will not enter.

ACKNOWLEDGMENTS

The author would like to acknowledge helpful discussions with Professor N. M. Kroll, Professor M. L. Goldberger, Professor R. Oehme, and Professor H. Lehmann. He also wishes to thank Columbia University for the award of a Boese Post-Doctoral Fellowship.

Analytic Properties of Transition Amplitudes in Perturbation Theory

S. Mandelstam

Phys. Rev. **115** (1959) 1741–1751

Abstract

The analytic properties of two-particle transition amplitudes as functions of both energy and momentum transfer are examined in perturbation theory. The modified Nambu representation previously proposed by the author for expressing these properties is discussed in a little more detail. It is shown that, as long as the masses do not satisfy certain inequalities connected with the existence of anomalous thresholds, the fourth order terms, calculated in the usual manner, satisfy the representation. The spectral functions are calculated explicitly for spinless particles. The proof can be extended to the sixth order, but is not worked out here. The modifications necessary when there exist anomalous thresholds are mentioned.

Two-Dimensional Representations of Scattering Amplitudes and Their Applications

S. Mandelstam

in *Quantum Theory of Fields. Proceedings of the Twelfth (1961) Solvay Conference on Physics*, October 1961, University of Brussels, Belgium, ed. R. Stoops (Interscience Publishers, 1962), pp. 209–233

Abstract

This report is divided into three sections. The first will be concerned with the foundations of two-dimensional representations and of analytic properties of transition amplitudes in general. The view will be taken that they are probably consequences of quantum field theory, though our mathematical tools are not yet sufficiently powerful to carry out the proof. In the second section an account will be given of the results so far obtained with the representations in conjunction with approximation schemes. Finally, recent proposals for overcoming some of the more serious difficulties of the present approximation scheme will be treated.

Theory of the Low-Energy Pion-Pion Interaction*

Geoffrey F. Chew and Stanley Mandelstam†
Lawrence Radiation Laboratory and Department of Physics, University of California, Berkeley, California
(Received January 18, 1960)

The double-dispersion representation is applied to the problem of pion-pion scattering, and it is shown that, if inelastic effects are important only at very high energies and S-wave scattering dominates at low energy, a set of integral equations for the low-energy amplitudes can be derived. The solution of these equations depends on only one arbitrary real parameter, which may be defined as the pion-pion coupling constant. The order of magnitude of the new constant is established, and a procedure for solving the integral equations by iteration is outlined. If P-wave scattering is large the equations become singular and must be modified. Such a modification can be performed, at the expense of introducing an extra parameter, but is not considered here.

I. INTRODUCTION

IT has become evident in recent times that no further substantial progress will be made in the theory of strong-interaction phenomena involving pions and nucleons until *something* is understood about the pion-pion interaction. Previous theoretical work on this problem has lacked a framework in which to make plausible approximations, so the results of past calculations are not considered reliable. Recently, however, one of us has proposed a generalization of dispersion relations that allows the simultaneous extension of energy and momentum transfer variables into the complex plane.[1] If the double-dispersion representation is accepted as correct, it becomes possible to formulate an approximation method for elementary-particle scattering at low energies that is extremely plausible. We propose in this paper to apply the new method to the pion-pion interaction.

The underlying motivation for the new approach is the property of an analytic function that its behavior in a limited region of the complex plane is dominated by nearby singularities. This circumstance is the basis of all "effective-range" theories for partial-wave scattering amplitudes. Effective-range theory leads to approximate formulas for partial amplitudes, valid in a small range of energies, that include nearby poles and branch points but ignore distant singularities. These formulas approximate the influence of the neglected singularities by arbitrary constants to be fitted by experiment. The content of the double-dispersion representation is essentially to give the location and character of *all* the singularities of a scattering amplitude as well as the behavior at infinity. Armed with this information, one may extend the usual "effective-range" approach so as to reduce drastically the number of free parameters. Of course one can never include all the distant singularities, but in the pion-pion problem the first difficult branch point occurs at such a high energy that we believe the omitted effects can to a good approximation be absorbed into one or two real parameters. If all phase shifts for $l>0$ are small, a single parameter suffices.

In the conventional Lagrangian formulation of field theory an independent constant appears in the pion-pion interaction, so one may be tempted to regard an effective-range approach with a single free parameter as the equivalent of a complete dynamical calculation. We prefer not to delve here into this very difficult question of principle but leave to the reader the theoretical interpretation of the constant λ that is to be introduced. Our definition of λ will be unambiguous from the experimental point of view.

As the price for including more of the nearby singularities than is usually attempted in effective-range theories, we shall have to solve nonlinear integral equations to find the pion-pion scattering amplitude. These equations will perhaps seem complicated, but they can be put into a form amenable to numerical solution. The results of the numerical solutions for various values of λ are given in the following paper. The equations are, strictly speaking, singular, and may possess a class of solutions which cannot be obtained by this direct procedure. We shall defer consideration of such solutions to a later paper.

II. SYMMETRIES AND KINEMATICS

Pion-pion elastic scattering may be represented by the diagram of Fig. 1, where the ingoing four-momenta and isotopic-spin indices are (p_1,α) and (p_2,β) and the outgoing are $(-p_3,\gamma)$ and $(-p_4,\delta)$.[2] It is convenient for

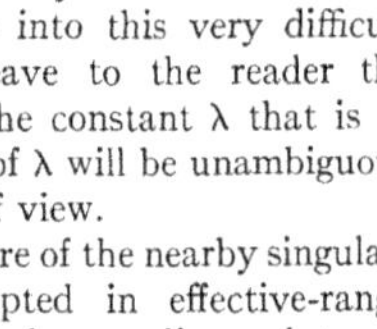

Fig. 1. The pion-pion interaction, $\pi+\pi \leftrightarrow \pi+\pi$.

---

* This work was supported in part by the U. S. Atomic Energy Commission and in part by the Air Force Office of Scientific Research of the Air Research and Development Command.
† Now at Department of Mathematical Physics, The University of Birmingham, Edgbaston, Birmingham 15, England.
[1] S. Mandelstam, Phys. Rev. **115**, 1741, 1752 (1959). Also Phys. Rev. **112**, 1344 (1958).
[2] The isotopic indices α, β, γ, and δ can each assume the values 1, 2, or 3. The value 3 corresponds to the neutral pion, while linear combinations of 1 and 2 correspond in the usual way to charged pions.

discussions of symmetry to use a notation in which all momenta are formally directed inward, although in the physical region p_1 and p_2 are positive timelike, with p_3 and p_4 negative timelike. The convenient invariant dynamical variables for the double-dispersion representation are the squares of the total center-of-mass energies for the three reactions:

$$\begin{align}
\text{I.} & \quad (p_1,\alpha) + (p_2,\beta) \to (-p_3,\gamma) + (-p_4,\delta), \\
\text{II.} & \quad (p_1,\alpha) + (p_4,\delta) \to (-p_2,\beta) + (-p_3,\gamma), \quad \text{(II.1)} \\
\text{III.} & \quad (p_1,\alpha) + (p_3,\gamma) \to (-p_2,\beta) + (-p_4,\delta).
\end{align}$$

Thus we define

$$s = (p_1+p_2)^2 = (p_3+p_4)^2 = 4(q^2+\mu^2),$$
$$u = (p_1+p_4)^2 = (p_2+p_3)^2 = -2q^2(1+\cos\theta), \quad \text{(II.2)}$$
and
$$t = (p_1+p_3)^2 = (p_2+p_4)^2 = -2q^2(1-\cos\theta),$$

where q is the magnitude of the three-momentum and θ the angle of scattering in the barycentric system. Note the important supplementary condition

$$s+t+u = 4\mu^2, \quad \text{(II.3)}$$

which means that only two of the three variables s, t, u are independent even when extensions are made into the complex plane.

Since isotopic spin is conserved and the three values $I=0, 1, 2$, can occur, we expect to have three independent invariant functions of s, t, u. These functions are conveniently introduced by writing the complete amplitude as

$$A(s,t,u)\delta_{\alpha\beta}\delta_{\gamma\delta} + B(s,t,u)\delta_{\alpha\gamma}\delta_{\beta\delta} + C(s,t,u)\delta_{\alpha\delta}\delta_{\beta\gamma}. \quad \text{(II.4)}$$

Crossing symmetry leads at once to the relations

$$\left.\begin{array}{c} A \to A \\ B \to C \end{array}\right\} t \to u, \quad s \to s, \quad \text{(II.5)}$$

and

$$\left.\begin{array}{c} A \to B \\ C \to C \end{array}\right\} s \to t, \quad u \to u, \quad \text{(II.6)}$$

$$\left.\begin{array}{c} A \to C \\ B \to B \end{array}\right\} s \to u, \quad t \to t. \quad \text{(II.7)}$$

The first of these relations simply expresses the Pauli principle, but the remaining two place a powerful new condition on the combined energy and angular dependence of the amplitude. Such a condition, even though it arises from very simple considerations, is not known outside field theory.

An elementary calculation gives the connection between A, B, C and the three amplitudes A^I corresponding to well-defined I spin:

$$\begin{align}
A^0 &= 3A + B + C, \\
A^1 &= B - C, \quad \text{(II.8)} \\
A^2 &= B + C.
\end{align}$$

At this point one may verify that (II.5), together with (II.2), means that only even powers of $\cos\theta$ appear in the amplitudes for $I=0, 2$ and only odd powers of $\cos\theta$ for $I=1$. The implications of (II.6) and (II.7) are much more subtle, as we shall see later.

The unitarity condition on the pion-pion amplitude is most usefully expressed in terms of the partial-wave expansion of the amplitudes A^I when these are considered as functions of q^2 and $\cos\theta$:

$$A^I(q^2, \cos\theta) = \sum_{\substack{l \text{ even}, I=0,2 \\ l \text{ odd}, I=1}} (2l+1) A^{(l)I}(q^2) P_l(\cos\theta). \quad \text{(II.9)}$$

Unitarity allows the partial amplitudes $A^{(l)I}(q^2)$ to be written in terms of phase shifts δ_l^I according to[8]

$$A^{(l)I}(q^2) = [(q^2+\mu^2)^{\frac{1}{2}}/q] \exp(i\delta_l^I) \sin\delta_l^I, \quad \text{(II.10)}$$

where the phase shifts are real for $q^2 < 3\mu^2$, the threshold for inelastic scattering with the production of two additional pions.[4] At higher energies the phase shifts are complex, but the content of (II.10) can generally be expressed by the relation

$$\text{Im} A^{(l)I} = [q/(q^2+\mu^2)^{\frac{1}{2}}] R^{(l)I} |A^{(l)I}(q^2)|^2,$$
or
$$\text{Im}(A^{(l)I})^{-1} = -[q/(q^2+\mu^2)^{\frac{1}{2}}] R_l^I, \quad \text{(II.11)}$$

where R_l^I is the ratio of the total to the elastic partial-wave cross section.

III. THE DOUBLE-DISPERSION REPRESENTATION

A prescription for extending the scattering amplitude to complex values of s, t, and u, subject to (II.3), has been given by one of us.[1] This rule is embodied by the representation[5]

$$A(s,t,u) = \frac{1}{\pi^2} \int\int ds'dt' \frac{A_{st}(s',t')}{(s'-s)(t'-t)}$$

$$+ \frac{1}{\pi^2} \int\int ds'du' \frac{A_{su}(u',s')}{(s'-s)(u'-u)}$$

$$+ \frac{1}{\pi^2} \int\int dt'du' \frac{A_{tu}(t',u')}{(u'-u)(t'-t)}, \quad \text{(III.1)}$$

where the integrations in the primed variables extend in each case over regions of the positive real axis extending to infinity, and the weight functions A_{ij} are real. The functions B and C have similar representations, but the crossing conditions tell us that only two out of the total of nine weight functions are independent, with one of these a symmetric function of its two

---

[3] The normalization of (II.10) is arbitrary, but the dependence on q follows from the Lorentz invariance of the S matrix.
[4] Production of any odd number of pions is forbidden by the G parity of T. D. Lee and C. N. Yang, Nuovo cimento 3, 749 (1956).
[5] As shown in reference 1, the correct π-π representation probably requires also single dispersion integrals and an over-all subtraction term. See the remarks below, following Eq. (III.5), in this connection, as well as those following (IV.7).

arguments. In particular, in order to satisfy (II.5), (II.6), and (II.7), we require

$$\rho(x,y) = A_{st}(x,y) = A_{su}(y,x) = B_{st}(y,x) = B_{tu}(x,y)$$
$$= C_{su}(x,y) = C_{tu}(y,x), \quad \text{(III.2)}$$
$$\rho_s(x,y) = \rho_s(y,x) = A_{tu}(x,y) = B_{su}(x,y) = C_{st}(x,y).$$

The region of the (x,y) plane in which the weight functions fail to vanish is bounded by $x=4\mu^2$ and $y=4\mu^2$, but the region is not rectangular. According to the rules developed by one of us on the basis of perturbation theory,[1] the boundary is given by the curves,

$$x = 16\mu^2 y/(y-4\mu^2), \quad \text{for } x > y,$$
and
$$y = 16\mu^2 x/(x-4\mu^2), \quad \text{for } y > x, \quad \text{(III.3)}$$

as shown in Fig. 2. The large distance to the boundary from the corner, $y = x = 4\mu^2$, is associated with the absence of a three-pion vertex and considerably simplifies our problem. The absence of a three-particle vertex also is responsible for the absence of poles in (III.1).[6]

A point of maximum symmetry in the s, t, u variables is the nonphysical point, $s = t = u = 4\mu^2/3$, where A, B, and C are all real and equal to each other. It is appropriate then to introduce the pion-pion coupling constant λ through the definition[7]

$$\lambda = -A(\tfrac{4}{3}\mu^2, \tfrac{4}{3}\mu^2, \tfrac{4}{3}\mu^2) = -B(\tfrac{4}{3}\mu^2, \tfrac{4}{3}\mu^2, \tfrac{4}{3}\mu^2)$$
$$= -C(\tfrac{4}{3}\mu^2, \tfrac{4}{3}\mu^2, \tfrac{4}{3}\mu^2). \quad \text{(III.4)}$$

It follows from (II.8) that at this symmetry point we have

$$A^0 = -5\lambda, \quad A^1 = 0, \quad A^2 = -2\lambda. \quad \text{(III.5)}$$

Simple relations involving the derivatives at this point are also easily obtained and will be discussed in a later paper.

Normally a coupling constant is defined through the residue of a pole, but here there are no poles. The new constant λ may be explicitly introduced into (III.1), if desired, by making a subtraction at the symmetry point. Subtractions are probably necessary to give a meaning to the double-dispersion representation (III.1), but we only need this expression in order to locate the singularities of the scattering amplitude. Thus we proceed at once to consider the analyticity properties of the partial-wave amplitudes $A^{(l)I}(q^2)$, which can be correctly obtained by inspection of (III.1).

IV. ANALYTICITY PROPERTIES OF THE PARTIAL-WAVE AMPLITUDES

In this paper we shall concentrate most of our attention on the low angular-momentum states. In

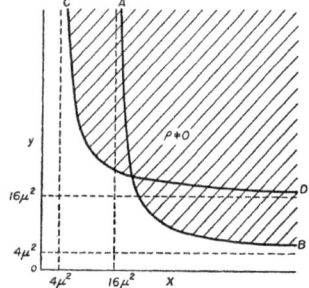

FIG. 2. The domain in which the spectral functions of the two-dimensional π-π representation are nonvanishing.

principle, the approximation scheme based on the double-dispersion representation does *not* consist of taking more and more angular-momentum states into account—such a procedure would be inadequate owing to the failure (to be discussed below) of the Legendre expansion to converge in the unphysical region. It will therefore ultimately be necessary to calculate the spectral functions in (III.1), so as to include effects of all the angular-momentum waves. An approximation scheme for calculating the spectral functions can be worked out and was outlined in reference 1. Even if the spectral functions were known, however, it still would be necessary to treat separately the low angular-momentum states. The reason is that, when the single-dispersion integrals are included in (III.1), the absorptive parts of the low angular-momentum states will no longer be determined by the spectral functions, as has been explained in reference 1. We shall see below that, because of special properties of the π-π system, the calculation in the lowest approximation can be based entirely on the low angular-momentum states.

From (II.9) it follows that

$$A^{(l)I}(q^2) = \frac{1}{2} \int_{-1}^{+1} d\cos\theta \, A^I(q^2, \cos\theta) P_l(\cos\theta), \quad \text{(IV.1)}$$

so that, in view of (II.2), the projection of a given partial wave amounts to an integration at fixed s over either dt or du. The two variables, t and u, each cover the range between 0 and $-4q^2$, moving in opposite directions. It is straightforward, then, by inspection of (III.1) to establish the nature and location of the singularities of $A^{(l)I}(q^2)$ in the q^2 complex plane.[8]

It is obvious, first of all, that all the singularities lie on the real axis.[9] Next it will be recognized that there are three sets of branch points. The first set is associated with the vanishing of denominators containing s, with

---

[6] We assume that there exist no strongly interacting particles with the same quantum numbers as a pair of pions. If such a particle should be found, corresponding poles must be added to (III.1), whether or not the new particle is interpreted as a two-pion bound state.

[7] This λ is, in conventional terminology, a *renormalized* unrationalized coupling constant. It corresponds to a term in the Lagrangian of the form $4\pi\lambda(\phi_\mu\phi_\mu)^2$.

[8] It is possible, of course, to carry out the integration over $d\cos\theta$ in (III.1) explicitly.

[9] With unequal masses, as in pion-nucleon scattering, the singularities in the partial-wave amplitudes do not all lie on the real axis, but they can be located without difficulty. See, for example, S. W. McDowell, Phys. Rev. 116, 774 (1960).

the lowest branch point occurring at $q^2=0$, the threshold of the physical region. The next branch point of this set will be at $q^2=3\mu^2$, the threshold for producing two additional pions, and so on. It is evidently appropriate to choose a cut running along the positive real axis from 0 to ∞. We shall refer to this as the "right hand" or "physical" cut.

The other two sets of branch points are associated with the vanishing of denominators containing t or u and are coincident, lying on the negative real axis. The first pair of branch points is at $q^2=-\mu^2$, the second at $q^2=-4\mu^2$, etc., the spacing being the same as on the positive axis. A second cut may then be chosen to run from $-\mu^2$ to $-\infty$; this will be called the "left hand" or "unphysical" cut.

Finally it should be recognized that our partial-wave amplitude is a *real* analytic function of q^2, whose boundary value as the physical cut is approached from above is the complex physical amplitude, but which is real in the gap between $-\mu^2$ and 0 on the real axis. The discontinuity in going across either cut is twice the imaginary part of the limit as the cut is approached. The required imaginary part is given for the right-hand cut by (II.11).

The calculation of the imaginary part on the left-hand cut is much more involved, as the unitarity condition cannot be used for negative values of q^2. We shall have to use crossing symmetry to obtain the imaginary part on the left-hand cut in terms of that on the right-hand cut, and the most convenient way to handle this problem is in terms of the absorptive parts for the three reactions, I, II, and III, defined in reference 1.

The absorptive parts A_s, B_s, and C_s may be identified with the imaginary parts of the corresponding amplitudes in the physical region of reaction I, $s>4\mu^2$ or $q^2>0$. Similarly, the absorptive parts with subscripts t and u are equal to imaginary parts in the physical regions of reactions II and III, respectively. These will be regions of negative q^2. It is possible to deduce from (III.2) the following crossing rules, which correspond to the relations (II.5) to (II.7):

and

$$\left.\begin{array}{c}A_s \to A_s \\ B_s \to C_s \\ C_s \to B_s\end{array}\right\} \text{ for } t \to u, \; s \to s,$$

$$\left.\begin{array}{c}A_t \to B_t \\ B_t \to A_s \\ C_t \to C_s\end{array}\right\} \text{ for } s \to t, \; u \to u,$$

$$\left.\begin{array}{c}A_u \to C_s \\ B_u \to B_s \\ C_u \to A_s\end{array}\right\} \text{ for } s \to u, \; t \to t. \quad \text{(IV.2)}$$

The other relation needed is that connecting the imaginary part of the amplitudes for $q^2<0$ with the absorptive parts for reactions II and III. By examination of (III.1) we find

$$\text{Im}A(q^2, \cos\theta) = -A_u(q^2, \cos\theta) - A_t(q^2, \cos\theta),$$
$$\text{for } q^2<0, \quad \text{(IV.3)}$$

with similar relations for $\text{Im}B$ and $\text{Im}C$.

If we now define

$$q'^2 = t/4 - \mu^2,$$
$$\bar{q}'^2 = u/4 - \mu^2,$$
$$\cos\theta' = 1 + s/2q'^2 = 1 + 2(q^2+\mu^2)/q'^2,$$

and

$$\cos\bar{\theta}' = -1 - s/2\bar{q}'^2 = -1 - 2(q^2+\mu^2)/\bar{q}'^2,$$

and recall from (II.2) that

$$q^2 = s/4 - \mu^2,$$

and

$$\cos\theta = 1 + 2(q'^2+\mu^2)/q^2 = -1 - 2(\bar{q}'^2+\mu^2)/q^2,$$

then the crossing rules (IV.2) allow us to write in place of (IV.3),

$$\text{Im}A(q^2, \cos\theta) = -C_s(\bar{q}'^2, \cos\bar{\theta}') - B_s(q'^2, \cos\theta'), \quad \text{(IV.3')}$$

where q'^2 ranges from $-q^2-\mu^2$ to $-\mu^2$ as $\cos\theta$ goes from -1 to $+1$, while $\bar{q}'^2$ covers the same range but in the opposite direction. It can be seen by inspection of (III.1) that B_s and C_s vanish in the range between 0 and $-\mu^2$, and so we have achieved our goal of expressing the imaginary part of the amplitude on the left cut in terms of absorptive parts on the right cut.

It remains now to project out the partial waves. From (IV.1) we have for $q^2<-\mu^2$,

$$\text{Im}A^{(l)}(q^2)$$
$$= \frac{1}{2}\int_{-1}^{+1} d\cos\theta \, \text{Im}A(q^2, \cos\theta) P_l(\cos\theta)$$
$$= \int_0^{-q^2-\mu^2} \frac{d\bar{q}'^2}{q^2} C_s\left(\bar{q}'^2, -1-2\frac{q^2+\mu^2}{\bar{q}'^2}\right)$$
$$\times P_l\left(-1-2\frac{\bar{q}'^2+\mu^2}{q^2}\right) + \int_0^{-q^2-\mu^2} \frac{dq'^2}{q^2}$$
$$\times B_s\left(q'^2, 1+2\frac{q^2+\mu^2}{q'^2}\right) P_l\left(1+2\frac{q'^2+\mu^2}{q^2}\right). \quad \text{(IV.4)}$$

The formulas for $\text{Im}B^{(l)}(q^2)$ and $\text{Im}C^{(l)}(q^2)$ are similar, and the corresponding expressions for amplitudes with well-defined isotopic spin are

$$\text{Im}A^{(l)I}(q^2) = \int_0^{-q^2-\mu^2} \frac{dq'^2}{q^2} P_l\left(1+2\frac{q'^2+\mu^2}{q^2}\right)$$
$$\times \sum_{I'=0,1,2} \alpha_{II'} A_s^{I'}\left(q'^2, 1+2\frac{q^2+\mu^2}{q'^2}\right), \quad \text{(IV.5)}$$

for $q^2<-\mu^2$, where

$$\alpha_{II'} = \begin{bmatrix} 2/3 & 2 & 10/3 \\ 2/3 & 1 & -5/3 \\ 2/3 & -1 & 1/3 \end{bmatrix}. \quad \text{(IV.6)}$$

Under the integrals in (IV.5) appear the absorptive parts of scattering amplitudes at values of $\cos\theta$ less than -1. From the boundary curve of Fig. 2 and Formula (III.3) it is possible to conclude that the Legendre polynomial expansion of $A_s{}^I(q'^2,\cos\theta)$ converges for the values of $\cos\theta$ required in (IV.5) so long as $q^2 > -9\mu^2$.[10] For the "effective-range" approach of this paper, such a limit might as well be $-\infty$ if all phase shifts for $l>0$ are small. On using the Legendre expansion, (IV.5) becomes

$$\mathrm{Im}A^{(l)}(q^2)$$
$$= \int_0^{-q^2-\mu^2} \frac{dq'^2}{q^2} P_l\left(1+2\frac{q'^2+\mu^2}{q^2}\right)$$
$$\times \sum_{l',I'=0,1,2} \alpha_{II'} A_s^{(l')I'}(q'^2) P_{l'}\left(1+2\frac{q^2+\mu^2}{q'^2}\right),$$
(IV.7)

for $(q^2 < -\mu^2)$. If the Legendre expansion is rapidly convergent, the sum over l' in (IV.7) can be terminated at an early stage.

The surprisingly large magnitude of the limit $q^2 = -9\mu^2$ is associated, as mentioned above, with the absence of a three-pion vertex. Crudely speaking, absence of a single-pion exchange mechanism reduces the range of the force to $\sim 1/2\mu$ and greatly improves the convergence of the partial wave expansion. Also it should be remembered, as emphasized by Lehmann,[10] that the expansion of the absorptive part of the amplitude always converges better than that of the real part.

It is possible to view in a slightly different way the approximation made in keeping only the first few terms of the polynomial expansion of the absorptive parts on the right of (IV.5). As shown in reference 1, the absorptive part can be written as a dispersion integral

$$A_s\{q^2(s),\cos\theta(s,t)\}$$
$$= a_0(s)+a_1(s)t+\cdots+(t-t_0)^n\frac{1}{\pi}\int dt'\frac{A_{st}(s,t')}{(t'-t_0)^n(t'-t)}$$
$$+(u-u_0)^n\frac{1}{\pi}\int du'\frac{A_{su}(s,u')}{(u'-u_0)^n(u'-u)}. \quad \text{(IV.8)}$$

The subtraction terms are here written explicitly, and the value of the exponent n is equal to the number of such terms. Perturbation theory prescribes that only one subtraction is necessary. However, further subtractions may be made either because one distrusts pertubation theory in this connection or to increase the accuracy of the calculation.[11] In this paper we make

---
[10] For a discussion of the convergence of the Legendre polynominal expansion of a scattering amplitude, see H. Lehmann, Nuovo cimento **10**, 579 (1958).
[11] Subtractions of this kind in one variable do not correspond to the introduction of new parameters. See reference 1.

two subtractions, as shown below explicitly in Formulas (IV.10) and (IV.11).

Let us examine the form of the region in which one of the spectral functions, A_{st} for instance, is nonzero. As explained in reference 1, this spectral function consists of a number of parts corresponding to different Feynman diagrams. The two parts extending to the lowest values of s and t are bounded by the curves AB and CD of Fig. 2. Now, the part bounded by AB begins at or above the value $s=(4\mu)^2$, the threshold for the production of two additional pions. In the following section we shall approximate the absorptive part in the physical region by neglecting inelastic processes in the unitarity condition. This is in line with the "effective-range" principle, which assumes that the behavior of the scattering amplitude at low momenta is dominated by the nearest singularities. The part of A_{st} bounded by AB is therefore zero in this approximation. Similar considerations will apply to the other spectral functions. If, further, we require the crossing relations (III.2) to be satisfied, we shall also have to assume that the part of A_{st} bounded by CD is zero in the lowest approximation, so that the spectral functions are to be neglected entirely. That is to say, all contributions to the spectral functions begin at values of s and t which are so far from the region of interest that they should be ignored in a consistent "effective-range" approach.

From Eq. (IV.8), the absorptive part A_s can then be approximated by an expression of the form

$$A_s\{q^2(s),\cos\theta(s,t)\} = a_0(s)+a_1(s)t\cdots,$$

which is terminated at an early stage. The absorptive parts are thus represented by taking a small number of angular-momentum states only. This conclusion bears out the statement made at the beginning of this section that, in the lowest approximation, the calculation can be based entirely on the low angular-momentum states.

The approach just outlined enables us to understand why the absence of a three-pion vertex is critical in allowing one to terminate the Legendre expansion of the absorptive part. Had there been such a vertex, the curve bounding the shaded area in Fig. 2 would have consisted of a single part which approached asymptotically the lines $x=4\mu^2$, $y=4\mu^2$. The neglect of the spectral functions would then not have been justified. It would have been necessary to insert them in some approximation into Eq. (IV.8), with the resulting expressions then substituted into the integrals of Eq. (IV.5). [Actually the fourth-order perturbation approximation could be used for the spectral functions, as all other contributions begin at values of either s or t greater than $(4\mu)^2$.] Even in the actual problem with no 3-pion vertex, if we were to go beyond the lowest approximation it would be necessary to calculate the spectral functions to an appropriate accuracy and then insert them into Eq. (IV.8).

It is worth emphasizing that we only assume the absorptive part of the scattering amplitude to be represented by its lowest angular-momentum waves. No such assumption regarding the real part is made. At the end of the calculation, the real part of the first angular-momentum state omitted can be computed; if its square turns out to be small at the energies under consideration, we were presumably justified in leaving out the absorptive part. There is thus a check on the number of angular states which it is necessary to include.

To illustrate the above considerations and for future reference, we now derive formulas that clearly show the difference in our treatment of low and high partial waves. With no subtractions, one could write the following momentum-transfer dispersion relation on the basis of (III.1):

$$A(q^2, \cos\theta)$$

$$= \frac{1}{\pi}\int_{4\mu^2}^{\infty} dt' \frac{A_t(q^2, 1+t'/2q^2)}{t'-t}$$

$$+ \int_{4\mu^2}^{\infty} du' \frac{A_u(q^2, -1-u'/2q^2)}{u'-u}$$

$$= \frac{1}{\pi}\int_{0}^{\infty} dq'^2 \frac{B_s(q'^2, 1+2(q^2+\mu^2)/q'^2)}{q'^2+\mu^2+\frac{1}{2}q^2(1-\cos\theta)}$$

$$+ \frac{1}{\pi}\int_{0}^{\infty} d\bar{q}'^2 \frac{C_s(\bar{q}'^2, -1-2(q^2+\mu^2)/\bar{q}'^2)}{\bar{q}'^2+\mu^2+\frac{1}{2}q^2(1+\cos\theta)}$$

$$= \frac{1}{\pi}\int_{0}^{\infty} dq'^2 B_s\left(q'^2, 1+2\frac{q^2+\mu^2}{q'^2}\right)$$

$$\times\left[\frac{1}{q'^2+\mu^2+\frac{1}{2}q^2(1-\cos\theta)}+\frac{1}{q'^2+\mu^2+\frac{1}{2}q^2(1+\cos\theta)}\right],$$

(IV.9)

with similar expressions for B and C. Now, the absorptive part B_s is, in general, complex, but the imaginary part of B_s vanishes in the lower range of the integral (IV.9) because from the equivalent of Eq. (IV.8), for $q^2>0$ and $q'^2>0$, we have

$$\text{Im}B_s(q'^2, 1+2(q^2+\mu^2)/q'^2) = B_{st}(4(q'^2+\mu^2), 4(q^2+\mu^2)),$$

which is zero outside the shaded region of Fig. 2. Thus, if we make a subtraction in the dispersion relation (IV.9) to suppress the high-energy part, the remainder will be almost entirely real for small q^2. Figure 2 shows, of course, that as q^2 becomes large, the imaginary part cannot be suppressed. These considerations are identical to those following Eq. (IV.8).

Let us make the subtraction by removing the S-wave part of Eq. (IV.9):

$$A(q^2, \cos\theta)$$

$$= A^{(0)}(q^2) + \frac{1}{\pi}\int_{0}^{\infty} dq'^2 B_s\left(q'^2, 1+2\frac{q^2+\mu^2}{q'^2}\right)$$

$$\times\left[\frac{1}{q'^2+\mu^2+\frac{1}{2}q^2(1-\cos\theta)}+\frac{1}{q'^2+\mu^2+\frac{1}{2}q^2(1+\cos\theta)}\right.$$

$$\left.-\frac{2}{q^2}\ln\left(1+\frac{q^2}{q'^2+\mu^2}\right)\right]. \quad \text{(IV.9')}$$

In the next section we shall determine $A^{(0)}(q^2)$, allowing it to be complex, but the residual amplitude (which starts here with the D wave) is real to a good approximation, for q^2 not too large. Furthermore, so long as the imaginary part of B_s is to be neglected, we are ignoring the singularities of this function in the variable q^2, and according to the arguments following Eq. (IV.8), we may consistently approximate it by a low-order polynomial. Since B_s when continued to the physical region of reaction I is the imaginary part of the amplitude, the appropriate procedure is to represent B_s in terms of precisely those partial waves that have been subtracted out, i.e., those that are allowed to be complex.

We now give the formulas for $I=0, 2$ that correspond to Eq. (IV.9'):

$$A^I(q^2, \cos\theta) = A^{(0)I}(q^2) + \frac{1}{\pi}\int_{0}^{\infty} dq'^2 \sum_{I'} \alpha_{II'} A_s^{I'}\left(q'^2, 1+2\frac{q^2+\mu^2}{q'^2}\right)$$

$$\times\left\{\frac{1}{2}\left[\frac{1}{q'^2+\mu^2+\frac{1}{2}q^2(1-\cos\theta)}+\frac{1}{q'^2+\mu^2+\frac{1}{2}q^2(1+\cos\theta)}\right]-\frac{1}{q^2}\ln\left(1+\frac{q^2}{q'^2+\mu^2}\right)\right\}. \quad \text{(IV.10)}$$

For $I=1$, we subtract the P wave:

$$A^1(q^2, \cos\theta) = 3\cos\theta A^{(1)1}(q^2) + \frac{1}{\pi}\int_{0}^{\infty} dq'^2 \sum_{I'} \alpha_{1I'} A_s^{I'}\left(q'^2, 1+2\frac{q^2+\mu^2}{q'^2}\right)$$

$$\times\left\{\frac{1}{2}\left[\frac{1}{q'^2+\mu^2+\frac{1}{2}q^2(1-\cos\theta)}-\frac{1}{q'^2+\mu^2+\frac{1}{2}q^2(1+\cos\theta)}\right]-\frac{3\cos\theta}{q^2}\left[\left(1+2\frac{q'^2+\mu^2}{q^2}\right)\ln\left(1+\frac{q^2}{q'^2+\mu^2}\right)-2\right]\right\}. \quad \text{(IV.11)}$$

These formulas are exact, but in practice under the integrals $A_s{}^{l'}(q'^2, \cos\theta')$ will be approximated by

$$A_s{}^{0,2}(q'^2, \cos\theta') \approx \mathrm{Im} A^{(0)0,2}(q'^2), \quad \text{for} \quad q'^2 > 0,$$

$$A_s{}^{1}(q'^2, \cos\theta') \approx 3\cos\theta' \, \mathrm{Im} A^{(1)1}(q'^2),$$

$$\text{for} \quad q'^2 > 0. \quad \text{(IV.12)}$$

It can be shown that such an approximation exactly satisfies the crossing conditions (II.5)–(II.7), provided $A^{(0)I}$ and $A^{(1)1}$ are calculated by the equations of the following section, in which a corresponding approximation is made.

V. FORMULATION OF INTEGRAL EQUATIONS

We now have the task of translating our knowledge about partial-wave amplitudes into integral equations. After introducing the variable $\nu = q^2/\mu^2$, the preceding statements about the location of singularities are equivalent to the dispersion relations,

$$A^{(l)I}(\nu) = \frac{1}{\pi}\int_{-\infty}^{-1} d\nu' \frac{\mathrm{Im} A^{(l)I}(\nu')}{\nu' - \nu}$$

$$+ \frac{1}{\pi}\int_0^\infty d\nu' \frac{\mathrm{Im} A^{(l)I}(\nu')}{\nu' - \nu}, \quad \text{(V.1)}$$

provided the functions in question behave properly at infinity. The unitarity condition (II.10) guarantees that the partial-wave amplitudes behave asymptotically no worse than like constants. In order to estimate the error in our approximation, it is plausible to assume that on the right-hand (physical) cut,

$$\mathrm{Im} A^{(l)I}(\nu) \to \tfrac{1}{2},$$

and

$$\mathrm{Re} A^{(l)I}(\nu) \to 0, \quad \text{(V.2)}$$

in other words, the limit of pure diffraction scattering.[12] Such behavior, i.e., the ratio of the real to the imaginary part going asymptotically to zero, can be consistent with Eq. (V.1) only if the limits on the left-hand cut are the same.[13]

A partial-wave amplitude of order l vanishes at the origin like ν^l, so we may consider new quantities

$$A'^{(l)I}(\nu) = (1/\nu^l) A^{(l)I}(\nu), \quad \text{(V.3)}$$

which also satisfy relations of the type (V.1) but whose imaginary parts, except for $l=0$, now vanish at infinity like ν^{-l}. It is clear that the higher the angular momentum, the smaller is the relative contribution from high values of ν' in the dispersion integrals (when ν is small). It is only for the S wave that distant contributions are expected to be important, so for the S wave we make a subtraction at the symmetry point

$$\nu_0 = -\tfrac{2}{3}, \quad \text{(V.4)}$$

to obtain[14]

$$A^{(0)I}(\nu) = a_I + \frac{\nu - \nu_0}{\pi}\int_{-\infty}^{-1} d\nu' \frac{\mathrm{Im} A^{(0)I}(\nu')}{(\nu' - \nu)(\nu' - \nu_0)}$$

$$+ \frac{\nu - \nu_0}{\pi}\int_0^\infty d\nu' \frac{\mathrm{Im} A^{(0)I}(\nu')}{(\nu' - \nu)(\nu' - \nu_0)}. \quad \text{(V.5)}$$

It is possible that even an S-wave subtraction is unnecessary in a treatment which includes in a serious way very-high-energy inelastic processes such as nucleon-antinucleon pair production. We do not believe, however, that such a treatment will be practical for a long time to come. Certainly nothing so ambitious will be attempted here.

Thus, either by dividing by ν^l or by subtracting we hope to suppress very high energies under the dispersion integrals. To obtain a rough indication of the contribution from high energies let us investigate the error made by cutting off the integral at $\nu = \pm L$. Equation (V.1) then becomes

$$\frac{\nu^l}{\pi}\int_{-L}^{-1} d\nu' \frac{\mathrm{Im} A^{(l)I}(\nu')}{\nu'^l(\nu' - \nu)} + \frac{\nu^l}{\pi}\int_0^L d\nu' \frac{\mathrm{Im} A^{(l)I}(\nu')}{\nu'^l(\nu' - \nu)},$$

or the corresponding subtracted expressions for S waves. Using Eq. (V.2) one can easily estimate the order of magnitude of the neglected contributions to be

$$\delta A^{(0)} \sim \frac{1}{\pi}\frac{\nu - \nu_0}{L},$$

and

$$\delta A^{(l)} \sim \frac{1}{\pi}\frac{1}{l}\left(\frac{\nu}{L}\right)^l, \quad \text{(V.6)}$$

which are small provided L can be made sufficiently large—in particular, if L is chosen in the range where inelastic scattering first becomes important. The inelastic threshold is at $\nu = 3$, but experience with pion-nucleon scattering suggests that double-pion production won't represent a substantial fraction of the cross section until $\nu \sim 10$. The important contributions to the integrals in (V.1) thus come from the region where only elastic scattering is important, and we may use the unitarity condition (II.11) with R_l set equal to unity:

$$\mathrm{Im}[A^{(l)I}(\nu)]^{-1} = -[\nu/(\nu+1)]^{\frac{1}{2}}. \quad \text{(V.7)}$$

Furthermore, as discussed in the preceding section, the imaginary parts on the left-hand cut as given by Eq. (IV.7) may be evaluated by taking only a few terms in the Legendre summation over l'. In particular

---

[12] Such behavior is expected because of the overwhelming competition from inelastic channels that sets in at very high energies.

[13] Considerations of this kind were first emphasized by I. Pomeranchuk, J. Exptl. Theoret. Phys. (U.S.S.R.) **34**, 725 (1958), in connection with forward-dispersion relations.

[14] The two subtraction constants a_0 and a_2 are not independent but are related to λ through Eq. (III.5). The relation is given below in Formula (V.18).

we shall keep only $l'=0$ and $l'=1$ terms in these integrals; the legitimacy of this approximation may be checked *a posteriori* by calculating the D waves that emerge from our system of equations. In terms of the variable ν the formula (IV.7) becomes, in this approximation:

$$\text{Im}A^{(l)I}(\nu) = \frac{1}{\nu} \int_0^{-\nu-1} d\nu' P_l\left(1+2\frac{\nu'+1}{\nu}\right)$$

$$\times \left[\alpha_{I0} \text{Im}A^{(0)0}(\nu') + \alpha_{I2} \text{Im}A^{(0)2}(\nu') \right.$$

$$\left. +3\left(1+2\frac{\nu+1}{\nu'}\right)\alpha_{I1} \text{Im}A^{(1)1}(\nu') \right]. \quad \text{(V.8)}$$

Now we put Eqs. (V.5), (V.7), and (V.8) together in order to obtain a procedure for calculating phase shifts in terms of the empirical constant λ.

Consider first the two S-wave amplitudes. We attempt to represent each of these by a quotient[15]

$$A^{(0)I}(\nu) = N_0^I(\nu)/D_0^I(\nu), \quad \text{(V.9)}$$

where $N_0^I(\nu)$ and $D_0^I(\nu)$ are both real analytic functions; the numerator contains the branch point at $\nu=-1$ with the left-hand cut, and the denominator contains the branch point at $\nu=0$ with the right-hand cut. It is also necessary, of course, that $D_0^I(\nu)$ have no zeros. By assumption, then, we have

$$\left.\begin{array}{l}\text{Im}N_0^I(\nu) = D_0^I(\nu) \text{ Im}A^{(0)I}(\nu), \\ \text{Im}D_0^I(\nu) = 0,\end{array}\right\} \text{ for } \nu<-1$$

$$\text{Im}N_0^I(\nu) = \text{Im}D_0^I(\nu) = 0, \text{ for } -1<\nu<0 \quad \text{(V.10)}$$

$$\left.\begin{array}{l}\text{Im}N_0^I(\nu) = 0, \\ \text{Im}D_0^I(\nu) = N_0^I(\nu) \text{ Im}\dfrac{1}{A^{(0)I}(\nu)}\end{array}\right\} \text{ for } \nu>0,$$

The subtracted dispersion relation (V.5) normalizes the S-wave amplitudes to a_I at the point $\nu=\nu_0$. We accomplish this normalization in our quotient by setting $N_0^I(\nu_0)=a_I$ and $D_0^I(\nu_0)=1$. Furthermore, the amplitudes $A^{(0)I}(\nu)$ at most approach constants at infinity, so we may assign constant asymptotic behavior to the numerator and require the denominator not to vanish. From (V.10), we are thus led first to write

$$N_0^I(\nu) = a_I + \frac{\nu-\nu_0}{\pi}\int_{-\infty}^{-1} d\nu' \frac{\text{Im}A^{(0)I}(\nu')D_0^I(\nu')}{(\nu'-\nu)(\nu'-\nu_0)}. \quad \text{(V.11)}$$

[15] To see that the N/D representation is always possible, observe that if

$$D(\nu) = \exp\left[-\frac{\nu-\nu_0}{\pi}\int_0^\infty d\nu' \frac{\delta(\nu')}{(\nu'-\nu_0)(\nu'-\nu)}\right],$$

then the denominator function is analytic, with the right-hand branch point only, and along the physical cut has the phase $e^{-i\delta}$. The numerator function is therefore real on the positive real axis and has only the left-hand branch point.

Second, remembering (V.7), we have

$$D_0^I(\nu) = 1 - \frac{\nu-\nu_0}{\pi}\int_0^\infty d\nu' \left(\frac{\nu'}{\nu'+1}\right)^{\frac{1}{2}}$$

$$\times \frac{N_0^I(\nu')}{(\nu'-\nu)(\nu'-\nu_0)}. \quad \text{(V.12)}$$

On defining

$$\omega = -\nu, \quad E_0^I(\omega) = D_0^I(\nu), \quad f_I^I(\omega) = \text{Im}A^{(1)I}(\nu), \quad \text{(V.13)}$$

the following integral equation is obtained:

$$E_0^I(\omega) = 1 + (\omega+\nu_0)K(\omega,-\nu_0)a_I$$

$$+ \frac{\omega+\nu_0}{\pi}\int_1^\infty d\omega' \frac{K(\omega,\omega')f_0^I(\omega')E_0^I(\omega')}{\omega'+\nu_0}, \quad \text{(V.14)}$$

with

$$K(\omega,\omega') = \frac{1}{\pi}\int_0^\infty d\nu'' \frac{[\nu''/(1+\nu'')]^{\frac{1}{2}}}{(\nu''+\omega)(\nu''+\omega')}$$

$$= \frac{2}{\pi(\omega-\omega')}\left\{\left(\frac{\omega}{\omega-1}\right)^{\frac{1}{2}} \ln[\sqrt{\omega}+(\omega-1)^{\frac{1}{2}}]\right.$$

$$\left. -\left(\frac{\omega'}{\omega'-1}\right)^{\frac{1}{2}} \ln[\sqrt{\omega'}+(\omega'-1)^{\frac{1}{2}}]\right\}. \quad \text{(V.15)}$$

If the function $f_0^I(\omega')$ were known, Eq. (V.14) would be a Fredholm equation, soluble by any number of standard methods. The question whether or not the equation is singular will be discussed below.

It is unfortunately true that $f_0^I(\omega')$ is not known in advance but is given only through Eqs. (V.13) and (V.8) in terms of the amplitudes we are looking for. Thus our system of equations is actually nonlinear. In the following paper, however, it will be shown that the problem can be solved by an iteration procedure in which at every stage the linear equations (V.14) are solved with f_0^I corresponding to the previous stage. We must, of course, also formulate an equation for the P amplitude since this is required in Eq. (V.8).

Before considering the P amplitude, however, a few general remarks about the S-wave problem are in order. First, an inspection of (V.14) with f_0^I set equal to zero shows that D_0^I will develop a zero for $\nu<\nu_0$ if a_I is negative. According to Eq. (V.8), both f_0^0 and f_0^2 will be negative if the S contributions under the integrals are dominant.[16] The zero will therefore not be removed when f_0^I is included, but if the zero appears sufficiently far out along the negative real axis—beyond the limit at which our calculation of $\text{Im}A^{(0)}$ ceases to be accurate—the associated pole in $A^{(0)}$ is of no physical significance and cannot be excluded. A crude estimate, based on Eq. (V.14) and neglecting f_0^I, indicates that for $-1.5 < a_I < 0$, the zero in $E_0^I(\omega)$ will occur for $\omega > 10$.

[16] Recall that the imaginary part of a partial-wave amplitude in the physical region is positive definite.

If a_I is positive, the requirement that there be no zero of $D_0^I(\nu)$ in the region $\nu_0<\nu<0$ (i.e., no bound state of the π-π system) puts an upper limit on a_I. As a_I increases, the zero will appear first at $\nu=0$, so we examine the condition that $D_0^I(0)$ be positive. Here the neglect of f_0^I is a good approximation, so one may deduce from Eq. (V.14) the requirement

$$1-\tfrac{2}{3}K(0,\tfrac{2}{3})a_I>0,$$

or, since we have

$$K(0,\tfrac{2}{3})=(3\sqrt{2}/\pi)\tan^{-1}(1/\sqrt{2}),$$

we can write

$$a_I<\frac{\pi}{2\sqrt{2}}\Big/\tan^{-1}\left(\frac{1}{\sqrt{2}}\right)=1.8.$$

One may inquire also about the possibility of zeros in $D_0^I(\nu)$ that are not on the real axis. Inspection of Eq. (V.13) shows that such zeros are impossible so long as $N_0^I(\nu)$ has no zeros on the positive real axis.[17] Should we find a solution that does have zeros in the physical region, this point would have to be investigated further.

Let us now determine the relation between a_I and λ and the consequent restrictions on λ that follow from the above limitations on a_I. According to (III.5), we have

$$A^0(\nu_0,0)=-5\lambda,$$
$$A^1(\nu_0,0)=0,\qquad\text{(V.16)}$$
$$A^2(\nu_0,0)=-2\lambda.$$

The second of these relations is identically satisfied, since A^1 contains only odd powers of $\cos\theta$. The first and the third, however, give us the required information about a_0 and a_2 which are defined by Eq. (V.5) to be

$$a_0=A^{(0)0}(\nu_0),$$
and
$$a_2=A^{(0)2}(\nu_0).\qquad\text{(V.17)}$$

Thus to a good approximation $a_0\approx-5\lambda$ and $a_2\approx-2\lambda$, since we expect D and higher partial-wave amplitudes to be small.

It is possible to correct for the higher waves within the approximation outlined at the end of Sec. IV. Formula (IV. 10), when evaluated at $\cos\theta=0$ and $\nu=\nu_0$, leads to the following result:

$$a_0=-5\lambda+\frac{1}{\pi}\int_0^\infty d\nu'\left\{\frac{1}{\nu_0}\ln\left(1+\frac{\nu_0}{\nu'+1}\right)-\frac{1}{\nu'+1+\nu_0/2}\right\}$$
$$\times\left\{\tfrac{2}{3}\operatorname{Im}A^{(0)0}(\nu')+\frac{10}{3}\operatorname{Im}A^{(0)2}(\nu')\right.$$
$$\left.+6\left(1+2\frac{\nu_0+1}{\nu'}\right)\operatorname{Im}A^{(1)1}(\nu')\right\},$$
(V.18)

$$a_2=-2\lambda+\frac{1}{\pi}\int_0^\infty d\nu'\left\{\frac{1}{\nu_0}\ln\left(1+\frac{\nu_0}{\nu'+1}\right)-\frac{1}{\nu'+1+\nu_0/2}\right\}$$
$$\left\{\tfrac{2}{3}\operatorname{Im}A^{(0)0}(\nu')+\tfrac{1}{3}\operatorname{Im}A^{(0)2}(\nu')\right.$$
$$\left.-3\left(1+2\frac{\nu_0+1}{\nu'}\right)\operatorname{Im}A^{(1)1}(\nu')\right\}.$$

The integral correction given by Eq. (V.18) to the simple relation between the a_I and λ is very small[18] and may be ignored except for highly refined considerations. The most restrictive conditions on λ are obtained by considering the $I=0$ state, for which $a_0\approx-5\lambda$. The absence of zeros on the negative real axis for $|\nu|<10$, as discussed above, then leads to the limits

$$-\tfrac{1}{5}(1.8)\lesssim\lambda\lesssim-\tfrac{1}{5}(-1.5),$$
or
$$-0.36\lesssim\lambda\lesssim0.3.\qquad\text{(V.19)}$$

A study of the formula for the cotangent of the S phase shifts reveals another interesting circumstance. We have for $\nu>0$

$$\left(\frac{\nu}{\nu+1}\right)^{\frac{1}{2}}\cot\delta_0^I=\frac{\operatorname{Re}D_0^I(\nu)}{N_0^I(\nu)}=\frac{1-(\nu-\nu_0)I(\nu,-\nu_0)a_I-\dfrac{\nu-\nu_0}{\pi}\displaystyle\int_1^\infty d\omega'\dfrac{I(\nu,\omega')f_0^I(\omega')E_0^I(\omega')}{\omega'+\nu_0}}{a_I+\dfrac{\nu-\nu_0}{\pi}\displaystyle\int_1^\infty d\omega'\dfrac{f_0^I(\omega')E_0^I(\omega')}{(\omega'+\nu)(\omega'+\nu_0)}},\qquad\text{(V.20)}$$

where

$$I(\nu,\omega')=\frac{P}{\pi}\int_0^\infty d\nu''\frac{[\nu''/(\nu''+1)]^{\frac{1}{2}}}{(\nu''-\nu)(\nu''+\omega')}$$
$$=\frac{2}{\pi(\nu+\omega')}\left\{\left(\frac{\omega'}{\omega'-1}\right)^{\frac{1}{2}}\ln[\sqrt{\omega'}+(\omega'-1)^{\frac{1}{2}}]-\left(\frac{\nu}{\nu+1}\right)^{\frac{1}{2}}\ln[\sqrt{\nu}+(\nu+1)^{\frac{1}{2}}]\right\}.\qquad\text{(V.21)}$$

---

[17] For $\nu=\nu_R+i\nu_I$, the imaginary part of $D_0^I(\nu)$ is given by

$$-\frac{\nu_I}{\pi}\int_0^\infty d\nu'\left(\frac{\nu'}{\nu'+1}\right)^{\frac{1}{2}}\frac{N_0^I(\nu')}{(\nu'-\nu_R)^2+\nu_I^2},$$

and therefore vanishes only for $\nu_I=0$ if $N_0^I(\nu')$ has a single sign.

[18] The smallness is due to the expression in the first curly bracket in the integrand of Eq. (V.18), which has a maximum value of 0.15 at $\nu'=0$ and falls rapidly to zero as ν' increases.

Again in the approximation where f_0^I is neglected we may study the possibility of a resonance developing, that is, $\cot\delta_0^I$ vanishing. We have

$$\left(\frac{\nu}{\nu+1}\right)^{\frac{1}{2}}\cot\delta_0^I \approx \frac{1}{a_I} - \frac{2}{\pi}\left\{\sqrt{2}\tan^{-1}\left(\frac{1}{\sqrt{2}}\right)\right.$$

$$\left. - \left(\frac{\nu}{\nu+1}\right)^{\frac{1}{2}}\ln[\sqrt{\nu}+(\nu+1)^{\frac{1}{2}}]\right\}, \quad (V.22)$$

an expression that does not vanish for $\nu>0$ if it is positive at $\nu=0$. The condition of being positive at $\nu=0$ for a_I positive is, however, exactly the condition that there shall be no bound state. Thus it seems unlikely that a resonance will develop in either S state for negative λ unless the effects of the f_0^I are very strong.

For positive λ and negative a_I, formula (V.22) has a zero but only for $\nu>10$ if the condition (V.19) is obeyed. Thus we tentatively conclude that there are no low-energy S-wave resonances in pion-pion scattering.[19] The results of the following paper confirm this conclusion.

We turn now to the P wave and again attempt to represent the amplitude by a ratio

$$(1/\nu)A^{(1)1}(\nu) = N_1(\nu)/D_1(\nu), \quad (V.23)$$

with the same division of singularities between the numerator and denominator as for the S wave. By arguments analogous to those used above, we may derive the equations

$$N_1(\nu) = \frac{1}{\pi}\int_{-\infty}^{-1} d\nu' \frac{\text{Im}A^{(1)1}(\nu')D(\nu')}{\nu'(\nu'-\nu)}, \quad (V.24)$$

and

$$D_1(\nu) = 1 - \frac{\nu}{\pi}\int_0^{\infty} d\nu' \left(\frac{\nu'}{\nu'+1}\right)^{\frac{1}{2}} \frac{N_1(\nu')}{\nu'-\nu}, \quad (V.25)$$

where N_1 has been assigned a $1/\nu$ behavior at infinity and D_1 required not to vanish. Introducing $E_1(\omega) = D_1(\nu)$, $f_1^1(\omega) = \text{Im}A^{(1)1}(\nu)$, the following integral equation is obtained by substituting Eq. (V.24) into Eq. (V.25):

$$E_1(\omega) = 1 + \frac{\omega}{\pi}\int_1^{\infty} d\omega' \frac{K(\omega,\omega')f_1^1(\omega')E_1(\omega')}{\omega'}. \quad (V.26)$$

The P phase shift in the physical region for $\nu>0$ is given by the formula

---

[19] The absence of S-state resonances in simple two-body systems is a very general circumstance and may be traced to the lack of a centrifugal barrier that can "confine" a positive energy state. The only way to get an S-wave resonance is to have the force sufficiently complicated so that a strong inner attraction is surrounded by an outer repulsion. P-wave resonances, in contrast, arise naturally whenever there is a sufficiently strong attraction.

$$\left(\frac{\nu^3}{\nu+1}\right)^{\frac{1}{2}}\cot\delta_1$$

$$= \frac{\text{Re}D_1(\nu)}{N_1(\nu)},$$

$$= \frac{1 - \frac{\nu}{\pi}\int_1^{\infty}d\omega'\frac{I(\nu,\omega')f_1^1(\omega')E_1(\omega')}{\omega'}}{\frac{1}{\pi}\int_1^{\infty}d\omega'\frac{f_1^1(\omega')E_1(\omega')}{\omega'(\omega'+\nu)}}. \quad (V.27)$$

If f_1^1 is predominantly positive and sufficiently large it is apparent that a resonance develops in the P wave.

Let us finally consider the question of a possible singularity in the integral Eqs. (V.14) and (V.26) at $\omega=\infty$. A careful examination shows that the equations are nonsingular if $f_0^I(\omega)$ and $f_1^1(\omega)$ tend to zero like any negative power of ω, or even like $(\log\omega)^{-1-\alpha}$ ($\alpha>0$), as ω tends to infinity. Now it follows from (V.8) and (V.13) that the contributions to $f_l^I(\omega)$ from $\text{Im}A^{(0)0}(\nu')$ and $\text{Im}A^{(0)2}(\nu')$ at a particular value of ν' tend to zero like ω^{-1} as ω tends to infinity. The contributions from indefinitely large values of ν' behave like the functions $\text{Im}A^{(0)0}(\nu)$ and $\text{Im}A^{(0)2}(\nu)$ themselves, but, as these functions tend to zero with increasing ν like $(\log\nu)^{-2}$, our equations are still nonsingular. However, the contribution to $f_l^I(\omega)$ from $\text{Im}A^{(1)1}(\nu)$, even for a particular finite value of ν, behaves like a constant at infinity, so that our integral equations are now just singular.

It is usually true of such marginally singular integral equations that, if the term in the kernel responsible for the singularity is less than some critical value, the equation still has a unique solution obtainable by standard methods. Such is the case with our equation, so that the equation will be singular if and only if $\text{Im}A^{(1)1}(\nu)$ is sufficiently large.

It turns out that the P phase shift cannot be at all appreciable before the transition to the singular case is reached. We shall see in the following paper that the equations do have a class of solutions for which all phase shifts other than in the S-waves are small, but that there may be in addition a class of solutions with large P-wave phase shifts. The methods described in this paper must be modified before they can be applied to this second class. Our basic scheme of approximation needs then to be reconsidered, since the polynomial expansion on the left is no longer reliable. A new calculation method has been developed for this situation, and it is hoped to describe it in a subsequent paper. Unfortunately the new method requires the introduction of a second independent parameter.

The sum of the higher partial-wave amplitudes is to be calculated from Eqs. (IV.10) and (IV.11). If individual phase shifts are desired, the appropriate projection from these formulas is straightforward.

VI. CONCLUSION

A set of coupled integral equations for the S- and P-wave pion-pion amplitudes has been formulated and in the following paper the numerical solution of these equations for various values of λ will be described. The D and higher phase shifts can consistently be calculated by integration over the left-hand cut only, where the discontinuity across this cut is expressed in terms of the S and P amplitudes. The complete amplitude generated by this method satisfies crossing symmetry exactly.

The physical meaning of our approximation in conventional language is that we consider explicitly only the exchange of pairs of virtual pions between the two physical pions being scattered, lumping 4-pion and higher multiplicity exchanges into the constant λ. Furthermore we only attempt to calculate accurately the exchanged pairs of lower energy—those which are mainly in S and P states. The higher energy pairs are included in λ along with all sorts of other high-energy exchanges. In terms of the range of various contributing mechanisms to the pion-pion force what we are trying to do, of course, is to calculate the longest-range effects in detail and to represent the short-range effects by an empirical constant. If there is an intrinsically incalculable zero-range force, as suggested by Lagrangian field theory, this also is included in λ.

Beside the solution discussed in the foregoing paragraphs, there are also an infinite number of other possible solutions, corresponding to the Castillejo, Dalitz, and Dyson (CDD) ambiguity.[20] We can add to the right-hand side of Eq. (V.12) any number of terms of the form $a_r/(\nu-\nu_r)$, since the only effect of such terms is to introduce zeros into the scattering amplitude. While a rigorous treatment of the CDD ambiguity has not been given for relativistic field theory, the problem has been solved for several models,[21] and there seems to be little doubt as to the meaning of the extra solutions. They correspond to theories in which, before the coupling is turned on, there are one or more particles with the same quantum numbers as two pions. Once the coupling is turned on, these particles become unstable, and appear experimentally as resonances. These "kinematical" resonances differ from "dynamical" resonances, such as that which we have suggested might appear in the P state of this problem, in that they occur for arbitrarily small values of the coupling constant. The absence of such unstable particles must be regarded as an additional postulate to be inserted into the theory.

FIG. 3. Diagram for the reactions, $\pi+N \leftrightarrow \pi+N$ and $\pi+\pi \leftrightarrow N+\bar{N}$.

A knowledge of the pion-pion scattering amplitude will allow a systematic calculation of many important properties of nucleons. The application to the nucleon electromagnetic structure has been emphasized already by Frazer and Fulco.[22] This application, however, actually requires a prior knowledge of the full amplitude for the graph shown in Fig. 3, which describes not only pion-nucleon scattering but also nucleon-antinucleon annihilation to form two pions. One of us has outlined a procedure for attacking this problem which is identical in spirit to that described here for the π-π problem.[23] The procedure requires a knowledge of π-π scattering and may now be implemented. It is hoped that a reasonably accurate description of the low-energy π-N phase shifts in terms of a single additional parameter, the pion-nucleon coupling constant, will result.

With an understanding of the graph of Fig. 3 one can proceed to a systematic calculation not only of nucleon electromagnetic structure but also of the two-pion exchange terms in the nuclear force. One can also, of course, make a solid theory of photopion production. All these problems are under investigation.

There is no reason why the generalized effective-range approach based on the double dispersion representation cannot be used in more complicated problems, such as those involving strange particles. As the structure of the nearby singularities becomes more complicated, of course, it becomes more and more difficult to include enough of them to constitute a good approximation. It is doubtful that any other problem can be found that is as favorable in this respect as π-π scattering.

---

[20] L. Castillejo, R. H. Dalitz, and F. J. Dyson, Phys. Rev. **101**, 453 (1956).
[21] N. G. Van Kampen, Physica **23**, 157 (1957).
[22] W. R. Frazer and J. R. Fulco, Phys. Rev. Letters **2**, 365 (1959) and Phys. Rev. **117**, (1960).
[23] S. Mandelstam, Phys. Rev. **112**, 1344 (1958).

DISPERSION RELATIONS IN STRONG-COUPLING PHYSICS

By S. MANDELSTAM

Department of Mathematical Physics, University of Birmingham

CONTENTS

	PAGE
§ 1. Introduction	100
§ 2. The role of dispersion relations in strong-coupling physics	102
§ 3. Causality conditions and dispersion relations	104
3.1. Outline of the procedure	104
3.2. Proof of the simple dispersion relation	105
§ 4. Dispersion relations in elementary-particle physics	108
§ 5. Unitarity and crossing relations	110
5.1. Unitarity	110
5.2. Poles and single-particle intermediate states	111
5.3. The crossing relations	113
§ 6. Phenomenological applications of dispersion relations	115
6.1. Experimental tests of forward dispersion relations	115
6.2. Effective-range formula for pion–nucleon scattering	116
6.3. Extrapolation to poles	117
§ 7. The approximation scheme for dynamical calculations	120
§ 8. Dispersion relations and analytic properties for non-forward scattering: rigorous results	122
8.1. Dispersion relations at fixed momentum transfer	122
8.2. Analytic properties in the momentum transfer	126
8.3. Analytic properties as a function of more than one complex variable	127
8.4. Summary of rigorous dispersion relations and analytic properties of scattering amplitudes	128
8.4.1. Dispersion relations for fixed momentum transfer	128
8.4.2. Analytic properties as functions of momentum transfer at fixed energy	128
8.4.3. Analytic properties as functions of energy and momentum transfer	128
§ 9. Outline of the proof of dispersion relations at fixed momentum transfer	128
§ 10. Dispersion relations and analytic properties for non-forward scattering: conjectured results	133
10.1. The principle of maximum analyticity	133
10.2. The double-dispersion representation	136
10.3. Dispersion relations for fixed angular-momentum states	142
10.4. Higher order thresholds	144
§ 11. Unstable elementary particles	146
§ 12. Applications of dispersion relations to dynamical calculations	149
12.1. Pion–pion scattering	149
12.2. Pion–nucleon scattering	151
12.3. Electromagnetic structure of nucleons	153
12.4. Nucleon–nucleon scattering	156
12.5. Other applications	157
§ 13. Concluding remarks	157
References	160

Abstract. A survey is given of the principles underlying dispersion relations in elementary-particle physics, and some of their phenomenological and dynamical applications are treated.

© IOP Publishing. Reproduced with permission from *Reports on Progress in Physics* **XXV** (1962) 99–162. All rights reserved.

S. Mandelstam

The general nature of dispersion relations and their connection with causality are examined, particular emphasis being placed on the role they have recently played in strong-coupling physics. An account is given of the approximation scheme obtained by combining dispersion relations with unitarity and which, it is hoped, will provide a basis for calculations. Proved and conjectured dispersion relations are written down and the plausibility of the conjectures is discussed. The calculations so far performed, and the results they have achieved, are summarized.

§ 1. INTRODUCTION

THE term ' dispersion relation ' is applied to a type of equation which occurs in a number of branches of physics and which is generally a consequence of some or other type of causal behaviour. Such relations are usually in the nature of very general restrictions placed on a physical theory, violations of which would lead to paradoxical consequences. In order to obtain a complete solution to any problem, therefore, one would have to introduce other information of a more specific kind.

In this article our interest will be centred mainly on the application of dispersion relations to the strong interaction of elementary particles, and here their scope is somewhat different. Originally they were introduced as general restrictions imposed by causality, but, with further work, it began to appear probable that they might actually be used to calculate observable quantities in terms of a small number of coupling constants. It may at first appear unreasonable that equations derived from causality can be so restrictive. However, in the theory of strong interactions one makes the causality assumption in conjunction with a number of other postulates, and the whole system can provide us with a complete formalism. We shall attempt to make this plausible to the reader without assuming a knowledge of quantum field theory.

In fact, attempts have been made in recent years to reformulate the theory of strong interactions so as to overcome certain inadequacies of the conventional formalism, and dispersion relations seem to provide a method of doing so. From a theoretical point of view the main advantage of the new formulation is that it operates with finite quantities throughout, no references being made to infinite ' renormalization constants '. Also, by working with quantities more directly related to experiment than the old formulation, it gives one a better idea of the validity of approximations made. The approach suggests a new approximation scheme, and, while one cannot be sure of its reliability, the signs appear to be fairly hopeful.

It should be emphasized that what we achieve is simply a reformulation of the conventional theory with nothing essentially new. The spectrum and symmetry properties of elementary particles remain unexplained. Further, it is practically certain that the masses of the particles and the coupling constants giving the strengths of their interactions will have to be inserted as parameters into the theory.

The first part of the review will be devoted to the general philosophy of the new approach. In the next section, we shall treat the desirability of having a new formulation and the part that dispersion relations would be expected to play. Following this, we shall in § 3 give a general discussion of dispersion relations and their connection with causality. The subject will be studied from the point of

Dispersion Relations in Strong-coupling Physics

view of the scattering of light by a dispersive medium. As their name implies, dispersion relations were first introduced in this problem, which provides a simple example enabling one to see essentially what is involved. We shall consider the specific case of elementary-particle physics in § 4, where we shall show how the causality principle is formulated and what form the relations take here. In § 5 we shall supplement the dispersion relations with other equations derived from general principles.

The applications of dispersion relations will be divided into two classes, 'phenomenological' applications which are useful in analysing experimental results and which may provide a means of measuring certain quantities, and 'dynamical' calculations where one is more ambitious. The phenomenological applications will form the subject of § 6. Included will be a discussion of extrapolation to poles, to which an appreciable amount of experimental effort has recently been devoted.

We shall then pass on to dynamical applications, and in § 7 shall outline the principles of the approximation scheme which will be necessary. The following three sections are devoted to a more detailed study of dispersion relations. A distinction is made between dispersion relations which have been proved and those which have only been made plausible. The proof itself is given in outline only.

Finally, in § 12 an account is given of the calculations that have so far been performed, and the results obtained are presented and evaluated.

Certain sections of the article may be omitted by various readers. In particular, the content of § 9, which outlines the steps involved in proving the relations, will not be required in subsequent sections. In that section, we have to assume a knowledge of quantum field theory. The reader who wants a general view of the subject or the experimentalist who does not wish to delve deeply into the theory may omit §§ 8 and 10, and perhaps also § 3.2. The section on extrapolation to poles (§ 6.3) may be of interest to some experimentalists but will not be required for an understanding of the rest of the article, and the reader who finds § 5.2 difficult is advised to pass on.

Most introductory treatments of dispersion relations that exist at present are in the form of lecture notes. Bogoliubov and Shirkov (1957) devote the final chapter of their book to the subject; it is based on a rather unusual formulation of quantum field theory and is quite sophisticated. The lectures given at the Scottish Universities' Summer School in 1960 have been published (Screaton 1961) and should prove useful. The 1960 les Houches lectures have also been published (Goldberger *et al.* 1960); the contribution by G. F. Chew† exists in preprint form. Chew's (1961) published lecture notes on 'Strong interactions' deal entirely with this subject. There is a monograph by Drell and Zachariasen (1961), devoted to the electromagnetic structure of nucleons, and a review article by Chew (1959). To turn to the lecture notes, a set by Gasiorowicz (1960) takes us to the end of 1959, and treats both theory and applications. Those by Symanzik (1959) were written about a year earlier and are more formal in outlook, though some applications are discussed. A recent set by Hagedorn (1961) concerns itself purely with formulating and proving dispersion relations. It should be useful to

---
† *Double Dispersion Relations and Unitarity as the Basis for a Dynamical Theory of Strong Interactions*, 1960, Lawrence Radiation Laboratory, preprint U.C.R.L. 9289.

the reader with an elementary knowledge of quantum field theory who wishes to reach the point where he can understand the proofs. Eden (1961 b) has written a detailed set of lectures on the use of perturbation methods in dispersion theory.

§2. The Role of Dispersion Relations in Strong-coupling Physics

The conventional formulation of the theory of elementary-particle interactions is obtained by quantizing classical field theory according to the standard procedure. Corresponding to the field variables of the unquantized theory it is assumed that there exist operators satisfying the usual commutation rules. As in ordinary quantum mechanics—or indeed ordinary classical mechanics—the postulates of the theory may be divided into two groups, the general principles of the theory and the specific Lagrangian (or Hamiltonian) which characterizes it. The first group, which includes postulates such as the relation between operators and observables, is common to all quantum field theories, whereas the second contains the details of the interaction and serves to distinguish one theory from another.

In writing down a Lagrangian for a particular theory, one is guided by certain requirements which it must satisfy, of which we may select (i) Lorentz invariance, (ii) causality, and (iii) unitarity of the resulting theory.

The first requirement simply demands that the theory conform to the principles of special relativity, and we need not dwell on it further. The second will be one of the principal themes of our discussion. It means essentially that the theory should not contain an instantaneous interaction at a distance; in other words, that the interaction must be *local*. As is well known, a violation of this principle, or, more generally, a violation of the postulate that a signal cannot propagate with a velocity faster than that of light, leads to paradoxical consequences.

In extending the causality requirement down to atomic dimensions one is really going beyond both the experimental evidence and the requirements of 'common sense'. The question whether such 'microscopic causality' is implied by ordinary macroscopic causality is one which has not been investigated in detail. Nevertheless, most current theories are microscopically causal, and we shall confine our attention to such theories.

The third requirement, unitarity, ensures that the total probability of something happening in an experiment is equal to unity. It is thus absolutely necessary to impose such a condition on the theory in order that it may be interpreted physically, and, strictly speaking, unitarity belongs to the first group of postulates. For our purposes, however, it is convenient to include it with the second group, since it is usually inserted into the theory by suitably restricting the Lagrangian—in mathematical terms, by demanding that the Lagrangian be 'Hermitian'.

As will probably be well known to many readers, it turns out that any Lagrangian satisfying the first two requirements mentioned above leads to a theory with infinite renormalization constants. These constants occur in the Lagrangian itself. Nevertheless, one can construct theories in which any observable quantity may be calculated in finite terms. As an example, the 'bare' mass of the particles is usually infinite. The interaction itself, however, leads to an additional contribution to the mass, which partly cancels the bare mass, so that the total mass, as measured in an experiment, is finite.

Dispersion Relations in Strong-coupling Physics

It should be mentioned that neither of the above conclusions—that the Lagrangian necessarily contains infinite constants and that any observable can be calculated in finite terms—has yet been given a proof which all physicists accept. It is certainly known, however, that solutions obtained as an expansion in powers of the coupling constant have these features and we shall assume that the same is true of the complete theory.

Whether or not one can be content with a theory involving infinite unobservable renormalization constants probably depends more on the temperament of the particular physicist than on anything else. One may take the view that it is the function of a physical theory to predict results of possible experiments, and that any theory that does so unambiguously is acceptable. The existence of infinite quantities at an intermediate stage is nevertheless, to say the least, aesthetically unattractive, and it would be preferable to avoid them if it were at all possible. It appears reasonable that, in a theory leading only to finite results, one should be able to circumvent the infinities.

The proposals which will be discussed in this article for reformulating the theory without infinities and, at the same time, for performing calculations depend on the observation that the conditions (i), (ii) and (iii) above are extremely restrictive on the Lagrangian—sufficiently so to determine it in terms of a small number of arbitrary constants. If, for instance, we have a theory containing no particles besides pions and nucleons, the only Lagrangian with the correct symmetry properties has the form

$$\mathscr{L} = \mathscr{L}_0 + ig\bar{\psi}\gamma_5\psi\tau_i\phi_i + \lambda\phi_i^2\phi_j^2, \qquad \ldots\ldots(2.1)$$

where $\mathscr{L}_0$ is the Lagrangian for non-interacting fields. Equation (2.1) contains the two arbitrary constants g and λ. If one writes down any other Lagrangian one finds that infinities appear in the calculation of observable quantities. Such a Lagrangian therefore leads to an inconsistent theory and cannot be accepted. Again, this is a result which has only been proved for power-series solutions, but it at least renders plausible the assumption that the requirements we have imposed on the Lagrangian determine it to within a few constants.

It therefore appears that, in strong-coupling physics, the arbitrariness corresponding to the choice of the Lagrangian is illusory. The general principles of field quantization, together with the requirements of Lorentz invariance, microscopic causality and unitarity, fully determine the theory once the symmetries, masses and coupling constants have been given. Though this may contradict our prejudices after thirty-five years of non-relativistic quantum mechanics it is by no means unreasonable, as the general principles supplemented by our requirements contain a large number of postulates. There is only one theory of interacting pions and nucleons, for instance, which does not lead to infinite observable results and is, therefore, acceptable. This has in fact been assumed in most work on strong interactions during the last ten years, when the Lagrangian (2.1) has been adopted in order to avoid the observable infinities.

If, then, we accept the conclusion that the theory does not contain an arbitrary Lagrangian, the question arises whether one can avoid using the Lagrangian at all in performing calculations. Before doing so it will be necessary to find some means of introducing directly the requirements (i), (ii) and (iii), as they had previously been introduced by suitably restricting the Lagrangian. Once this has been achieved, we need not be surprised if we have enough equations to calculate observable quantities in terms of a small number of coupling constants. The equations of motion obtained from the Lagrangian will be superfluous, as, according to our reasoning, they are determined in principle from the conditions we are imposing.

The formulation of the theory based on these principles will be equivalent to the conventional formulation; no new physical postulate is introduced. However, as we avoid the intermediary of a Lagrangian, we do not make any reference to infinite constants. The only place where the infinite constants occur in the conventional formulation is in the Lagrangian itself. As will be seen later, the coupling constants enter into the new formulation as finite quantities fairly closely related to experiment.

The problem facing us is thus to incorporate Lorentz-invariance, causality and unitarity directly into the theory of interacting particles. The conditions for Lorentz-invariance are well known and need not detain us. We shall be concerned with the unitarity requirements in subsequent sections, but, again, to write down the equations does not present any problems. It remains to consider the consequences of the causality condition, and this will be done in the next section.

To summarize the principles underlying the new approach: conventional quantum field theory requires the specification of a Lagrangian. It turns out, however, that there is only one possible Lagrangian which satisfies all the necessary conditions and which leads to finite experimental results. Any other Lagrangian, e.g. a Lagrangian with derivative coupling, is 'unrenormalizable' and gives infinite results for measurable quantities; such a Lagrangian is therefore inadmissible in a consistent theory. The Lagrangian not being arbitrary, one is tempted to reformulate the theory without introducing it at all and, in doing so, one would avoid having to deal with infinite constants. It is then necessary to incorporate certain conditions, and, in particular, the causality condition, directly, as they had previously been introduced by suitably restricting the Lagrangian. The equations derived from the causality condition will be the so-called 'dispersion relations' and they will play a fundamental role in the new formulation.

The arguments leading to the conclusion that the Lagrangian is not arbitrary are based on perturbation theory and are therefore, not rigorous. They are, nevertheless, very plausible. It will turn out that the equations of the new formulation do appear to be sufficient for calculating observables in terms of a small number of coupling constants, and the conclusion appears to be confirmed.

An approach along the lines proposed was first suggested by Lehmann, Symanzik and Zimmermann (1955) though its actual realization took a form rather different from that originally proposed by them.

§3. Causality Conditions and Dispersion Relations

3.1. Outline of the Procedure

The causality condition has been used in a number of branches of physics, and the type of equations so obtained are very similar. The first application was made independently by Kramers (1927) and Kronig (1926) to the scattering of light by a

Dispersion Relations in Strong-coupling Physics

dispersive medium. It will therefore be instructive to begin by examining this simple problem as an illustration of the principles involved. The name 'dispersion relation' derives from this particular case. It has since been given to similar equations in quite different fields and does not really have any descriptive significance.

Kramers and Kronig showed that one must impose a restriction on the refractive index of a dispersive medium if gross violations of causality are to be avoided. We shall adopt the causality principle in the form that the polarization of a medium due to a light signal should not begin until the signal reaches the medium.

To set up the problem, let us imagine an instantaneous light signal approaching the medium (in one dimension) and reaching it at a time $t = 0$. The polarization at a time t will be denoted by $P(t)$. According to our causality postulate, the polarization is zero before the signal arrives, so that

$$P(t) = 0 \text{ for } t < 0. \qquad \ldots\ldots(3.1)$$

Equation (3.1) is actually the complete expression of the causality postulate. However, for practical purposes, it is convenient to state it in terms of the refractive index rather than in terms of the polarization produced by an instantaneous light signal, as this quantity is never observed experimentally. Now, if $n(\nu)$ is the refractive index at a frequency ν, it is easy to show that the quantity $E(\nu) = n(\nu) - 1$ is proportional to the Fourier transform of $P(t)$, i.e.

$$E(\nu) = 4\pi \int_{-\infty}^{\infty} dt\, P(t)\, e^{i\nu t}. \qquad \ldots\ldots(3.2)$$

Thus, given that $P(t)$ satisfies the limitation (2.1), we require to find the corresponding limitation on its Fourier transform (3.2). This is quite a simple problem, and the result is most easily stated as a relation between the real and imaginary parts of E:

$$\mathscr{R}E(\nu) = P\frac{1}{\pi}\int_{-\infty}^{\infty} d\nu' \frac{\mathscr{I}E(\nu')}{\nu' - \nu}. \qquad \ldots\ldots(3.3)$$

The symbol P refers to the integration about the point where the integrand becomes infinite. It implies that one excludes a small interval centred about $\nu = \nu'$ from the range of integration. This procedure is known as taking the principal value of the integral.

Equation (3.3) is the prototype of dispersion relation which occurs in many branches of physics. The real and imaginary parts of E correspond respectively to the dispersion and absorption of light by the medium, and will be referred to as the dispersive and absorptive parts. The dispersion relation enables the dispersive part to be calculated once the absorptive part is known at all frequencies, so that a real limitation is imposed on the refractive index by simple causality conditions.

In deriving (3.3) we have assumed that $E(\nu)$ approaches zero for large values of ν. This is the case in practice, but it is an additional assumption which does not follow from causality.

3.2. Proof of the Simple Dispersion Relation

We shall now show how (3.3) follows from (3.2), as the derivation provides additional insight into the nature of a dispersion relation. The essential point is that the definition (3.2) has a meaning when ν is complex with a positive imaginary part. Thus, if $\nu = \nu_1 + i\nu_2$ ($\nu_2 > 0$),

$$E(\nu) = 4\pi \int_{-\infty}^{\infty} dt\, P(t) \exp[i\nu t]$$

$$= 4\pi \int_{-\infty}^{\infty} dt\, P(t) \exp[i\nu_1 t - \nu_2 t]. \qquad \ldots\ldots(3.4)$$

As this is an infinite integral, the convergence at $t = \infty$ and $t = -\infty$ must be investigated. At $t = \infty$ there is no difficulty, since the exponential $\exp(-\nu_2 t)$

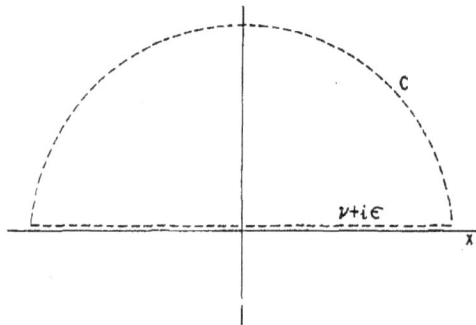

Figure 1. Cauchy integral corresponding to a dispersion relation.

tends strongly to zero. At $t = -\infty$, on the other hand, $P(t) = 0$ from our causality postulate, so that there is again no difficulty. The integral (3.4) thus has a meaning if $\nu_2 > 0$.

We can go further. Equation (3.4) has the form $\int f(\nu, t)\, dt$, where f is an analytic function of ν. According to a well-known theorem in complex-variable theory, such an integral is itself an analytic function of ν, subject to certain convergence conditions. The theorem requires no assumption regarding the analytic properties of f as a function of t. Thus $E(\nu)$ defined by (3.4) for complex ν (with $\nu_2 > 0$) is an analytic function of ν.

The analytic properties of E will be used to perform a contour integration, and to derive (3.3), but it will be necessary to make a postulate regarding the behaviour of E as $\nu \to \infty$. It can be shown from the causality condition that $E(\nu)$ tends to infinity less rapidly than ν^n for some n, and also that the value of n in a complex direction is no larger than along the real axis. We shall assume first that $E(\nu)$ in fact tends to zero as $\nu \to \infty$.

By applying Cauchy's theorem to the contour C of figure 1, it follows that

$$E(\nu + i\epsilon) = \frac{1}{2\pi i} \int_C d\nu' \frac{E(\nu')}{\nu' - \nu - i\epsilon}, \qquad \ldots\ldots(3.5)$$

since we have seen that $E(\nu)$ is analytic if $\mathscr{I}\nu > 0$. According to our assumption regarding the behaviour of E as $\nu \to \infty$, the integration over the semicircular portion of the contour is zero, so that

$$E(\nu + i\epsilon) = \frac{1}{2\pi i} \int_{-\infty}^{\infty} d\nu' \frac{E(\nu')}{\nu' - \nu - i\epsilon}. \qquad \ldots\ldots(3.6)$$

Dispersion Relations in Strong-coupling Physics

The dispersion relation is really nothing more than a slightly disguised form of this Cauchy integral. To perform the transformation, we observe that

$$0 = \frac{1}{2\pi i} \int_{-\infty}^{\infty} d\nu' \frac{E(\nu')}{\nu' - \nu + i\epsilon}, \quad \ldots\ldots(3.7)$$

as the integrand now contains no pole in the semicircle. Taking the conjugate complex gives the equation

$$0 = \frac{1}{2\pi i} \int_{-\infty}^{\infty} d\nu' \frac{E^*(\nu')}{\nu' - \nu - i\epsilon}. \quad \ldots\ldots(3.8)$$

Adding (3.8) to (3.6) and letting $t \to 0$, we obtain the relation

$$E(\nu) = \lim_{\epsilon \to 0} \frac{1}{\pi} \int d\nu' \frac{\mathscr{I}E(\nu')}{\nu' - \nu - i\epsilon}. \quad \ldots\ldots(3.9)$$

This is one form of the dispersion relation which is often useful. It is not difficult to show that the real part of the right-hand side is obtained by omitting the $i\epsilon$ and taking the principal value, so that

$$\mathscr{R}E(\nu) = P \frac{1}{\pi} \int d\nu' \frac{\mathscr{I}E(\nu')}{\nu' - \nu}$$

which is just the dispersion relation (3.3).

The assumption that $E(\nu)$ approaches zero as ν approaches infinity, which had to be made in the derivation of (3.3), will now be weakened to require that $E(\nu)/\nu$ approach zero. In place of (3.5) we take the difference between the contour integrals for two different values of ν:

$$E(\nu + i\epsilon) - E(\nu_0 + i\epsilon) = \frac{1}{2\pi i} \int_C d\nu' \left[\frac{E(\nu')}{\nu' - \nu - i\epsilon} - \frac{E(\nu')}{\nu' - \nu_0 - i\epsilon} \right].$$

The integration over the semicircular part of the contour is still zero, although this would not be the case for the two terms of the integrand taken separately. By repeating the steps in the derivation of (3.3) precisely as before, we arrive at the equation

$$\mathscr{R}E(\nu) - \mathscr{R}E(\nu_0) = P \frac{1}{\pi} \int_{-\infty}^{\infty} d\nu' \left[\frac{\mathscr{I}E(\nu')}{\nu' - \nu} - \frac{\mathscr{I}E(\nu')}{\nu' - \nu_0} \right]. \quad \ldots\ldots(3.10)$$

Thus, under the weaker assumption regarding the behaviour of $E(\nu)$ at large values of ν, (3.3) does not hold as it stands, but the equation obtained by taking the difference between (3.3) at two values of ν is still true. Corresponding to the weaker assumption, the amount of information provided by (3.10) is slightly less than that provided by (3.3). In order to calculate the dispersive part of E one would now have to know, besides the absorptive part at all values of ν, the dispersive part at some value ν_0 of ν.

If one were to make a still weaker assumption about the behaviour of $E(\nu)$ as ν tends to ∞, and require that $E(\nu)/\nu^2$ approached zero, one would have to perform two subtractions on (3.3). A calculation of the quantity

$$E(\nu) - E(\nu_0) - \frac{\nu - \nu_0}{\nu_1 - \nu_0} \{E(\nu_1) - E(\nu_0)\} \quad \ldots\ldots(3.11)$$

according to (3.3) would still be valid, even though the equation for the individual terms is not. In general, if $E(\nu)/\nu^n$ approaches zero as ν approaches infinity,

n subtractions are necessary to obtain a valid equation, and we have remarked that there certainly exists a value of n for which this assumption is true. It is easiest to consider (3.3) as the fundamental form of the dispersion relation, always bearing in mind that subtractions may be necessary. The number of subtractions cannot be determined from causality alone. When an equation such as (3.3) is written down in this article, the possibility of having to perform subtractions will be implied.

Essentially the consequence of the causality postulate is that $E(\nu)$ must be an analytic function of ν in the upper half-plane. Equation (3.3), with subtractions if necessary, is equivalent to this assumption, as we have seen by using Cauchy's theorem.

§4. Dispersion Relations in Elementary-particle Physics

The derivation of equations of the form (3.3) in other branches of physics is relatively recent. The principles underlying them are much the same; causality implies that a certain function vanishes for some values of the time and as a result the Fourier transform, which is the quality determined by experiment, obeys a dispersion relation. Applications to non-relativistic quantum mechanics have been given in different forms by van Kampen (1953 a, b), Wong (1957), Khuri (1957), Khuri and Treiman (1958), and Klein and Zemach (1959). In all these instances the dispersion relations were general limitations imposed by causality which restricted the solutions to the problems but by no means determined them.

That dispersion relations may also be applicable in elementary-particle physics was first suggested by Kronig (1946). His suggestion was carried through by Gell-Mann, Goldberger and Thirring (1954), Goldberger (1955), Karplus and Ruderman (1955), and others. In several of these papers the conclusions are supported by plausibility arguments rather than proofs. What was common to all of them was that dispersion relations were still viewed as broad limitations common to several theories. As the subject was developed the power of the relations became more and more evident, until Gell-Mann (1956) suggested the possibility that they may in fact determine the theory. In the second section we argued that such a conclusion is by no means unreasonable. Before we examine it further, however, let us look at the equations from the phenomenological point of view from which they were first introduced.

Many current experiments in high-energy physics are concerned with the scattering of one particle by another. The scattering is characterized by an amplitude representing the transition from a given incoming state to a given outgoing state. The square of this transition amplitude, when multiplied by suitable kinematic forces, gives the differential scattering cross section and is directly observable. The transition amplitude is a function of two variables, the energy of the particles and the angle of scattering. One can also construct more complicated transition amplitudes, whose square gives the cross section for production of particles in collisions. The mathematics of dispersion relations for functions of more than one variable are very much more complicated than for functions of one variable, and, in order to illustrate the principles involved, we shall begin by limiting ourselves to forward elastic scattering of a pair of particles. The transition

Dispersion Relations in Strong-coupling Physics

amplitude is then a function of one variable only, the energy. Such forward dispersion relations are useful for phenomenological examination of experimental results. Before we obtain a complete dynamical theory we shall certainly require more general dispersion relations, but let us pass over this point for the moment.

As in the dispersion of light, the relations we require are deduced from a causality postulate, which is now expressed in terms of the operators of quantized fields. We assume that at each point of space there exist operators corresponding to the ordinary field variables in the classical theory. These operators will in general not commute, the lack of commutativity being a consequence of the interference of the measurements associated with them. If, however, two points x_1 and x_2 are space-like separated, so that a signal cannot pass between them without exceeding the velocity of light, a measurement of ϕ_1 at x_1 cannot interfere with a measurement of ϕ_2 at x_2. In such a case, therefore, the two operators would be expected to commute. The causality postulate is thus

$$[\phi_1(x_1), \phi_2(x_2)] = 0 \quad \text{for } (x_1 - x_2) \text{ space-like.} \quad \ldots\ldots(4.1)$$

There is a striking difference between the intuitively obvious causality postulate of the previous section and the much more detailed equation (4.1). When Kronig proposed applying dispersion relations to elementary-particle physics he had in mind a causality postulate of the former type, such as that a signal cannot make itself felt before it reaches an observer, and attempts have been made to prove dispersion relations in this branch of physics from similar postulates. In the author's opinion the attempts have been unsuccessful and it is doubtful whether the dispersion relations are a consequence of such general reasoning.

The existence of local field operators—which is what (4.1) implies—is probably justified down to distances of the order of 10^{-13} cm, since one has probes in the form of nucleons which can go to such distances. Whether such concepts are valid for arbitrarily small distances is a question which can only be answered by comparing their consequences with experiment. Perhaps our whole idea of the structure of space needs drastic revision at these distances. However, our aim at the moment is something much less radical; we are trying to reformulate the current assumptions of field theory in such a way that they provide equations which can be applied in practice and which involve no infinite quantities. For this purpose, we can assume the validity of (4.1). As this is not a self-evident postulate, conclusions derived from it are not obvious *a priori* in the absence of experimental evidence. On the other hand, it is not excluded that the equations derived from it will be sufficient to determine the theory.

Working from the causality postulate (4.1), Gell-Mann, Goldberger and Thirring proved that the amplitude for the forward scattering of photons by nucleons satisfies a dispersion relation

$$\mathscr{R}A(\nu) = \frac{1}{\pi}\int_{-\infty}^{\infty} d\nu' \frac{\mathscr{I}A(\nu')}{\nu' - \nu}. \quad \ldots\ldots(4.2)$$

$A(\nu)$ is the amplitude in question and ν the laboratory energy, i.e. the energy in the frame where the nucleon is at rest.† We shall ignore the spinor and vector subscripts which complicate the equations without introducing any new principles.

---

† In this article the laboratory energy is defined to include the rest mass of the colliding particle only, whereas the centre-of-mass energy is defined to include both rest masses.

Goldberger then generalized such dispersion relations to the scattering of particles with mass, such as pions, by nucleons. The equation has the same form and need not be written down again. Goldberger's proofs of his relations were mathematically unsound, as they involved integrations over divergent exponentials, but rigorous proofs were later given by Symanzik (1957) and Bogoliubov, Medvedev and Polivanov (Bogoliubov and Shirkov 1957). We shall give a more complete list of references when discussing dispersion relations for non-forward scattering, but we may remark that the proofs draw on the detailed postulates of quantum field theory, including the causality condition (4.1). Nevertheless, the mathematics involved is not too different from that of the last section; again functions vanish for certain values of the time t, so that their Fourier transforms satisfy the dispersion relation (4.2).

Before equation (4.2) can be applied, either phenomenologically or to dynamical calculations, it is necessary to obtain more information about the function $\mathcal{I}A(\nu')$ which appears in the integrand. In a phenomenological application one would want to obtain $\mathcal{I}A(\nu')$ from experiment. For a detailed calculation, a second equation between the two quantities $\mathcal{R}A$ and $\mathcal{I}A$ is required to supplement (4.2). In either case the requisite physical information is provided by unitarity which, as we have emphasized, now has to be introduced directly into the theory.

§5. Unitarity and Crossing Relations

5.1. *Unitarity*

The unitarity equation for the transition amplitude from an incoming state E to an outgoing state O is

$$\mathcal{I}A_{E \to O}(\nu) = k \sum_I A_{E \to I}{}^*(\nu) A_{I \to O}(\nu). \qquad \ldots\ldots(5.1)$$

The constant k is a kinematical factor depending on the precise manner in which the amplitudes A are defined and on the spins of the particles involved. For spinless particles and with the definition of A in most current literature,† $k = q/32\pi^2 W$, q being the centre-of-mass momentum and W the total centre-of-mass energy.

Some readers may be more familiar with the unitarity condition in the form of $S^*S = 1$. On writing $S = 1 - iT$ and denoting the matrix elements of T by A to within kinematic factors, we arrive at equation (5.1).

The symbol I in (5.1) denotes any state into which the incoming state can go without violating any conservation laws. Suppose for instance we are interested in the transition amplitude for pion–nucleon scattering. E and O then both represent states with one pion and one nucleon. Part of the summation (5.1) will be over states I which are also pion–nucleon states, so that $A_{E \to I}$ and $A_{I \to O}$ are again amplitudes for pion–nucleon scattering. There will be further terms from states I with a nucleon and two pions, in which case A will be the amplitude for single pion production in pion–nucleon collisions, i.e. for the process $\pi + N \to \pi + \pi + N$. More generally, I could be a state containing a nucleon and any number of pions

---

† Unfortunately the normalization of the scattering amplitude has not become standard. We shall define A to be minus the conventional T-matrix element, with a factor $(E/M)^{\frac{1}{2}}$ for each incoming or outgoing fermion line and a factor $(2E)^{\frac{1}{2}}$ for each incoming or outgoing boson line to make it covariant. With this definition, a Feynman diagram without external lines is $-(2\pi)^4 A\delta^4(p)$, the δ-function giving overall conservation of energy and momentum. The scattering amplitudes used by Goldberger, Miyazawa and Oehme (1955), and by Chew and Mandelstam (1960) differ from these amplitudes by factors of $1/4\pi$ and $1/16\pi$ respectively.

Dispersion Relations in Strong-coupling Physics

and, in addition, any number of nucleon–anti-nucleon pairs; $A_{E\to I}$ or $A_{I\to O}$ will be the corresponding production amplitudes. Strange particles will provide additional states. Some states I are excluded, since the excess number of nucleons over anti-nucleons must be equal to one and the electric charge must be the same as in the states E or O. Nevertheless, (5.1) represents a complicated summation and, if it is used in a dynamical calculation, approximations will have to be made (§ 7).

In the particular case of forward elastic scattering, equation (5.1) simplifies greatly. The states E and O are then the same state, so that the equation becomes

$$\mathscr{I}A_{E\to O}(\nu) = k \sum_I |A_{E\to I}(\nu)|^2.$$

Now the square of a transition amplitude is proportional to the cross section for the process, so that the right-hand side is proportional to the sum of the cross sections for the reactions leading from E all possible states, or to the total cross section. Such a quantity can easily be measured experimentally. Thus

$$\mathscr{I}A_{E\to O}(\nu) = K\sigma_{\mathrm{tot}}(\nu) \quad \quad \ldots\ldots(5.2)$$

for forward scattering. K is again a kinematic factor. For the scattering of spinless particles and with the usual definition of A, $K = 2Wq$ and, if one of the particles has spin, $K = Wq/M$. Equation (5.2) is known as the optical theorem and, with its aid, forward dispersion relations may easily be applied phenomenologically. Probably because of equation (5.2) the phrase 'absorptive part' is retained in field theory to denote $\mathscr{I}A$.

5.2. Poles and Single-particle Intermediate States

The one state I which we have avoided mentioning in discussing the summation (5.1) for pion–nucleon scattering is the single-nucleon state. Nevertheless, it has the same quantum numbers as the pion–nucleon state and must be included in the summation. The amplitudes $A_{E\to I}$ and $A_{I\to O}$ will now be amplitudes for the reaction $\pi + N \to N$ and $N \to \pi + N$. Such a reaction cannot occur unless the momenta of the particles can take complex values. From the equations of conservation of energy and momentum, one finds easily that the process can only take place at one value of ν, the laboratory energy of the pion, namely at $\nu = -\mu^2/2M$, μ being the pion mass and M the nucleon mass. As this value is less than μ, the momentum of the pion will be imaginary. To give a meaning to the amplitude $A_{\pi+N\to N}$ one must therefore turn to the hypotheses of quantum field theory and to the mathematics of the proof of dispersion relations. It is then seen that the amplitude does have a precise meaning and that the one-nucleon state must be included in the summation (5.1) when calculating the function $\mathscr{I}A(\nu)$ on the right of (4.2). The quantity $A_{\pi+N\to N}$ is just a single number, since both the energy and angle of scattering are fixed by the equations of conservation of energy and momentum. We shall denote $k|A_{\pi+N\to N}|^2$, the contribution to $\mathscr{I}A$ from the one-nucleon state according to (5.1), by†

$$\frac{\pi g^2}{2M} \delta\left(\nu + \frac{\mu^2}{2M}\right). \quad \quad \ldots\ldots(5.3)$$

† As usual we are neglecting spin, which would bring in an extra factor $\mu^2/2M$. For forward scattering we could just as easily have included spin, but we omit it to obtain uniformity with the rest of the article.

The factor $\pi/2M$ is inserted to conform to the conventional notation,† and the δ-function indicates that the term is only non-zero when the energy condition $\nu + (\mu^2/2M) = 0$ is satisfied. The contribution (5.3) to $\mathscr{I}A(\nu)$ can now be inserted into (4.2), so that $\mathscr{R}A(\nu)$ has a term

$$\frac{g^2}{2M} \frac{1}{-(\mu^2/2M) - \nu}. \qquad \ldots\ldots(5.4)$$

(As $\mathscr{I}A(\nu')$ given by (5.3) is zero unless $\nu' = -(\mu^2/2M)$, we replaced ν' by this value in the integrand of (4.2).) If the term (5.4) is inserted explicitly on the right of (4.2) the remaining integral will only contain ordinary physical contributions to $\mathscr{I}A(\nu')$ from states with two or more particles.

The term (5.4) in the expression for $\mathscr{R}A$ shows that A, considered as a function of ν, has a pole when $\nu = -(\mu^2/2M)$. This is the energy at which the incoming pion–nucleon state can go into an intermediate single-nucleon state with conservation of energy and momentum. The result is quite generally true and is not confined to pion–nucleon scattering. If, at a certain energy, a given reaction can go via a one-particle intermediate state and conserve energy and momentum, the amplitude for the reaction will have a pole at that energy. Except for scattering of photons such a pole will always be in an unphysical region, for example, at a negative kinetic energy as in the present instance. However, when the term corresponding to it, such as (5.3) is inserted into the dispersion relation (4.2), it will contribute for all values of the energy and it is, therefore, important in the physical region.

In the conventional Lagrangian formulation of quantum field theory, all reactions are regarded as a consequence of the fundamental reaction $N \to \pi + N$ and the constant g characterizing it is the fundamental constant of the theory. It is the coupling constant appearing in the Lagrangian. The constant g appearing in our formulae is not quite the same, as the amplitude for the reaction $N \to \pi + N$ contains correction terms from more complicated interactions. We may therefore write

$$g = g_u + g_c$$

where g_u is the fundamental coupling constant in the Lagrangian, g_c the correction to the reaction $N \to \pi + N$, and g the resultant amplitude for this reaction. The constant g_u is known as the 'unrenormalized' coupling constant while g is the renormalized coupling constant which appears in experimentally observable quantities. In the present approach g_u and g_c do not occur separately. In fact, they are the infinite constants which we are trying to avoid.

For the benefit of readers with a knowledge of quantum field theory, we may be more precise regarding the definition of g. The pion–nucleon vertex function depends on the squares of the momenta of the two nucleons and of the pion. The renormalized coupling constant is defined as the value of this function for specified values of the momenta. Dyson originally suggested taking two nucleons on the mass shell ($p^2 = -M^2$) but putting the pion momentum equal to zero. The present definition puts the pion momentum on the mass shell as well ($p^2 = -\mu^2$), a convention suggested by Lepore and Watson (1949). It has the apparent

---

† See footnote to previous page.

disadvantage that the momenta must be taken to be complex. The vertex function can, however, be continued to these complex values, as is proved rigorously at an intermediate stage in the derivation of dispersion relations. The Lepore–Watson renormalized coupling constant then appears in the dispersion relations and can be related to experiment, whereas the Dyson coupling constant is very far removed from experiment. It is thus common practice now to define the renormalized coupling constant using the Lepore–Watson convention.

5.3. The Crossing Relations

Even with unitarity we do not possess enough information to apply the dispersion relation (4.2). The reason is that unitarity only gives the imaginary part of A for positive kinetic energies, whereas the integration in (4.2) is over all energies ν'.

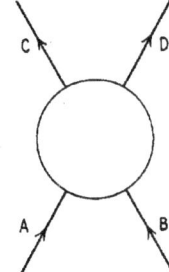

Figure 2. Diagram for the reaction $A + B \to C + D$.

In fact, dispersion relations are true for some types of potential scattering and unitarity is certainly always true, but it is only in relativistic quantum field theory that we can use these concepts for dynamical calculations. The dispersion relations and the unitarity condition must be supplemented by an additional equation to calculate $\mathscr{I}A(\nu)$ for negative ν, and such an equation is peculiar to field theory.

The reaction

$$A + B \to C + D \qquad \text{I}$$

is sometimes represented by the diagram in figure 2. The same diagram also represents the reactions

$$A + \bar{D} \to \bar{B} + C \qquad \text{II}$$

and

$$A + \bar{C} \to \bar{B} + D \qquad \text{III}$$

where a bar denotes the anti-particle (for a pion the bar simply reverses the sign of the charge). In perturbation theory, for instance, the same formula which serves to calculate the amplitude for the reaction I may be used to calculate it for the other two reactions. It can be proved independently of perturbation theory that the amplitudes for the three reactions are given by the same analytic function. This does not mean that the amplitudes for all of the reactions are known once one is known, because they will only represent physical processes when the kinematic variables have certain values. To take the example with which we are dealing, the amplitude for forward pion–nucleon scattering will only be of direct physical interest if the laboratory energy is greater than μ. The three ranges of values of the variables for which our amplitude characterizes the physical reactions I, II and III will always be mutually disjoint, so that, if the amplitude for one of the reactions is known in its physical region, the amplitude for the other two will only be known in certain unphysical regions. But this is just what we require at the moment; we want the amplitude for a reaction (or, rather, its imaginary part) at negative unphysical values of the energy in order to be able to insert it into the dispersion relation (4.2). The foregoing discussion enables us to equate it to the imaginary part of the amplitude for one of the other two reactions at a point in the physical region, the precise point depending on the details of the kinematics.

As the diagrams for the reactions II and III may be obtained from the diagram for the reaction I by crossing a pair of external lines, the relation between the

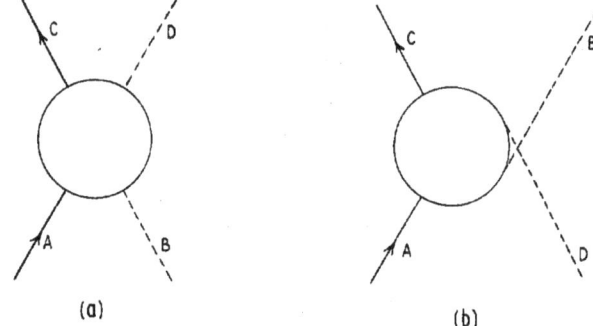

Figure 3. Uncrossed and crossed diagrams for pion–nucleon scattering.

amplitudes for the three processes is known as the crossing relation. It provides us with our additional equation and gives us $\mathscr{I}A(\nu)$, $\nu < 0$, in terms of the imaginary parts at positive energies of the other two reactions.

Like the unitarity condition, the crossing condition takes a particularly simple form for forward scattering. Suppose that A and C are nucleons, B and D pions. The diagram for the reaction I is now figure 3(a) while that for the crossed reaction II is figure 3(b). With neutral pions the reactions I and II are in fact identical. However, when the directions of the lines B and D are reversed the sign of their four-momenta must be changed. As the laboratory energy is the fourth component of the momentum associated with the lines B or D, the crossing relation tells us that the amplitude for the reaction I at an energy ν is equal to that for the reaction II, the same reaction, at an energy $-\nu$. Since the definition of the imaginary part of a scattering amplitude in the unphysical region is not unambiguous, the relation can only be applied to the real part, so that

$$\mathscr{R}A(-\nu) = \mathscr{R}A(\nu) \qquad \ldots\ldots(5.5)$$

for forward π^0–nucleon scattering. It then follows from (5.5) and (4.2) that

$$\mathscr{I}A(-\nu) = -\mathscr{I}A(\nu) \qquad \ldots\ldots(5.6)$$

for forward π^0–nucleon scattering. With this equation we have sufficient information to apply our dispersion relation.

For charged pion–nucleon scattering, the crossing of the external lines to give figure 3(b) from figure 3(a) reverses the sign of the charge, so that

$$\mathscr{I}A_+(-\nu) = -\mathscr{I}A_-(\nu), \quad \quad \quad (5.7a)$$

and

$$\mathscr{I}A_-(-\nu) = -\mathscr{I}A_+(\nu), \quad \quad \quad (5.7b)$$

where A_+ and A_- are the amplitudes for π^+-proton and π^--proton scattering. Defining

$$A^{(+)} = \tfrac{1}{2}(A_- + A_+), \quad \quad \quad (5.8a)$$

$$A^{(-)} = \tfrac{1}{2}(A_- - A_+), \quad \quad \quad (5.8b)$$

the crossing relations become

$$\mathscr{I}A^{(\pm)}(-\nu) = \mp \mathscr{I}A^{(\pm)}. \quad \quad \quad (5.9)$$

The slightly confusing notation has unfortunately become standard. $A^{(+)}$ and $A^{(-)}$ will both have pole terms such as (5.4). To maintain the crossing relation one has to add an additional pole term

$$\pm \frac{g^2}{2M} \frac{1}{-(\mu^2/2M) + \nu}. \quad \quad \quad (5.10)$$

For energies ν between μ and $-\mu$ the unitarity equation does not, strictly speaking, give us information about $\mathscr{I}A(\nu)$, as we are in the unphysical region, the laboratory energy being below the rest mass of the pions. The crossing relations (5.6) or (5.8) clearly do not help us. A formal application of the unitarity condition (5.1) gives zero for this imaginary part, as there are no intermediate states with energies in this range except the one-nucleon state which we have already considered. From the rigorous derivation of forward dispersion relations it turns out that $\mathscr{I}A(\nu)$ is indeed zero for $-\mu < \nu < \mu$. The integration in (4.2) therefore extends only over the physical regions from $-\infty$ to $-\mu$ and from μ to ∞, provided the pole terms (5.4) and (5.10) are included separately.

§6. Phenomenological Applications of Dispersion Relations

6.1. Experimental Tests of Forward Dispersion Relations

We can now collect together the information provided by dispersion relations, unitarity and crossing for forward pion–nucleon scattering. On inserting the equations

$$\mathscr{I}A^{(\pm)}(\nu) = 0, \quad -\mu < \nu < \mu,$$

$$\mathscr{I}A^{(\pm)}(-\nu) = \mp \mathscr{I}A^{(\pm)}(\nu),$$

into (4.2) and adding the pole terms (5.4) and (5.9), we arrive at the equation

$$\mathscr{R}A^{(\pm)}(\nu) = \frac{g^2 \mu^2}{4M^2}\left\{-\frac{1}{(\mu^2/2M) - \nu} \pm \frac{1}{-(\mu^2/2M) + \nu}\right\}$$
$$+ \frac{1}{\pi}\int_\mu^\infty d\nu' \mathscr{I}A^{(\pm)}(\nu')\left\{\frac{1}{\nu' - \nu} \pm \frac{1}{\nu' + \nu}\right\}. \quad \quad (6.1)$$

We have multiplied the pole terms by the factor $\mu^2/2M$ to give the correct spin kinematics. The function $\mathscr{I}A^{(+)}(\nu')$ in the integrand is given in terms of the total cross section by the optical theorem while $|A(\nu)|^2$, or $\{\mathscr{R}A(\nu)\}^2 + \{\mathscr{I}A(\nu)\}^2$, is the differential cross section for scattering through very small angles ('forward scattering'). The dispersion relation thus enables the differential forward cross section to be calculated once the total cross section has been measured at all energies. It can either be compared directly with experiment or used to resolve ambiguities when the experimental data are insufficient.

Equations (6.1) have been put in a form suitable for direct comparison with experiment by Goldberger, Miyazawa and Oehme (1955)† and have been tested by Anderson, Davidon and Kruse (1955), Puppi and Stanghellini (1957), Hamilton (1958) and Chiu (1958). They appear to be well verified for π^+–proton scattering and are at least consistent with the experimental data for π^-–proton scattering, the original discrepancy noted by Puppi and Stanghellini having gradually been reduced.

The equations (6.1) involve the renormalized coupling constant g, and the value which must be inserted to give agreement with experiment turns out to be $g^2/4\pi = 14-15$. This gives one of the principal methods of measuring the coupling constant.

Unfortunately the agreement of the dispersion relations for π^+–p scattering with the data does not provide a critical check on the validity of the assumptions underlying them. For energies below 250 MeV the scattering is due principally to the pion–nucleon resonance, and a resonance formula satisfies the dispersion relations in any case. At higher energies the imaginary part of the scattering amplitude is considerably larger than its real part, so that the experiments, which measure $\mathscr{I}A$ and $\{\mathscr{R}A\}^2+\{\mathscr{I}A\}^2$, cannot provide a good determination of $\mathscr{R}A$. One is therefore not able to check the dispersion relations accurately for these larger values of ν.

6.2. Effective-range Formula for Pion–Nucleon Scattering

A relatively simple method which can give some information on non-forward scattering from dispersion relations is to construct such relations for the derivatives of the forward scattering amplitude, with respect to the angle, as well as for the forward scattering amplitude itself. If only a few partial waves are important at low energies, one can then use these relations to derive others relating to the partial-wave amplitudes. Such an approach has been followed by Chew, Goldberger, Low and Nambu (1957 a). Their paper also indicates how dispersion relations can be used for particles with spin. For purely forward scattering the spin–flip amplitude was zero and the spin essentially played no part.

The main result of Chew *et al.* was the derivation of an 'effective-range' formula for P-wave pion–nucleon scattering, which expressed the phase shift in terms of two phenomenological parameters without detailed calculations. The result took the form

$$q^3 \cot \delta_{33}/\omega = \frac{4}{3}\left(\frac{\mu}{2M}\right)^2 \frac{g^2}{4\pi}(1-r_a\omega), \qquad \ldots\ldots(6.2)$$

where q is the centre-of-mass momentum, ω the centre-of-mass energy of the pion, δ_{33} the phase shift in the resonant 3–3 state and r_a a parameter. Equation (6.2) had previously been derived by Chew and Low (1956) on the static model. It is found to fit the experimental data satisfactorily and provides another method of measuring g.

† These authors use a subtracted form of dispersion relation; they subtract (6.3) at $\nu = \mu$ from the equation at a general value of ν. Doing so gives equations (2.5) and (2.6) of their paper.

Dispersion Relations in Strong-coupling Physics

Chew, Low, Goldberger and Nambu also applied similar phenomenological considerations to photo-production of pions.

Another use to which dispersion relations for derivatives of forward scattering amplitudes have been put is the elimination of the Yang and Minami phase shifts (Davidon and Goldberger 1956, Lindenbaum and Sternheimer 1958).

6.3. Extrapolation to Poles

In § 5.2 we observed that the pion–nucleon forward scattering amplitude, considered as a function of the laboratory energy ν, has a pole at the energy for which the pion–nucleon state can go into a single-nucleon state ($\nu = \mu^2/2M$), and that the residue at the pole gives the pion–nucleon coupling constant g. We remarked that the result was quite general. When such considerations are combined

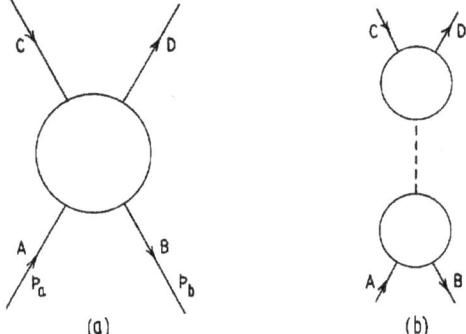

Figure 4. One-pion intermediate state in $N\bar{N}$ scattering.

with the crossing relation one can obtain formulae which are sometimes helpful in analysing experimental results and which may possibly lead to the indirect measurement of otherwise elusive quantities.

Our first application will be to nucleon–nucleon scattering, but it will be introduced by looking at nucleon–anti-nucleon scattering. Thus, in figure 4(a), A and D will be nucleons, B and C anti-nucleons. (In general an arrow pointing in the direction of a reaction corresponds to a nucleon, while an arrow pointing in the opposite direction corresponds to an anti-nucleon.) There exists a one-particle state having the same quantum numbers as the incoming nucleon–anti-nucleon state, namely the one-pion state. A nucleon–anti-nucleon state can go into a single pion provided that the laboratory energy of the incoming particle is $(\mu^2/2M) - M$—a highly unphysical energy—so that we would expect the amplitude for nucleon–anti-nucleon scattering to have a term of the form

$$\frac{g^2}{4\pi} \frac{1}{2M} \frac{1}{(\mu^2/2M) - M - \nu}. \qquad \ldots\ldots(6.3)$$

The actual formula will be more involved due to charge and spin, but we shall ignore such complications. The constant g is the coupling constant for the reaction $N + \bar{N} \to \pi$; according to the crossing relation this is also the constant for the reaction $N \to \pi + N$. The same constant g thus appears in the pole (6.3) as in the pion–nucleon pole (5.3).

The existence of the pole may be represented diagrammatically by figure 4(b). The diagram shows that the nucleon–anti-nucleon state can go, via a one-pion state, to another nucleon–anti-nucleon state, while conserving all the quantum numbers.

It is usually more convenient to express the pole term in terms of the centre-of-mass energy rather than the laboratory energy. We represent the square of the centre-of-mass energy, including both rest masses, by t. Then

$$t = 2M(\nu+M), \qquad \ldots\ldots(6.4)$$

and, on putting this into (6.3), the pole term becomes

$$\frac{g^2}{4\pi} \frac{1}{\mu^2-t}. \qquad \ldots\ldots(6.5)$$

The value of the centre-of-mass energy which the incoming state must have to be able to transform into the one-pion state is of course the mass of that state, so that the pole will occur at $t = \mu^2$, as in (6.5). Thus centre-of-mass energies are simpler, as regards their kinematics, than laboratory energies for this purpose.

In terms of the momenta of figure 4(a)

$$t = -(p_a+p_b)^2. \qquad \ldots\ldots(6.6)$$

The momenta are regarded as four-vectors, with the convention

$$x^2 = -x_0^2 + x_1^2 + x_2^2 + x_3^2.$$

The nucleon–anti-nucleon scattering amplitude can now be related to the nucleon–nucleon scattering amplitude by crossing. Figure 4(a), read sideways, corresponds to the reaction $A+\bar{C} \to \bar{B}+D$, which is nucleon–nucleon scattering. (Note that the arrows now always point forward.) Since the nucleon–nucleon and nucleon–anti-nucleon scattering amplitudes are given by the same analytic function, the nucleon–nucleon amplitude will have a term of the form (6.5).† The quantity t must now, however, be related to the kinematics of our diagram read sideways. We still have a formula like (6.6)

$$t = (p_a-p_b)^2; \qquad \ldots\ldots(6.7)$$

the sign of p_b is reversed since B was an incoming particle in nucleon–anti-nucleon scattering but is now an outgoing particle. Thus t is minus the square of the difference of the momentum between the incoming and the outgoing nucleon, or minus the square of the momentum transfer of the nucleon. It can be related to the angle of scattering by the formula

$$t = -2q^2(1-\cos\theta), \qquad \ldots\ldots(6.8)$$

q being the centre-of-mass momentum of the nucleons.

Inserting (6.8) into (6.5) we obtain finally the expression

$$\frac{g^2}{4\pi} \frac{1}{\mu^2+2q^2(1-\cos\theta)}. \qquad \ldots\ldots(6.9)$$

The existence of a term (6.9) in the nucleon–nucleon scattering amplitude was first suggested from this point of view by Chew (1958). It is not a rigorous result

† We are here anticipating later results (§§ 8 and 10), and are assuming that the term (6.5) occurs quite generally in the amplitude for nucleon–anti-nucleon scattering, not only for forward scattering.

Dispersion Relations in Strong-coupling Physics

because, from the existence of the term (6.5) in forward nucleon–anti-nucleon scattering, we cannot strictly infer its existence in the quite different kinematic region that now interests us. Further, dispersion relations for nucleon–anti-nucleon scattering have not been proved rigorously up till now. Nevertheless, the reasoning given does make the existence of the term (6.5) very plausible, and it is certainly present in perturbation theory. The plausibility is strengthened by the fact that the result can be proved rigorously for 'nucleons' whose mass is sufficiently small (§ 8.3).

The expression (6.9) has a pole when

$$\cos\theta = 1 + (\mu^2/2q^2). \qquad \ldots\ldots(6.10)$$

For q^2 sufficiently large, this value of $\cos\theta$ is only just greater than 1. The pole should therefore make itself felt in experimental results, and applications have been considered by Cziffra and Moravcsik (1959), Cziffra, MacGregor, Moravcsik and Stapp (1959), and MacGregor and Moravcsik (1960). Their work is summarized by Moravcsik (Screaton 1961). It was found that the value of g in (6.9) needed to fit the experiments is the same as that found from pion–nucleon scattering, but the experimental errors are fairly large.

It is of interest to observe that figure 4(b), read sideways, may be thought of as the exchange of a single pion. One of the incoming nucleons emits a 'virtual' pion and goes out in a different direction, the other nucleon then absorbs the pion. This was the mechanism for nuclear forces originally suggested by Yukawa in 1935, and the existence of the term (6.9) in the amplitude was known before dispersion relations were used. Applications to experimental analysis were not, however, considered too seriously, mainly because it was difficult to judge the relative importance of the pole term and the remaining part of the amplitude. Dispersion theory helps us to do this by writing the latter part as an integral (4.2); for further details, we refer the reader to the above-mentioned articles (but see also Landshoff and Treiman 1961).

Another application of poles in transition amplitudes has been suggested by Chew and Low (1959), who give a procedure for the determination of the cross section for pion–pion scattering from experimental observations on the reaction $\pi + N \to 2\pi + N$.

In the reaction $N + \bar{N} \to 3\pi$ (figure 5 read upwards), there is a single-particle state, the one-pion state, with the same quantum numbers as the initial and final states. There will accordingly be a term

$$\frac{gf\{(p_d+p_e)^2, (p_c-p_d)^2\}}{4\pi(\mu^2-t)} \qquad \ldots\ldots(6.11)$$

in the transition amplitude, where t is the energy of the process and is equal to $(p_a+p_b)^2$. The lower half of the diagram represents once more the process $N + \bar{N} \to \pi$, and is given by the constant g. The upper half, however, is now the process $\pi + \pi \to \pi + \pi$. It will be a function f of two kinematic variables, which we have arbitrarily taken to be $(p_d+p_e)^2$ and $(p_c-p_d)^2$, and the cross section for the process will be related to $|f|^2$ by kinematic factors.

The diagram can now be read sideways, when it represents the reaction $\pi + N \to 2\pi + N$. As in the case of nucleon–nucleon scattering, t is now minus the square of the momentum transfer of the nucleon, and it may be linearly related to the cosine of the angle through which the nucleon is scattered. On inserting the formula relating t and $\cos\theta$ into (6.11), we find that the amplitude has a pole at a value of $\cos\theta$ a little greater than 1. The constant g being known, a measurement of the strength of the pole will yield the value of f. Depending on which of the momenta p_c, p_d and p_e are measured, one would get the total or the differential cross section.

The extrapolation of the cross section from physical values of $\cos\theta$ to the value, slightly larger than 1, where the pole occurs always entails a very substantial loss of accuracy. Such a procedure can probably never yield any results with absolute

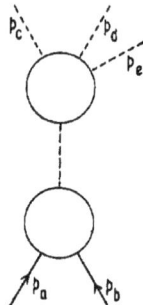

Figure 5. The one-pion pole in pion production.

certainty, though it may very well provide evidence for some qualitative features of pion–pion scattering and may even give numerical values. Preliminary results by Anderson, Bang, Burke, Carmony and Schmitz (1961) indicate that the cross section is not too large below about 500 MeV centre-of-mass energy (including the rest masses), but that it may increase after that.

Another application, to find the electric form factor of the pion from experiments on electro-production of pions, has been considered by Frazer (1959). Unfortunately, the experiment does not appear to be feasible at the present moment.

Dowker (1961) has proposed a method to obtain more information on hyperon–nucleon interactions from experiments on hyperon–deuteron scattering, but again the experiment appears very difficult. In general it is easy to find the pole terms in a particular reaction by drawing any possible diagram of the type of figure 4(b) or figure 5. All relevant quantities must be conserved at the vertices.

§7. The Approximation Scheme for Dynamical Calculations

We shall now pass on from the phenomenological applications of the previous section to treat the approximation scheme for calculating observable quantities in terms of a small number of parameters. Before applying the scheme one certainly requires more dispersion relations than those for forward scattering presented thus far, but the principles of the scheme may be understood without such complications.

In the previous sections we have introduced three types of equation:

(i) The dispersion relation (4.2) which gives an equation connecting the real and the imaginary parts of the scattering amplitude.

Dispersion Relations in Strong-coupling Physics

(ii) The unitarity condition (5.1) which provides another equation between the real and imaginary parts of the scattering amplitude for positive energies.

(iii) The crossing relation, which expresses the imaginary part of the scattering amplitude at negative energies in terms of the imaginary part of the scattering amplitude for the two 'crossed reactions' at positive energies. The crossed reactions are those represented by the same diagram, figure 2.

The three equations should in principle be sufficient to solve the problem. As we have pointed out, however, the unitarity condition (5.1) involves a summation over an infinite number of states, and approximations will have to be made if we are to obtain a soluble system of equations.

The scheme as we shall present it here only enables calculations to be made for fairly low values of the energy. A very interesting proposal for extending it to certain high-energy calculations has been given by Chew and Frautschi (1960, 1961) but we shall not be able to examine it in detail.

As the integral in (4.2) is over all values of ν', one requires to know the scattering amplitude at high values of ν' to perform a calculation for low ν. The fundamental principle of the approximation scheme is that, owing to the energy denominator $\nu' - \nu$, the integral in (4.2) will be dominated by not too high values of ν' if we require it at low values of ν. If this is in fact true, one would be able to use an equation for $\mathscr{I}A(\nu')$ in the integrand which fails at large values of ν'. The precise value of ν' up to which one requires an accurate knowledge of $\mathscr{I}A(\nu')$ depends of course on the details of the problem and on the accuracy required. Expressed in terms of the analytic properties of the scattering amplitude described in § 3.2, the assumption takes the form that the amplitude at a particular value of ν is dominated by the near singularities. Distant singularities may be approximated or neglected.

To illustrate how our fundamental approximation may be used for simplifying the unitarity condition we shall again refer to pion–nucleon scattering. The unitarity condition is then

$$\mathscr{I}A_{\pi,N\to\pi,N}(\nu) = k \sum_{I} A^*_{\pi,N\to I}(\nu) A_{I\to\pi,N}(\nu). \qquad \ldots\ldots(7.1)$$

The summation is over all possible intermediate states I and is in general very involved. If, however, ν is greater than μ but below the threshold for pion production, the only intermediate state I whose energy is low enough for the reaction $\pi + N \to I$ to take place is the pion–nucleon state. Hence the A's appearing on the right of (7.1) will all be of the form $A_{\pi,N\to\pi,N}$, the amplitude for pion–nucleon scattering. These are just the amplitudes under consideration. When ν rises above the threshold for pion production, terms such as $A_{\pi,N\to 2\pi,N}$ enter into (7.1). Until we are considerably above threshold we know from experience that the amplitude for pion production is very small, so that it is a good approximation to restrict the summation in (7.1) to intermediate pion–nucleon states. Once the energy becomes too high, above about 600 MeV, the approximation fails. Since, however, we are prepared to accept an approximation that is accurate at fairly low energies only, it satisfies our requirements. We therefore take the sum in (7.1) over the pion–nucleon states I, so that only the amplitudes for pion–nucleon scattering appear on the left and on the right.

The one-nucleon intermediate state must also be included, and it gives the pole term which we have discussed in detail.

In general, for other processes, the approximation to the unitarity condition (5.1) will consist in restricting the summation to low-energy states I.

It may well be the case in practice that we cannot include a sufficient number of intermediate states in the unitarity condition to give a good approximation without doing a prohibitive amount of work. This would not force us to give up, as we could represent the neglected contributions to $\mathcal{I}A(\nu)$ by one or more empirical constants. For instance, we could approximate the integral

$$\frac{1}{\pi}\int d\nu' \frac{\{\mathcal{I}A(\nu')\}_n}{\nu'-\nu},$$

the subscript n denoting the neglected part, by $a/(\nu_a-\nu)$, where a and ν_a are phenomenological parameters. Such parameters are on a different footing from the coupling constants g and λ, which could probably not be determined in principle with present-day theory. The new parameters might be able to be determined by including a sufficient number of states in the summation I, but must be taken from experiment in the calculations actually performed.

The approximation of neglecting high-energy contributions to the dispersion integral (4.2) may be rephrased as neglecting interactions which take place at short distances. If such interactions are important they have to be represented by the phenomenological parameters just mentioned. The interactions at larger distance should be amenable to treatment by dispersion-relation methods.

If the number of phenomenological parameters which must be introduced is less than the number of parameters given by experiment, one can perform useful calculations with the tools described in this article. In many cases one may introduce more parameters than is necessary in order to lessen the amount of computational labour. By using sufficient parameters to reduce the calculations to a minimum one can obtain some general formulae; the effective-range formula of the last section is an example. There is thus no sharp dividing line between a phenomenological and a dynamical application of dispersion relations, but in practice one finds a difference between the types of calculations which characterize the two approaches.

§ 8. Dispersion Relations and Analytic Properties for Non-forward Scattering: Rigorous Results

8.1. Dispersion Relations at Fixed Momentum Transfer

We turn now to examine dispersion relations more general than those for forward scattering presented so far. As usual we shall consider the problem from the point of view of pion–nucleon scattering and we shall ignore spin and charge. The reaction is represented by figure 6 read upwards. There are two variables involved, the energy of scattering ν (in the laboratory), and the angle of scattering θ. It is usual, however, to use instead the invariants from figure 6

$$\left.\begin{aligned} s &= (p_a+p_b)^2 = (p_c+p_d)^2 \\ t &= (p_a+p_c)^2 = (p_b+p_d)^2 \\ u &= (p_a+p_d)^2 = (p_b+p_c)^2 \end{aligned}\right\} \quad \ldots\ldots(8.1)$$

Dispersion Relations in Strong-coupling Physics

Notice that we have defined the outgoing momenta with a minus sign, so that the equation of conservation of momentum is $p_a + p_b + p_c + p_d = 0$. From this equation it follows that s, t and u are not independent, but satisfy the equation

$$s + t + u = 2M^2 + 2\mu^2. \qquad \ldots\ldots(8.2)$$

The quantity s is the square of the centre-of-mass energy of the system, including both rest masses. It is given in terms of the laboratory energy by

$$s = M^2 + 2M\nu + \mu^2. \qquad \ldots\ldots(8.3)$$

t will be minus the square of the momentum transferred from the initial to the final pion, or from the initial to the final nucleon. Its relation to the centre-of-mass scattering angle is

$$t = -2q^2(1 - \cos\theta) \qquad \ldots\ldots(8.4)$$

where q, the centre-of-mass momentum, can be calculated from s by the formula

$$q^2 = \{s^2 - 2s(M^2 + \mu^2) + (M^2 - \mu^2)^2\}/4s. \qquad \ldots\ldots(8.5)$$

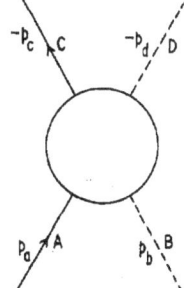

Figure 6. Diagram for pion–nucleon scattering.

To begin let us rewrite the forward dispersion relation (4.2) with A expressed as a function of s rather than of ν. As s is linearly related to ν by (8.3), it is not difficult to see that the dispersion relation preserves its form

$$\mathscr{R}A(s) = \frac{1}{\pi}\int_{-\infty}^{\infty} ds' \frac{\mathscr{I}A(s')}{s' - s}. \qquad \ldots\ldots(8.6)$$

(In order to avoid confusion, we emphasize that we have changed our variable and $A(s)$ no longer has the same meaning as before; in our new notation the functions appearing in (4.2) would be $A(M^2 + 2M\nu + \mu^2)$.) In this equation t and $\cos\theta$ are set equal to zero. From several points of view the simplest generalization of (8.4) is to put t equal to a constant, and the new relation takes the form

$$\mathscr{R}A(s, t) = P\frac{1}{\pi}\int ds' \frac{\mathscr{I}A(s', t)}{s' - s}. \qquad \ldots\ldots(8.7)$$

For positive kinetic energies the function $\mathscr{I}A(s', t)$ in the integrand is given by the unitarity condition, while for negative kinetic energies the crossing relation must be used. As before, the crossed reaction

$$A + D \to B + C$$

is also pion–nucleon scattering, but the roles of the pion momenta p_b and p_d become interchanged. From (8.1) it follows that s and u are interchanged but t is unaltered,

so that the crossing relation is
$$\mathcal{I}A(s,t) = -\mathcal{I}A(u,t), \qquad \ldots\ldots(8.8)$$
or, using (8.2),
$$\mathcal{I}A(s,t) = -\mathcal{I}A(2M^2+2\mu^2-s-t,t). \qquad \ldots\ldots(8.9)$$

For values of s below $(M-\mu)^2-t$ the quantity $2M^2+2\mu^2-s-t$, which represents the square of the energy in the amplitude on the right, will be greater than $(M+\mu)^2$, and $\mathcal{I}A(s,t)$ may be found from (8.9). For values of s between $(M-\mu)^2-t$ and $(M+\mu)^2$, both s and $2M^2+2\mu^2-s-t$ are less than $(M+\mu)^2$, so that no intermediate states occur in the summation (5.1) for $\mathcal{I}A(s,t)$ or $\mathcal{I}A(2M^2+2\mu^2-s-t,t)$, and $\mathcal{I}A(s,t)$ is zero. We must of course include the pole term (5.3), which in terms of s is simply $g^2/(M^2-s)$, and the corresponding term $g^2/(-M^2-2\mu^2+s+t)$ obtained by crossing. Equation (8.7) therefore becomes

$$\mathcal{R}A(s,t) = P\frac{1}{\pi}\int_{(M+\mu)^2}^{\infty}ds'\frac{\mathcal{I}A(s',t)}{s'-s} - P\frac{1}{\pi}\int_{-\infty}^{(M-\mu)^2-t}ds'\frac{\mathcal{I}A(2M^2+2\mu^2-s'-t,t)}{s'-s}$$
$$+\frac{g^2}{M^2-s}+\frac{g^2}{s-M^2-2\mu^2+t}, \qquad \ldots\ldots(8.10)$$

or
$$\mathcal{R}A(s,t) = P\frac{1}{\pi}\int_{(M+\mu)^2}^{\infty}ds'\,\mathcal{I}A(s',t)\left\{\frac{1}{s'-s}+\frac{1}{s'+s-2M^2-2\mu^2+t}\right\}$$
$$+g^2\left\{\frac{1}{M^2-s}+\frac{1}{s-M^2-2\mu^2+t}\right\}. \qquad \ldots\ldots(8.11)$$

As t is to be kept constant in these equations, it follows from (8.4) that $\cos\theta$ will vary along the range of integration as s' and therefore $q(s')$ vary. We assume t to be negative so that $1-\cos\theta$ is positive. Even then, however, $\cos\theta$ will be less than -1 if q is sufficiently small. As q^2 is zero at the lower limit of the region of integration $(s'=(M+\mu)^2)$, one will be integrating over unphysical angles for a range of values of s' above the lower limit. The length of this range increases from zero as $|t|$ increases from zero. One therefore requires information about the function $\mathcal{I}A(s',t)$ for these unphysical angles before the integration can be performed. As we shall see below, it turns out that, with values of t for which the dispersion relation can be proved, one can use the Legendre expansion into partial waves at all relevant angles.

Owing to the gap in the range of integration in (8.10), one can extend the analytic properties of A treated in § 3. We have already pointed out that the right-hand side of (8.10) is an analytic function of s in the upper half-plane, i.e. for $\mathcal{I}s>0$. In fact, the dispersion relation was deduced as a consequence of this property. However, the right-hand side of (8.10) involves s only in the denominators, and will therefore be analytic in s at all values for which the denominators never vanish as s' varies over its range of integration. Thus, if s is between $(M-\mu)^2-t$ and $(M+\mu)^2$, our function (excluding the pole) will be analytic, even for $\mathcal{I}s=0$. If s is in the lower half-plane, the denominators certainly never vanish as s' is always real. Hence A will be an analytic function of s in the whole plane except for cuts from $-\infty$ to $(M-\mu)^2$ and from $(M+\mu)^2$ to ∞, and poles at M^2 and $M^2+2\mu^2-t$. It follows from (8.10) that the discontinuity of A across the cut is $2\mathcal{I}A(s,t)$.

By differentiating (8.10) any number of times with respect to t and then setting t equal to zero, one can obtain dispersion relations for the derivative of the forward

Dispersion Relations in Strong-coupling Physics

scattering amplitude with respect to the angle. The range of integration does not involve unphysical angles. These relations have already been mentioned in § 6 and have been used in phenomenological calculations. Efremov, Meshcheryakov, Shirkov and Tzu (1960) have suggested using such an approach for dynamical calculations, but it is doubtful whether this is possible. For these more ambitious calculations genuine non-forward dispersion relations are needed.

Though equation (8.6) appears to be a natural generalization of (4.2) and was written down independently by several workers (Gell-Mann, Goldberger, Nambu and Oehme (unpublished), Salam (1956), Capps and Takeda (1956)), it took some time before a proof from field theory was furnished. As a matter of fact, even the

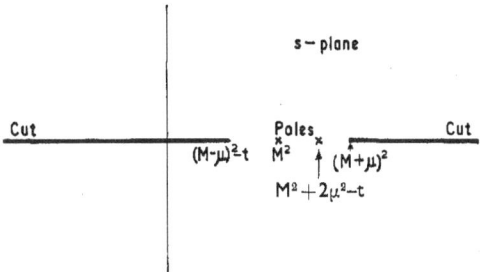

Figure 7. Analytic properties of A as a function of s.

proofs of the forward dispersion relations are only straightforward if the scattered particles are massless. Bogoliubov, Medvedev and Polivanov (Bogoliubov and Shirkov 1957) finally gave a rigorous proof of dispersion relations for non-forward pion–nucleon scattering provided $|t|$ is not too large. The most complicated step in their proof was later simplified considerably by Bremermann, Oehme and Taylor (1958) and especially by Lehmann (1958), who showed that the dispersion relations are valid provided that

$$0 < -t < \frac{32}{3}\frac{2M+\mu}{2M-\mu}\mu^2. \qquad \ldots\ldots(8.12)$$

The limitation on the value of t is almost certainly due to the method of proof and not to a breakdown in the validity of the dispersion relations. An article by Lehmann (1959) combines the ideas of the papers just mentioned to give what is probably the most economical proof of the dispersion relations.

The difficulty in the proofs arises from the contribution of the unphysical region to the dispersion integral. For forward scattering of massless particles there is no unphysical region and the proof is straightforward. For forward scattering of pions by nucleons the intermediate one-nucleon state is in the unphysical region (as it occurs at a value of s below $(M+\mu)^2$) and more work is involved. On extending the relations to non-forward scattering one also encounters a continuum in the unphysical region, as there is a range of values of s for which the cosine of the angle is less than -1. When $|t|$ increases sufficiently this range is large enough to cause the breakdown of the proofs we have at present.

Dispersion relations for nucleon–nucleon scattering have not yet been proved rigorously, even in the forward direction. The difficulty is once more due to a large unphysical region in the dispersion integral. As usual, one has to determine the imaginary part of the scattering amplitude at negative kinetic energies by means of the crossing relations. The crossed reaction is nucleon–anti-nucleon scattering, whose imaginary part will be non-zero above the energy of the lowest intermediate with the same quantum numbers. This is the two-pion state, even the one-pion state will contribute a pole term. However, the physical region only begins at an energy of two nucleon masses. There is thus a large intermediate unphysical region which contributes to the dispersion integral.

The dispersion relations for nucleon–nucleon scattering have been proved in all orders of perturbation theory if $|t|$ is sufficiently small (Symanzik 1958), and have been examined by Goldberger, Nambu and Oehme (1957) and Goldberger and Oehme (1960). One cannot perform the dispersion integral and compare the result with experiment, as one did with pion–nucleon scattering, since the contribution to the integral from the unphysical region cannot be determined directly.

In the hypothetical case where the mass of the nucleon is less than $(1+\sqrt{2})$ times the mass of the pion, the dispersion relations for nucleon–nucleon scattering have been proved rigorously by Bremermann, Oehme and Taylor (1958). Proofs of dispersion relations for scattering of strange particles on nucleons are subject to the same limitations as for nucleon–nucleon scattering.

8.2. Analytic Properties in the Momentum Transfer

The dispersion relations show that the amplitude $A(s,t)$ for pion–nucleon scattering is an analytic function of s, the square of the energy, in the cut plane when the momentum transfer t is fixed. Lehmann (1958), besides carrying out a step in the proof of these relations, also showed that the amplitude has analytic properties as a function of t when s is kept fixed. In other words, it has analytic properties as a function of $\cos\theta$, which is linearly related to t by (8.4). A knowledge of such properties is essential if one is to make use of non-forward dispersion relations. We have remarked that a use of the Legendre expansion for unphysical values of $\cos\theta$ is necessary even to give a meaning to the integrand in the relation, and the validity of this expansion depends on the analytic properties in $\cos\theta$.

We shall state Lehmann's results in terms of $\cos\theta$, which we shall denote by z, the transformation from z to t being trivial by (8.4). The physical region is the interval $-1 < z < 1$. One cannot obtain rigorously analytic properties in the momentum transfer which are anything like as extensive as those in the energy, where the whole cut plane can be included. Lehmann was able to show, however, that the scattering amplitude is an analytic function of z in an ellipse surrounding the physical region in the complex z-plane. The foci are at ± 1, and the major axis z_0 is given by

$$z_0 = \left[1 + \frac{2\mu^3(2M+\mu)}{q^2\{s-(M-2\mu)^2\}}\right]^{1/2}. \qquad \ldots\ldots(8.13)$$

Notice that, as s approaches infinity, the ellipse closes down on the physical region.

If A is analytic within the ellipse, the same must be true about its absorptive part, but Lehmann showed that the absorptive part is in fact an analytic function

Dispersion Relations in Strong-coupling Physics

of z in a larger ellipse. The major axis of this ellipse is now of length $2z_0^2 - 1$, z_0 being again given by (8.13). Of course, it is only on the real axis that the absorptive part of A is equal to its imaginary part. Within the remainder of the ellipse it is defined by analytic continuation.

According to a well-known theorem in analysis, the Legendre expansion for a function may be used within its ellipse of analyticity. As points in the integral (8.8) are always in the large Lehmann ellipse if the inequality (8.10) is true, we may use the Legendre expansion to calculate $\mathscr{I}A$, even in the unphysical region.

Results similar to these can be proved for all scattering processes, and one can even prove certain analytic properties of production amplitudes (Ascoli 1960).

We have already used the analytic properties in z in § 6.3, when we assumed that, in certain processes, there were poles in z just outside the physical region. Unfortunately, these poles are always outside the small Lehmann ellipse, so that the existence of the poles is not proved by these results.

8.3. Analytic Properties as a Function of more than One Complex Variable

Applications of analytic properties of scattering amplitudes to dynamical calculations, at any rate in their present form, require a knowledge of these properties as a function of two variables, s and t, both complex. In general such properties have not been proved rigorously. The one exception is pion–pion scattering, where certain rigorous properties have been obtained (Mandelstam 1960 b). We shall discuss more fully the conjectured analytic properties in § 10, and shall merely state here that, for pion–pion scattering, these properties have actually been proved within a certain region of the complex $s-t$ space.

For the case of 'neutral scalar pions' where the reaction $\pi + \pi \to \pi$ could take place, an interesting result is that the domain in which analyticity can be proved is widened by using the unitarity condition. (Previously the unitarity condition had been used for applying analytic properties but not for proving them.) As the transition amplitudes for production processes have not seriously been considered, except in perturbation theory, the unitarity condition can only be used at values of the energy below the threshold for production; since the treatment is rigorous, approximations cannot be made. Had we been able to use the unitarity condition elsewhere, the domain of analyticity would have been larger still. This strongly suggests that unitarity will be a powerful tool for proving analytic properties once production processes are taken into account.

Another result, obtained by considering the amplitude for the scattering of neutral scalar pions as a function of two complex variables, was that the pole in $\cos \theta$, analogous to the pole in nucleon–nucleon scattering discussed in § 6.3, was now within the domain where analytic properties could be proved. The pole was shown to exist. Finally, the range of values of t for which dispersion relations can be proved—0 to $8\mu^2$ by Lehmann's method—is now extended to $9\mu^2$. For these larger values of t it is no longer always true that $\mathscr{I}A(s,t)$ can be obtained from its Legendre expansion in the unphysical region. More sophisticated methods of continuation, based on the analytic properties of $\mathscr{I}A$ as a function of t or z, have to be used.

8.4. Summary of Rigorous Dispersion Relations and Analytic Properties of Scattering Amplitudes

Let us summarize here the results that have actually been proved.

8.4.1. Dispersion relations for fixed momentum transfer.

Dispersion relations have been proved for pion–nucleon scattering

$$\left(0 < t < \frac{32}{3}\frac{2M+\mu}{2M-\mu}\mu^2\right)$$

(Lehmann 1958), and for pion–pion scattering ($0 < t < 28\mu^2$). In general, dispersion relations have been proved, or can easily be proved, for the scattering of pions or photons off any 'elementary' particle, or for the photo-production of pions (Oehme and Taylor 1959, Logunov and Isaev 1958, Logunov 1959). One inelastic process—the double Compton effect—has also been treated (Logunov, Belenkii and Tavkhelidze 1958). For other cases dispersion relations have not been proved, but they could be proved if certain mass-ratios were less than the experimental mass-ratio. For instance, dispersion relations for nucleon–nucleon scattering could be proved if M/μ were less than $\sqrt{2}+1$.

8.4.2. Analytic properties as functions of momentum transfer at fixed energy.

The scattering amplitude for any reaction is an analytic function of z within a certain ellipse centred about the physical region (Lehmann 1958).

8.4.3. Analytic properties as functions of energy and momentum transfer.

It has been shown that the pion–pion scattering amplitude has certain analytic properties (to be described more fully in § 10) within a region in (s, t) space (Mandelstam 1960 b). For the scattering of scalar pions this region can be extended by unitarity, and includes the poles in $\cos\theta$.

These are the only analytic properties of scattering amplitudes that have been proved at the time of writing (May 1961).

§ 9. Outline of the Proof of Dispersion Relations at Fixed Momentum Transfer

While we shall not be able to present the complete proof of dispersion relations here, we shall outline the method in order to show the type of reasoning involved. We shall have to presuppose a knowledge of quantum field theory on the part of the reader.

The coordinate system used in the proof of non-forward dispersion relations is neither the laboratory nor the centre-of-mass system, but the 'Breit' system in which the nucleon's momentum both before and after the scattering is independent of the energy. The kinematics are illustrated in figure 8, ω being the pion energy; the broken line represents the pion and the solid line the nucleon. In terms of our usual variables

$$\omega = \tfrac{1}{2}E_\Delta^{-1/2}(s - M^2 - \mu^2 - 2\Delta^2), \quad \ldots\ldots(9.1a)$$
$$= E_\Delta^{-1/2}(M\nu - \Delta^2), \quad \ldots\ldots(9.1b)$$
$$t = -4\Delta^2, \quad \ldots\ldots(9.1c)$$

where $\quad E_\Delta^2 = M^2 + \Delta^2.$

Dispersion Relations in Strong-coupling Physics 129

According to the fundamental reduction formula of Lehmann, Symanzik and Zimmermann (1955) and Low (1955), the scattering amplitude has the following form:

$$(2\pi)^4 A \delta^4(p_A+p_B+p_C+p_D) = i(2\pi)^3 2E_\Delta \int d^4x\, d^4y \exp\{i(p_B x+p_D y)\}$$
$$\times (\Box_x^2-\mu^2)(\Box_y^2-\mu^2) \langle N(-p_C)| T\{\phi(x),\phi(y)\}|N(p_A)\rangle. \quad \ldots\ldots(9.2)$$

As before, p_A and $-p_C$ are the momenta of the incoming and outgoing nucleons, p_B and $-p_D$ those of the incoming and outgoing pions. The state $\langle N(-p_C)|$ is a state with one nucleon having momentum p_C and similarly for $|N(p_A)\rangle$, while T

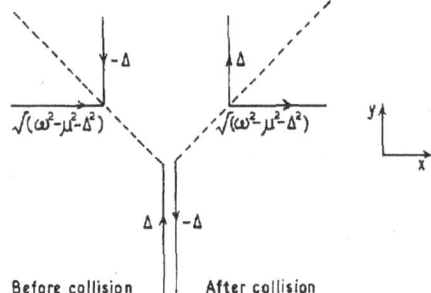

Figure 8. Breit system of coordinates for pion–nucleon scattering.

is a time-ordered product of the pion field operators $\phi(x)$ and $\phi(y)$. It is possible to subtract from this time-ordered product the unordered product $\phi(x)\phi(y)$ without changing the expression. For, writing $\phi(x)\phi(y)$ as $\Sigma_I \phi(x)|I\rangle\langle I|\phi(y)$ and inserting it in place of $T\{\phi(x),\phi(y)\}$ in (9.2), we obtain

$$i(2\pi)^3 2E_\Delta \int d^4x\, d^4y \exp\{i(p_B x+p_D y)\}$$
$$\times(\Box_x^2-\mu^2)(\Box_y^2-\mu^2)\sum_I \langle N(-p_C)|\phi(x)|I\rangle\langle I|\phi(y)|N(p_A)\rangle. \quad \ldots\ldots(9.3)$$

Now consider the factor

$$\int dy \exp(ip_D y)\langle I|\phi(y)|N(p_A)\rangle,$$

or

$$\langle I|\int dy \exp(ip_D y)\phi(y)|N(p_A)\rangle.$$

The state $|N(p_A)\rangle$ has four-momentum p_A, and the operator between the two states has four-momentum p_D, so that I must have four-momentum p_A+p_D. As p_A is the four-momentum of a nucleon and p_D is minus the four-momentum of a pion, no such state exists. The factor, and therefore the expression (9.3), is zero. Accordingly we may subtract the product $\phi(x)\phi(y)$ from the time-ordered product in (9.2).

Now
$$T\{\phi(x),\phi(y)\}-\phi(y)\phi(y) = R\{\phi(y),\phi(x)\}, \quad \ldots\ldots(9.4)$$
where, by definition,
$$R\{\phi(y),\phi(x)\} = \theta(y_0-x_0)[\phi(y),\phi(x)], \quad \ldots\ldots(9.5)$$

$\theta(y_0-x_0)$ being the function which is 1 for $y_0>x_0$, 0 for $y_0<x_0$. We are thus permitted to replace $T\{\phi(x),\phi(y)\}$ by $R\{\phi(y),\phi(x)\}$ in (9.3). Making this substitution, noting that the integral over $(x+y)$ for fixed $(x-y)$ just gives $(2\pi)^4 \delta(p_A+p_B+p_C+p_D)$ and putting in the relevant values of p_B and p_D from figure 8, we find that

$$A = i(2\pi)^3 2E_\Delta \int d^4(x-y) \exp\{i[\omega(y_0-x_0) - \sqrt{(\omega^2-\mu^2-\Delta^2)}(y_s-x_s)]\}$$
$$\times (\Box_x^2 - \mu^2)(\Box_y^2 - \mu^2) \langle N(-p_C)| R\{\phi(y),\phi(x)\}| N(p_A)\rangle. \quad \ldots\ldots(9.6)$$

We have written y_0-x_0 for the time component of $y-x$ and y_s-x_s for the component in the x-direction of figure 8.

The reason why we replaced the T-product by the R-product is that the latter possesses simple causality properties. According to our fundamental assumption, the commutator in (9.5) vanishes if $(y-x)$ is space-like. Together with the property of the θ-function, this means that $R\{\phi(y),\phi(x)\}$ vanishes unless $(y-x)$ is time-like with $y_0-x_0>0$. Thus the integrand in (9.6) will certainly vanish unless $y_0-x_0>y_s-x_s$.

Now take the simple case of forward scattering of massless particles ($\mu=0, \Delta^2=0$). The exponential in (9.6) is then $\exp i\omega\{(y_0-x_0)-(y_s-x_s)\}$. If ω is given a positive imaginary part ($\omega=\omega_1+i\omega_2$), this expression becomes $\exp\{(i\omega_1-\omega_2)[(y_0-x_0)-(y_s-x_s)]\}$. The integrand vanishes unless the factor in x and y is positive, so that the exponential is less in magnitude than unity. It therefore follows exactly as in the case of dispersion of light that A is an analytic function of ω in the upper half-plane. Since ω is linearly related to s or ν, A is an analytic function of these variables in the upper half-plane, and the dispersion relation is proved.

This is the situation with the forward scattering of massless particles. The simple proof breaks down when either μ^2 or Δ^2 is greater than zero, as the exponent in (9.6) is then not necessarily negative if $\omega_2>0$. In such cases the dispersion relations are not a consequence of equation (9.6) unless we put in more physical information. Bogoliubov, Medvedev and Polivanov proceeded further by defining the following function:

$$A(s,t,\gamma) = i(2\pi)^3 2E_\Delta \int d^4(x-y) \exp\{i[\omega_\gamma(y_0-x_0) - \sqrt{(\omega_\gamma^2-\gamma-\Delta^2)}(y_s-x_s)]\}$$
$$\times (\Box_x^2 - \mu^2)(\Box_y^2 - \mu^2) \langle N(-p_C)| R\{\phi(y),\phi(x)\}| N(p_A)\rangle, \quad \ldots\ldots(9.7)$$

where

$$\omega_\gamma = \tfrac{1}{2} E_\Delta^{-1/2}(s-M^2-\gamma-2\Delta^2), \quad \Delta^2 = -(t/4).$$

The mass μ^2 has simply been replaced by γ in (9.6) and (9.1) so that $A(s,t,\mu^2)$ is our ordinary scattering amplitude. Apart from the factors $\Box_x^2-\mu^2$ and $\Box_y^2-\mu^2$, (9.7) is just the pion–nucleon Green's function with the pions 'off the mass shell'. If γ is less than $-\Delta^2$, it is easy to show that the exponent

$$i[\omega_\gamma(y_0-x_0) - \sqrt{(\omega_\gamma^2-\gamma-\Delta^2)}(y_s-x_s)]$$

has a negative real part when $\mathscr{I}\omega_\gamma>0$, $(y_0-x_0)>(y_s-x_s)$. For this case we can

Dispersion Relations in Strong-coupling Physics

therefore prove the dispersion relation as before:

$$\mathscr{R}A(s,t,\gamma) = P\frac{1}{\pi}\int_{-\infty}^{\infty} ds' \frac{\mathscr{I}A(s',t,\gamma)}{s'-s}, \quad (\gamma < -\Delta^2). \quad \ldots\ldots(9.8)$$

Owing to the restriction on γ, equation (9.8) is not of direct significance. Applying the crossing relation and putting in the pole term explicitly, one can write (9.8) in a form analogous to (8.9)

$$\mathscr{R}A(s,t,\gamma) = P\frac{1}{\pi}\int_{(M+\mu)^2}^{\infty} ds'\,\mathscr{I}A(s',t,\gamma)\left\{\frac{1}{s'-s}+\frac{1}{s'+s-2M^2-2\gamma+t}\right\}$$

$$+\{g(\gamma)\}^2\left\{\frac{1}{M^2-s}+\frac{1}{s-M^2-2\gamma+t}\right\}. \quad \ldots\ldots(9.9)$$

The method of Bogoliubov, Medvedev and Polivanov was to show that both sides of (9.9) could be continued analytically from $\gamma < -\Delta^2$ to $\gamma = \mu^2$. Once this is done the physical dispersion relation is proved, since $A(s,t,\mu^2)$ is the scattering amplitude required. To do the analytic continuation, $\mathscr{I}A(s,t,\gamma)$ must be expressed as the Fourier transform of an amplitude with causal properties. We can write the R-product in (9.7) as

$$R\{\phi(y),\phi(x)\} = \tfrac{1}{2}[\phi(y)\phi(x)] + \tfrac{1}{2}\epsilon(y_0-x_0)[\phi(y)\phi(x)] \quad \ldots\ldots(9.10)$$

where $\epsilon(y_0-x_0)$ is ± 1 according as y_0-x_0 is positive or negative. By Hermitian conjugation and reflection in the y-direction (figure 8), one can easily show that the real and imaginary parts of (9.7) are obtained by replacing the R-product by the second and first terms of (9.10) respectively. Thus

$$\mathscr{I}A(s,t,\gamma) = \tfrac{1}{2}(2\pi)^3\,2E_\Delta\int d^4(x-y)\exp\{i[\omega_\gamma(y_0-x_0)-\sqrt{(\omega_\gamma^2-\gamma-\Delta^2)}(y_s-x_s)]\}$$

$$\times(\Box_x^2-\mu^2)(\Box_y^2-\mu^2)\langle N(-p_C)|[\phi(y),\phi(x)]|N(p_A)\rangle.$$
$$\ldots\ldots(9.11)$$

On using the reduction formulae and performing manipulations similar to those leading to the derivation of (9.7) we can rewrite this as

$$\mathscr{I}A(s,t,\gamma) = -\tfrac{1}{2}\int d^4(x-z)\,d^4(y-w)\,d^4(z-w)$$

$$\times \exp\{i[p_B(x-z)+p_D(y-w)+(p_A+p_B)(z-w)]\}$$

$$\times(\Box_x^2-\mu^2)(\Box_y^2-\mu^2)(\Box_z^2-M^2)(\Box_w^2-M^2)$$

$$\times \langle 0|R\{\phi(y),\psi(w)\}\,R\{\phi(x),\bar{\psi}(z)\}|0\rangle. \quad \ldots\ldots(9.12)$$

The state $\langle 0|$ is the vacuum state; $\psi(w)$ and $\bar\psi(z)$ are nucleon operators. The momenta p_A, p_B, p_C and p_D in the exponential are now referred to the centre-of-mass system. They are of course functions of s, t and γ. (In the kinematical relations, μ^2 is still replaced by γ.)

The two retarded products possess causality properties in $x-z$ and $y-w$, so that the Fourier transform will have analytic properties in p_B and p_D. Bogoliubov was able to show that the expression could be continued from $\gamma < -\Delta^2$ to $\gamma = \mu^2$. The right-hand side of (9.9), except possibly for the pole terms, could therefore be continued to $\gamma = \mu^2$. From this fact and from the expression (9.7) for $A(s,t,\gamma)$, Bogoliubov et al. showed that the left-hand side of (9.9) and the pole terms could also be continued analytically to $\gamma = \mu^2$ and thus completed the proof of the dispersion relations.

S. Mandelstam

By far the most difficult part of Bogoliubov's proof was the demonstration that $\mathscr{I}A(s, t, \gamma)$, given by (9.9), could be continued to $\gamma = \mu^2$. Bremermann, Oehme and Taylor (1958) gave a rather simpler alternative proof of this step using the theory of functions of several complex variables. A year before this, Jost and Lehmann (1957) had proposed a representation which exhibited the analytic properties of causal commutators such as occur in (9.9). Their representation was only valid when the two particles had equal mass, but Dyson (1958) extended it to the case of non-equal masses. This enabled Lehmann to give yet another proof that $\mathscr{I}A(s, t, \gamma)$ could be continued to $\gamma = \mu^2$, and at the same time to obtain analytic properties of $\mathscr{I}A(s, t, \gamma)$ as a function of t. By using an expression for A different from (9.7), Lehmann showed that A also possesses analytic properties as a function of t.

The analytic continuation of (9.9) to $\gamma = \mu^2$ confirmed that, in the unphysical region $(M-\mu)^2 - t < s < (M+\mu)^2$, $A(s, t, \mu^2)$ was equal to zero except at the poles. It also enabled the pole terms themselves to be placed on a rigorous footing. Between the operators $\phi(y)$ and $\phi(x)$ in (9.11) one can introduce a sum over states $\Sigma_I |I\rangle\langle I|$ and the pole term is obtained by taking only the one-nucleon state. We therefore obtain for the pole terms

$$(2\pi)^4 \mathscr{I}_{\text{Pole}} A(s, t, \gamma) \delta^4(p_A + p_B + p_C + p_D) = \tfrac{1}{2}(2\pi)^3 2E_\Delta \int dx\, dy \exp\{i(p_B x + p_D y)\}$$
$$\times (\Box_x^2 - \mu^2)(\Box_y^2 - \mu^2) \sum_{N_I} \{\langle N(-p_C)|\phi(y)|N_I\rangle \langle N_I|\phi(x)|N(p_A)\rangle$$
$$- \langle N(-p_C)|\phi(x)|N_I\rangle \langle N_I|\phi(y)|N(p_A)\rangle\}. \quad\quad\ldots\ldots(9.13)$$

The momenta p_B and p_D are the pion momenta in figure 8 expressed in terms of s and t, with μ^2 replaced by γ. We have rewritten the integral as one over x and y separately. In the summation over N_I only the one-nucleon states are considered and these will be restricted by energy conservation. Now, according to the Lehmann–Symanzik–Zimmermann reduction formulae, the factor

$$(2\pi)^{3/2}(2E_\Delta)^{1/2}\int dx \exp(ip_B x)(\Box_x^2 - \mu^2)\langle N_I|\phi(x)|N(p_A)\rangle$$

is just equal to

$$(2\pi)^{-3/2}(2p_{0I})^{-1/2} g(\gamma)(2\pi)^4 \delta(p_I + p_A + p_B),$$

where $g(\gamma)$ is the vertex function with the nucleons on the mass shell but with p_B^2, the square of the pion momentum, equal to $-\gamma$. For the negative values of γ with which we are dealing at the moment all the momenta can be made real and the vertex function has a well-defined meaning. Performing the integration over the intermediate states, we find for the first term of (9.13)

$$\mathscr{I}_{\text{Pole},1} A(s, t, \gamma) = (\pi/2p_{0I})\{g(\gamma)\}^2 \delta(p_{0I} + p_A + p_B),$$

or, in the centre-of-mass system,

$$\mathscr{I}_{\text{Pole},1} A(s, t, \gamma) = \pi\{g(\gamma)\}^2 \delta(s - M^2). \quad\quad\ldots\ldots(9.14)$$

Substitution of (9.14) in the dispersion integral gives the first pole term in (9.9). The other term of (9.13) gives similarly the second pole term.

Thus, for negative values of γ, the function of $g(\gamma)$ in (9.9) is the vertex function. The proof of dispersion relations shows that $\{g(\gamma)\}^2$ can be continued analytically

Dispersion Relations in Strong-coupling Physics

to $\gamma = \mu^2$. It follows that the vertex function can be continued to the point where all three particles are on the mass shell, and that its value there is the coupling constant which occurs in the dispersion relations.

§ 10. DISPERSION RELATIONS AND ANALYTIC PROPERTIES FOR NON-FORWARD SCATTERING: CONJECTURED RESULTS

10.1. The Principle of Maximum Analyticity

The analytic properties we have described in § 8 do not appear to be sufficient to carry out the programme of calculating observable quantities in terms of a small number of coupling constants. We pointed out that the known properties of scattering amplitudes and Green's function—causality and unitarity —have not nearly been fully exploited yet, and that the analyticity properties which could in principle be obtained with our present methods are probably much more extensive than those that have been obtained to date. If calculations are to be performed at present or in the near future, one will have to find some way of judging what results, as yet unproved, follow from our hypotheses. Fortunately, the position is that the physicist who is prepared to work along non-rigorous lines is not prevented from doing calculations by this particular gap in our knowledge.

A powerful guide to the analytic properties of scattering amplitudes is furnished by perturbation theory. But first let it be emphasized that, in appealing to perturbation theory, we need not introduce the Lagrangian formulation with its infinite constants which we are trying to avoid. It appears that our present perturbation series can be reproduced without referring to the Lagrangian, but by using instead the principles of causality and unitarity discussed in this article. To our knowledge this has not been proved to all orders, but sub-sets of the perturbation series have been constructed by Mandelstam (1959 b) and Muraskin and Nishijima (1961, see also Nishijima 1960). As with all calculations performed on this basis, the series appear from the beginning in renormalized form without infinite constants. It is necessary in these approaches to use proved analyticity properties only but, if one wishes to do so, one must examine the 'Green's functions', which are generalizations of scattering amplitudes. Once the extra analyticity properties have been obtained, one need work only with the scattering amplitudes.

The fact that the perturbation theory can be constructed without the Lagrangian formalism is itself progress. In electrodynamics perturbation theory is the basis of our calculations, so that all results of electrodynamics are now justified by a theory which contains no infinite constants. In strong-coupling physics perturbation theory is worse than useless for actual calculations and is to be used only as a guide for finding analytic properties.

The assumption we make is essentially that, if a scattering amplitude has any analytic property in all orders of perturbation theory, it has that property generally. This does not amount to an assumption that the perturbation series converges. It frequently happens that information can be obtained about a function from a series expansion even if the expansion diverges, and this is the situation we are envisaging here. Perturbation theory is obtained by treating the equations of the theory in a certain way, and one should be able to obtain a guide to the theory from its perturbation series.†

We may restate the assumption we have made as the assumption of *maximum analyticity*. The scattering amplitude is analytic in all its variables except at those points where singularities arise as a consequence of the unitarity condition. For instance, we know that the scattering amplitude is an analytic function of the square of the energy for fixed momentum transfer, with cuts along the real axis. The fact that these cuts exist is forced upon us by unitarity, as the discontinuity across the cuts is equal to twice the imaginary part of the scattering amplitude—the quantity given by the unitarity equation. When considered as a function of more than one complex variable the scattering amplitude is found to have other types of singularities as a consequence of the unitarity condition.‡

The two ways of stating our assumption are almost certainly equivalent. If a singularity arises from the unitarity condition it must occur in perturbation theory, which satisfies this condition. Conversely, if perturbation theory can be generated from causality and unitarity, the singularities of the perturbation terms must be a result of features of the unitarity equation.

General formulae for finding the singularities of transition amplitudes in perturbation theory have been given by J. D. Bjorkén (1960),§ Landau (1959), Tarski (1960) and Polkinghorne and Screaton (1960). Usually these formulae are difficult to apply in the general case. One therefore makes a supplementary assumption, which gives one sufficient analytic properties to set up equations on the basis of the approximation scheme outlined in § 7. Suppose that one had written down the unitarity equations with neglect of intermediate states above a certain energy. From the equations one would find singularities in the transition amplitude, but one would certainly miss some of the singularities owing to the neglect of the higher intermediate states. It is very plausible to assume that these extra singularities occur at energies comparable with those of the neglected states. According to our fundamental approximation, these singularities can be ignored, and we have all the analytic properties we require.

Assuming that perturbation theory gives correctly the singularities, Eden (1960 a, b, 1961 a) and Landshoff, Polkinghorne and Taylor (1961) have proposed a method of proving that, in two-particle transition amplitudes, singularities due to high intermediate states do occur at correspondingly large distances away; we shall refer again to their work in § 10.2. Their proof is at the moment incomplete, however (Eden, Landshoff, Polkinghorne and Taylor 1961 b). All perturbation terms so far examined have this property.

The precise analytic properties of the transition amplitude, then, will depend on the process studied and perhaps on the approximation used. Since,

---

† We really mean a generalized perturbation theory in which elementary particles can have any mass for which there exist states, simple or composite, with the corresponding quantum numbers. The analytic properties of generalized perturbation theory are all that are used in calculations. The difference between ordinary and generalized perturbation theory is that the latter tells us nothing about ' unphysical sheets '.

‡ If the principle of maximum analyticity is to give unambiguous results it should be applied to the Green's functions instead of to the scattering amplitudes; the analytic properties of the latter then follow as a special case of those of the former.

§ Stanford University preprint.

Dispersion Relations in Strong-coupling Physics

however, one is to find them from the equations which will later have to be solved, their determination as such should not prove the main stumbling block in a problem.

What are our reasons for believing that the principle of maximum analyticity is a reasonable working hypothesis? As the rigorous analytic properties contain sufficient information to construct the perturbation series, one would expect that they probably contain sufficient information in principle to solve the theory—always on the assumption that the theory has a consistent solution. The principle of maximum analyticity gives a solution possessing at least some of the correct properties and this should be the solution required, as it seems to reproduce the perturbation series when expanded in powers of the coupling constant. Another reason for believing our principle was given in § 8. In the case of pion–pion scattering or scalar pion–pion scattering one can prove rigorously that, within a certain domain in the (s, t) space, the only singularities of the scattering amplitude are those given by the unitarity condition. The limitation on the size of the domain was due to the fact that we were not considering production amplitudes and could therefore use the unitarity condition only below production thresholds. We would therefore hope that, the more processes we considered, the larger would be the domain within which we could prove that the only singularities were those given by the unitarity condition.

The foregoing discussion is based on the assumption that the principles of field theory are in fact valid. We shall make some remarks on this, and also on the suggestion that the assumption of maximum analyticity should be taken as a postulate in place of field theory, in the final section.

Though we cannot of course give firm reasons why our conjecture should be true, we may at least dispose of one objection which is sometimes made. In a problem with bound states, there will be the usual poles in the scattering amplitude associated with them as single-particle intermediate states. As an example, the triplet nucleon–nucleon scattering amplitude will have a pole at the deuteron energy. The poles do not occur in perturbation theory, where the bound state is never seen. Thus there are at least some singularities in the scattering amplitude which are not given by perturbation theory. More generally, it is usually possible to continue scattering amplitudes through the cuts along the real axis on to the so-called 'unphysical sheet' though we shall make no use of such continuation in our article. One then finds poles corresponding to resonances, poles which again do not occur in perturbation theory. Having established that there are non-perturbation singularities in the scattering amplitude, one may ask why there cannot be others. But the manner in which these singularities arise is known: poles that do exist on the unphysical sheet in perturbation theory begin to move about as the coupling increases. Such a phenomenon does not happen on the physical sheets, except for the movement due to the change in energy of a bound state, since all the poles which exist on the physical sheet when the coupling constant is zero correspond to particles. Moreover, poles cannot cross from the unphysical sheet to the physical sheet without violating the unitarity condition, except in the case where a state just becomes bound. Thus, if non-perturbation singularities do develop on the physical sheet, the mechanism is something other than this.

10.2. The Double-dispersion Representation

For most two-particle scattering amplitudes, including pion–pion, pion–nucleon and nucleon–nucleon scattering, the domain of analyticity obtained under the assumptions of § 10.1 assumes a particularly simple form when states with more than two particles are neglected in the unitarity condition. The result is most easily visualized by recalling that the diagram, figure 6, represents three reactions and by treating them simultaneously and on an equal footing. From the point of view of pion–pion scattering, s (equation (8.1)) is the square of the energy and t minus the square of the momentum transfer between the pions or between the nucleons. The variable u will similarly be the square of the momentum transfer between an incoming pion and an outgoing nucleon or vice versa. With the other two reactions the variables will be interchanged and we may draw up the following table:

I	$A+B \to C+D$	$\pi+N \to \pi+N$	s energy	t, u momentum transfer
II	$A+\bar{D} \to C+\bar{B}$	$\pi+N \to \pi+N$	u energy	s, t momentum transfer
III	$A+\bar{C} \to \bar{B}+D$	$N+\bar{N} \to \pi+\pi$	t energy	s, u momentum transfer.

There are only two independent variables, s, t and u being related by (4.2).

The dispersion relation (8.9) may be rewritten

$$\mathscr{R}A(s,t) = P\frac{1}{\pi}\int_{(M+\mu)^2}^{\infty} ds' \frac{A_1(s',t)}{s'-s} + P\frac{1}{\pi}\int_{(M+\mu)^2}^{\infty} du' \frac{A_2(u',t)}{u'-u} + \frac{g^2}{M^2-s} + \frac{g^2}{M^2-u}.$$
$$\dots\dots(10.1a)$$

We have written $A_1(s',t)$ for $\mathscr{I}A(s',t)$. The ranges of integration in the two terms are the physical regions for the first and second reactions respectively. As we have seen, the expression A_2 in the second integrand is obtained by using the unitarity condition for the second reaction. In this particular case the first and second reactions are identical and $A_1(s,t) = A_2(s,t)$ but equation (10.1a) is true generally, even if all four particles are different.

The dispersion relation (10.1a) shows that A is analytic in s' or u' for fixed t, except for poles at $s = M^2$, $u = M^2$ and cuts along the real axis when s or u is greater than $(M+\mu)^2$. If multi-particle states are neglected in the unitarity equation, the analytic properties obtained as a function of two variables are the simplest possible generalization of these properties (Mandelstam 1959 a, b). A is analytic in its variables except for cuts when s, t or u is real and greater respectively than $(M+\mu)^2, (M+\mu)^2$ and $(2\mu)^2$, the energies of the lowest intermediate states in the corresponding reactions. A has also the usual poles when $s = M^2$, $u = M^2$. From these analytic properties one can use Cauchy's theorem to derive the 'double dispersion relation' analogous to ordinary dispersion relations.

$$\mathscr{R}A(s,t) = P\frac{1}{\pi^2}\int ds'dt' \frac{A_{13}(s',t)}{(s'-s)(t'-t)} + P\frac{1}{\pi^2}\int du'dt' \frac{A_{23}(u',t')}{(u'-u)(t'-t)}$$
$$+ P\frac{1}{\pi^2}\int ds'dt' \frac{A_{12}(s',u')}{(s'-s)(u'-u)} + \frac{g^2}{s-M^2} - \frac{g^2}{u-M^2}.$$
$$\dots\dots(10.2)$$

As usual we have assumed that 'subtractions' are unnecessary. This assumption is certainly incorrect and we shall write down modified relations below. We leave for the moment the discussion of the range of integration.

Dispersion Relations in Strong-coupling Physics

An equation similar to (10.2) but much more general had been postulated earlier by Nambu. This representation, however, is known to be invalid when in the lowest order of perturbation theory, whereas (10.2) is certainly valid for a large sub-set of the perturbation series at least.

The analytic properties of A as a function of two variables imply that, considered as a function of a single variable, it should be analytic in the cut place, and we would expect to be able to derive ordinary dispersion relations from (10.2). This is in fact the case. Let us define

$$A_1(s,t) = \frac{1}{\pi}\int dt' \frac{A_{13}(s,t')}{t'-t} + \frac{1}{\pi}\int du' \frac{A_{12}(s,u')}{u'-u}, \quad \ldots\ldots(10.3a)$$

$$A_2(u,t) = \frac{1}{\pi}\int dt' \frac{A_{23}(s,t')}{t'-t} + \frac{1}{\pi}\int ds' \frac{A_{12}(s',u)}{s'-s}, \quad \ldots\ldots(10.3b)$$

$$A_3(s,t) = \frac{1}{\pi}\int ds' \frac{A_{13}(s',t)}{s'-s} + \frac{1}{\pi}\int du' \frac{A_{23}(t,u')}{u'-u}. \quad \ldots\ldots(10.3c)$$

It is unfortunately impossible to adopt a notation symmetrical in s, t and u for the arguments of the functions without using three variables. We can then prove by direct substitution that $\mathscr{R}A$, defined by (10.2), is given by (10.1a) and also by the following alternative expressions

$$\mathscr{R}A(s,t) = \frac{1}{\pi}\int_{(M+\mu)^2}^{\infty} ds' \frac{A_1\{s',t(s',u)\}}{s'-s} + \frac{1}{\pi}\int_{(2\mu)^2}^{\infty} dt' \frac{A_3\{s(u,t'),t'\}}{t'-t} + \frac{g^2}{M^2-s},$$
$$\ldots\ldots(10.1b)$$

$$\mathscr{R}A(s,t) = \frac{1}{\pi}\int_{(M+\mu)^2}^{\infty} du' \frac{A_2\{u',t(s,u')\}}{u'-u} + \frac{1}{\pi}\int_{(2\mu)^2}^{\infty} dt' \frac{A_3(s,t')}{t'-t} + \frac{g^2}{M^2-u},$$
$$\ldots\ldots(10.1c)$$

where, for instance, $t(s',u)$ is defined by $s'+t(s',u)+u = 2M^2+2\mu^2$. This implies that, in the integral (10.1b), u is kept constant while in (10.1c) s is kept constant.

Let us first consider equations (10.1) when the constant parameter—respectively t, u and s—is negative. These are our ordinary dispersion relations for pion–nucleon scattering. In (10.1b), the momentum transfer between the incoming pion and the outgoing nucleon, rather than between the incoming and outgoing pions, is kept constant. The lower part of the range of integration then takes us into the physical region, not for reaction II ($u > (M+\mu)^2$) but for reaction III ($t > (2M)^2$). The imaginary part in this region has therefore been written A_3. It will not be equal to A_1 and A_2, but must be found from the unitarity condition for the annihilation of a nucleon–anti-nucleon pair into two pions. The lowest intermediate state is the two-pion state, so that the value $(2\mu)^2$ has been assigned to the lower limit of the t-integration.

The dispersion relations (10.1a), (10.1b) and (10.1c) are thus all on the same footing except for the altered kinematics resulting from the pion–nucleon mass difference. Owing to the kinematics a large part of the range of integration in (10.1b) and (10.1c) is in the unphysical region, in fact. Between s' (or u') = $(M+\mu)^2$ and s' (or u') = $2M^2+2\mu^2-u$, the cosine of the angle of pion–nucleon scattering is negative while, for the t'-integration, the range $(2\mu)^2 < t' < (2M)^2$ corresponds to a negative pair kinetic energy in the reaction $N+\bar{N} \to 2\pi$. For t' immediately above $(2M)^2$, the angle of scattering is in general unphysical. Thus these dispersion relations are not of direct use and have not been proved rigorously. For pion–pion scattering all three dispersion relations are really on an equal footing.

Before going further we can say something about the range of integration of the variables in (10.2). The absorptive parts A_1, A_2 and A_3, defined by (10.3), must have their lower limits at $(M+\mu)^2, (M+\mu)^2$ and $(2\mu)^2$. Thus the region where the function A_{13}, say, is non-zero must lie entirely within the domain $s > (M+\mu)^2, t > (2\mu)^2$, and the curve bounding it must touch these lines at least

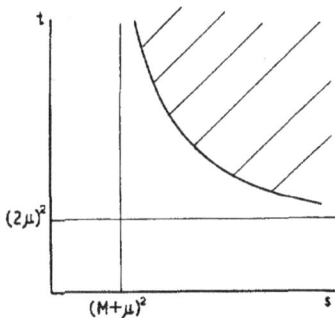

Figure 9. Region where the function A_{13} is non-zero.

at one point. Actually it turns out that this point is at infinity, so that the region has the form shown in figure 9. The equation for the bounding curve can be obtained from the unitarity condition (Mandelstam 1959 b). We shall see that, in this particular case, the region is bounded by two different curves.

Now let us return to consider equations (10.1) when the constant parameter is positive. To take (10.1c) with $s > (M+\mu)^2$, we are in the physical region for the reaction I, so that the equation expresses the analytic properties of the scattering amplitude as a function of the momentum transfer with the energy kept fixed. It implies that A is analytic in t except for cuts from $(2\mu)^2$ to ∞ and from $-\infty$ to $-s+(M-\mu)^2$, and a pole at $-s+M^2+2\mu^2$. (The last two values are obtained by putting $u = (M+\mu)^2$ and $u = (2\mu)^2$ in (8.2).) The physical region $-4q^2 < t < 0$ lies entirely between the two cuts. It is evident that the analytic properties in t given by the double dispersion representation are much more extensive than those that have been proved rigorously, where we only have analyticity in the Lehmann ellipse.

Equations (10.3) show that the absorptive part is also analytic as a function of the momentum transfer in the cut plane. The discontinuities across the cuts are given by $2i$ times the 'double spectral functions', A_{13}, A_{12}, A_{23}. In this case the cuts do not extend right to the limits $(2\mu)^2$ and $-s+(M-\mu)^2$ as, according to figure 8, there will be a region above the limits where the double spectral functions are zero. Also there are no poles in the absorptive parts. Thus both the conjectured and the rigorous domains of analyticity in the momentum transfer are greater for the absorptive part than for the whole scattering amplitude.

While the scattering amplitude is analytic in all its variables, the absorptive part is only analytic in the momentum transfer. It cannot be analytic in the energy, as it is zero for some values and non-zero for others.

Dispersion Relations in Strong-coupling Physics

Equation (10.2) has been obtained from the postulated analyticity properties on the assumption that the scattering amplitude tends to zero when either s or t tends to infinity. We were not really very consistent in writing it down, as the pole terms violate this assumption. If we allow the scattering amplitude to tend to a constant, the equation becomes

$$\mathcal{R}A(s,t) = P\frac{1}{\pi^2}\int ds'dt' A_{13}(s',t')\left\{\frac{1}{s'-s}-\frac{1}{s'-s_0}\right\}\left\{\frac{1}{t'-t}-\frac{1}{t'-t_0}\right\}$$
$$+P\frac{1}{\pi^2}\int du'dt' A_{23}(u',t)\left\{\frac{1}{u'-u}-\frac{1}{u'-u_0}\right\}\left\{\frac{1}{t'-t}-\frac{1}{t'-t_0}\right\}$$
$$+P\frac{1}{\pi^2}\int ds'du' A_{13}(s',u')\left\{\frac{1}{s'-s}-\frac{1}{s'-s_0}\right\}\left\{\frac{1}{u'-u}-\frac{1}{u'-u_0}\right\}$$
$$+P\frac{1}{\pi}\int ds' f_1(s')\left\{\frac{1}{s'-s}-\frac{1}{s'-s_0}\right\}+P\frac{1}{\pi}\int du' f_2(u')\left\{\frac{1}{u'-u}-\frac{1}{u'-u_0}\right\}$$
$$+P\frac{1}{\pi}\int dt' f_3(t')\left\{\frac{1}{t'-t}-\frac{1}{t'-t_0}\right\}+\frac{g^2}{M^2-s}+\frac{g^2}{M^2-u}+\lambda. \quad\ldots\ldots(10.4)$$

The last term λ is the subtraction constant analogous to the term $\mathcal{R}E(\nu_0)$ in (3.10), while the fourth, fifth and sixth terms may be regarded as subtraction constants in one variable but dispersion integrals over the other. The pole terms are actually special cases of these 'single dispersion integrals'.

If (4.10) holds, the corresponding subtracted dispersion relation for the absorptive part A_1 will be

$$A_1(s,t) = f_1(s) + \frac{1}{\pi}\int dt' A_{13}(s,t')\left\{\frac{1}{t'-t}-\frac{1}{t'-t_0}\right\}$$
$$+\frac{1}{\pi}\int du' A_{23}(s,u')\left\{\frac{1}{u'-u}-\frac{1}{u'-u_0}\right\}. \quad\ldots\ldots(10.5)$$

Thus knowledge of the absorptive parts is equivalent to knowledge of the double spectral functions and of the f's in (10.4), and it might be expected that one could rewrite the unitarity equations for the absorptive part as equations for the f's and the double spectral functions. This can in fact be done (Mandelstam 1959 b, also 1958). However, the overall constant λ cannot be obtained from unitarity, and is a fundamental constant of the theory.

The representation corresponding to (10.4) will take on slightly different forms depending on the assumptions made regarding the asymptotic behaviour of A. Now the asymptotic behaviour of A as both s and t tend to infinity is limited by unitarity, since the amplitude for each partial wave has the form $8\pi W e^{i\delta}\sin\delta/q$, with $\mathcal{I}\delta > 0$. From this it can be shown that λ is the only overall subtraction that can be allowed. Representations with subtraction terms such as λs or λt, which would result if one assumed that A became infinite as s and t become infinite, are inconsistent with the unitarity limitation. Further, if either of the particles has spin, no overall subtraction constant independent of s and t is possible at all.

We should emphasize that simple unitarity arguments alone do not provide a limitation on the behaviour of A as either s or t becomes infinite. Thus it may be that, in the correct form of the representation, terms such as

$$tP\frac{1}{\pi}\int ds' f_4(s')\left\{\frac{1}{s'-s}-\frac{1}{s'-s_0}\right\}$$

also appear. This would be the case if two subtractions had to be performed in the t-variable, as in (3.11). The number of subtractions necessary in single-variable dispersion relations is likewise not limited in any obvious way by unitarity.

The constant λ which has to be introduced in (10.4) corresponds to the constant in the $\lambda\phi^4$ term of the Lagrangian formulation. As a matter of fact, in that formulation one has to define the renormalized λ as the value of the pion–pion scattering

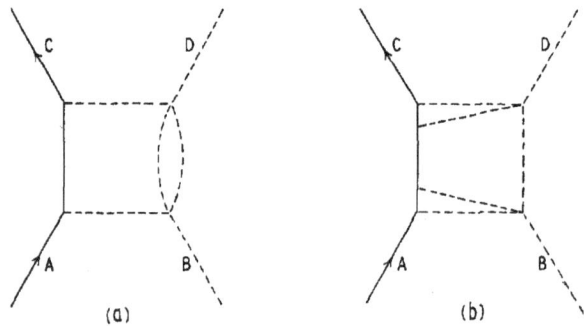

Figure 10. Typical Feynman diagrams for pion–nucleon scattering.

amplitude at a particular point, and this is just how it appears in (10.4). Four-field point interactions of particles with spin are not renormalizable; this fits in with the absence of overall subtraction constants in the double-dispersion representation for their scattering amplitudes.

We have seen that the curve bounding the region where the double spectral function A_{13} is non-zero approaches asymptotically the lines $s = (M+\mu)^2$,

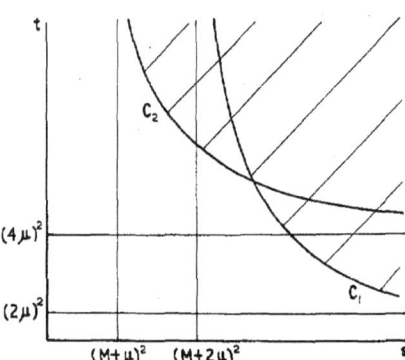

Figure 11. Double spectral function for π–N scattering.

$t = (2\mu)^2$. The corresponding curve for individual Feynman diagrams may approach higher thresholds. In figure 10(a) read upwards, the pion–nucleon scattering goes through an intermediate two-pion–nucleon state, whereas, read from the side, the pair-annihilation reaction goes through an intermediate two-pion state. The expected thresholds would accordingly be $s = (M+2\mu)^2$, $t = (2\mu)^2$, and the double spectral function would be non-zero in the region bounded by the curve

Dispersion Relations in Strong-coupling Physics

C_1 of figure 11. Similarly the spectral function from figure 10(b) would be bounded by C_2. Owing to the absence of a three-pion vertex it is impossible to draw a single diagram for which pion–nucleon scattering goes through an intermediate pion–nucleon state and the pair-annihilation relation goes through a two-pion state. The curves C_1 and C_2 therefore bound the region in which the complete spectral function is non-zero.

For pion–pion scattering, too, the spectral function will always be bounded by two curves. It is only in nucleon–nucleon scattering that there exists a diagram which, read both ways, goes through the lowest intermediate states $(2M)^2$ and

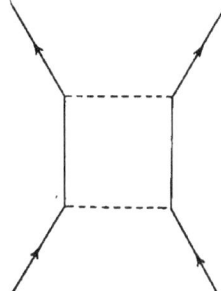

Figure 12. Feynman diagram for nucleon–nucleon scattering.

$(2\mu)^2$ (figure 12). The equations for the bounding curves for the three reactions have been given by Mandelstam (1959 b).

The double-dispersion representation has until now only been proved in perturbation theory (or on the assumption of maximum analyticity), and only with neglect of multi-particle intermediate states in the unitarity condition. Attempts to prove it quite generally in perturbation theory have been made by Eden (1960 a, b, 1961 a), and by Landshoff, Polkinghorne and Taylor (1961) (also Eden, Landshoff, Polkinghorne and Taylor 1961 a). These authors later found a gap in their proof which has not yet been filled (Eden, Landshoff, Polkinghorne and Taylor 1961 b). For non-relativistic scattering by superpositions of Yukawa potentials, rigorous proofs have been given by Blankenbecler, Goldberger, Khuri and Treiman (1960), Klein (1960) and Regge (1959, 1960); the last author also treats the asymptotic properties. We should stress, however, that the applications which have been made or which will be made with the double representation do not depend on the assumption that all the perturbation terms satisfy it, but only on the much weaker assumption that any other singularities, due to multi-particle intermediate states, are sufficiently far from the low-energy physical region. For certain strange-particle processes such as Σ–Σ scattering (Karplus, Sommerfield and Wichmann 1959) and Λ–N scattering (Eden, Landshoff, Polkinghorne and Taylor 1961 b) the double-dispersion relation is certainly invalid, for Σ–Σ scattering it fails even in fourth-order perturbation theory. Nevertheless, one can deal with the extra singularities using the unitarity condition, just as one can with the singularities given by the double dispersion representation, and their presence does not prevent one from doing calculations.

10.3. *Dispersion Relations for Fixed Angular-momentum States*

One technique for combining the analyticity properties with unitarity consists in treating the different partial waves separately. The reason why it is convenient to perform the calculations with a fixed partial wave, rather than, say, at fixed momentum transfer, is merely that the unitarity condition is so much simpler in the first case. Due to conservation of angular momentum the unitarity condition does not mix the different partial waves whereas, for fixed momentum transfer, the sum over intermediate states would be a sum over all other momentum transfers and would lead to a complicated integral equation.

Dispersion relations for fixed angular-momentum states will therefore be very useful. Though they are single-variable dispersion relations, they have not been proved rigorously, except that, for pion–pion scattering, the methods mentioned in § 8.4 enable analyticity to be proved in a limited region of the cut plane. In general, one must use the assumptions of this section, and the dispersion relations may easily be proved simply by expressing the partial waves in terms of the spectral functions using the double-dispersion relation (MacDowell 1959, Chew and Mandelstam 1960). As the kinematics is very much simplified by taking the equal mass case, we shall use pion–pion scattering instead of pion–nucleon scattering as our example. The relation is then

$$A^{(l)}(s) = A^{(l)}(s_0) + \frac{1}{\pi}\int_{(2\mu)^2}^{\infty} ds'\, \mathscr{I}A^{(l)}(s') \left\{\frac{1}{s'-s} - \frac{1}{s'-s_0}\right\}$$
$$+ \frac{1}{\pi}\int_{-\infty}^{0} ds'\, \mathscr{I}A^{(l)}(s') \left\{\frac{1}{s'-s} - \frac{1}{s'-s_0}\right\}. \quad \ldots\ldots(10.6)$$

Equation (10.6) is, of course, just a single-variable dispersion relation with one subtraction. With fixed partial waves, as distinct from fixed momentum transfer, we know that there is at most one subtraction, as we have seen that unitarity gives $A^{(l)}(s)$ a constant upper bound at infinity.

For the equal mass case, Bjorkén (1959) has shown that all terms in the perturbation series satisfy fixed partial-wave dispersion relations.

With non-equal masses, and in particular for pion–nucleon scattering, the range of integration also includes a circle in the complex s-plane with its centre at the origin (MacDowell 1959). However, the fact that the range of integration is not everywhere real alters nothing in principle.

As we have seen several times already, the unitarity condition only determines the function $\mathscr{I}A^{(l)}(s)$ occurring in the first integral of (10.6); the imaginary part occurring in the second must be determined by crossing. While the unitarity equation is much simpler for fixed partial waves than for fixed momentum transfer, the crossing relation is unfortunately more complicated. We start from the crossing condition (for neutral pions)

$$A(s,t) = A(t,s) \quad \ldots\ldots(10.7)$$

which is evident from the fact that changing from the reaction I to the reaction III, which is the same reaction, changes the role of s and t. From (10.7) the crossing relation for fixed partial waves is found to be (Chew and Mandelstam 1960)

$$\mathscr{I}A^{(l)}(s) = \frac{2}{s-4\mu^2}\int_{4\mu^2}^{-s+4\mu^2} ds'\, P_l\!\left(1+\frac{2s'}{s-4\mu^2}\right) \sum_{l'} \mathscr{I}A^{(l')}(s')\, P_{l'}\!\left(1+\frac{2s}{s'-4\mu^2}\right)$$
$$(s<0,\, l \text{ even}). \quad \ldots\ldots(10.8)$$

Dispersion Relations in Strong-coupling Physics

The equation gives us a formula for $\mathscr{I}A^{(l)}$ at negative values of s in terms of $\mathscr{I}A^{(l)}$ for positive s, as is required, but instead of being a point-to-point relation as was the case at fixed momentum transfer, the right-hand side is an integral over s'. More important, it contains the sum over all angular-momentum waves l'. In fact, the argument of the last Legendre function is sometimes less than -1 and it is not certain that the series converges. We have remarked that the question of convergence of a Legendre series depends on the analytic properties, which in this particular case can be determined from the double-dispersion representation. The

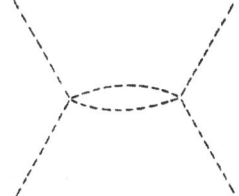

Figure 13. Exchange of two pions in pion–pion scattering.

expansion is then found to converge over the entire range of the s'-integration in (10.8) provided that $s > -32\mu^2$. As s goes below this point the series begins to diverge for a short section of the s'-integration, and the length of the section increases as s decreases further.

In our approximation scheme, we are prepared to use an expression for $\mathscr{I}A^{(l)}(s)$ which is invalid at large values of s, and we neglect, in any case, four-pion states in the unitarity condition, which come in at $s = (4\mu)^2$. Owing to the fairly high value of s at which the series begins to diverge, one would hope to get an adequate approximation by restricting the summation in (10.8) to a few small values of l'. The dispersion relations (10.6), the unitarity equation (5.1), and the crossing relation (10.8) then provide us with coupled integral equations for the problem. The difficulties of solution are not as great as they may seem and numerical results have been obtained without too much labour or computer time (Chew, Mandelstam and Noyes 1960, Chew and Mandelstam 1961, Desai 1961).

The termination of the partial-wave expansion in (10.8) has been discussed from a slightly different but essentially equivalent point of view by Chew and Mandelstam (1960) and by Cini and Fubini (1960). In this treatment the connection between the approximation and the neglect of distant singularities is perhaps more evident.

With processes other than pion–pion scattering where the crossed reactions are not the same as the direct reaction, the equation corresponding to (10.8) would contain, in the integral on the right, the imaginary part for the two crossed reactions.

The reason why the angular-momentum expansion in (10.8) converges to such a high value of s is that there are no three-pion vertices. In conventional terminology, pion–pion scattering must proceed by the exchange of two pions (figure 13); exchange of one pion is impossible. The range is then one-half of a pion Compton wavelength instead of one pion wavelength. In processes for which there is not this reduction in range—such as nucleon–nucleon scattering—the convergence of the series is much worse. Also, if we were to go to higher approximations, we would not be able to use a series which began to diverge at $-36\mu^2$. One therefore requires methods which do not depend on a partial-wave expansion. Moreover, it turns out that, if values of l' greater than zero are considered in (10.8), the integral equation becomes singular and a cut-off must be introduced. There are reasons for believing that, in the exact solution, no cut-off would be necessary.

A partial-wave expansion is clearly not adequate for a complete calculation, then. As formulae have been given for calculating the double-spectral functions (Mandelstam 1958, 1959 b, Cutkosky 1960 a, b, 1961), it is not difficult to improve on the partial-wave expansion. No calculations have yet been completed with the better approximation and we shall not expand upon this point. Let us only remark here that the neglect of higher partial waves in (10.8) is the first stage of a scheme, subsequent stages of which contain terms which are not partial-wave expansions.

As far as the lowest approximation is concerned (for pion–pion and pion–nucleon scattering), the double-dispersion representation is not used explicitly, but is referred to in the derivation of the partial-wave dispersion relations, and in finding the range of convergence of the partial-wave expansion in (10.8).

10.4. Higher Order Thresholds

The simplest types of singularities of transition amplitudes are the poles at the energies of single-particle intermediate states, and the cuts that begin at the lowest energies of the continuum of intermediate states. There may also be other

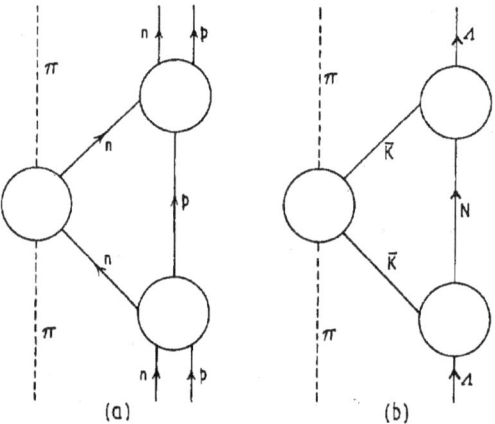

Figure 14. Reactions with second-order thresholds.

singularities of increasing orders of complexity, and the positions at which they occur can also be seen without much difficulty from the unitarity condition. Our treatment follows that of Cutkosky (1960 a, b, 1961).

The nature of the new singularities is most easily understood by examining a three-particle to three-particle reaction such as $\pi+p+n \to \pi+p+n$. The succession of reactions depicted in figure 14(a) can only take place (through real intermediate states) with conservation of energy and momentum if the momentum transfer of the pion is less than half the momentum of the neutron in the system where the proton is at rest. In terms of the invariants the condition is

$$t < -(1/M_p)^2 \{E_{pn}^2 - (M_p + M_n)^2\}\{E_{pn}^2 - (M_p - M_n)^2\}, \quad \ldots\ldots(10.9)$$

Dispersion Relations in Strong-coupling Physics

where E_{pn} is the relative energy of the proton and neutron. When 'side-reactions' such as the scattering of two of the particles can take place the unitary condition must be modified to take them into account. There will thus be a singularity at the point

$$t = -(1/M_p)^2 \{E_{pn}^2 - (M_p + M_n)^2\} \{E_{pn}^2 - (M_p - M_n)^2\}, \quad \ldots\ldots(10.10)$$

which is the limit of the region where the reactions of figure 14 can occur.

Now let us consider pion–deuteron scattering. If E_{pn} in (10.10) is replaced by M_D, the value of t becomes positive and the singularity moves into the unphysical region. Nevertheless, pion–deuteron scattering will show the same features as the scattering of a pion off an unbound nucleon pair, and the singularity will still be there. We are by now familiar with singularities in the unphysical region, as ordinary pole terms are examples of them.

If the binding energy of the particle increases, the singularity continues to exist. Thus, for pion–lambda scattering, the graph drawn in figure 14(b) has a singularity at

$$t = (1/M_N)^2 \{M_\Lambda^2 - (M_K + M_N)^2\} \{M_\Lambda^2 - (M_K - M_N)^2\}. \quad \ldots\ldots(10.11)$$

This value of t is less than $(2M_K)^2$, the beginning of the cut due to the intermediate $K\bar{K}$ state in the crossed reaction $\pi + \pi \to \Lambda + \bar{\Lambda}$. If one were to imagine the mass of the lambda particle decreasing to the point where $M_\Lambda^2 = M_K^2 + M_N^2$, the singularity given by (10.11) would occur at $t = (2M_K)^2$ and would coincide with the beginning of the cut. If the mass of the lambda particle decreased still further, the singularity given by (10.11) would no longer exist.

One could write a dispersion relation in t for pion–lambda scattering. As the scattering amplitude has a singularity—which turns out to be a branch point—at the value of t given by (10.11), it is necessary to introduce a cut along the real axis from this value of t upwards. The lower limit of the dispersion integral is therefore given by (10.11) instead of by $t = (2M_K)^2$ (if the intermediate states containing pions are ignored). This limit may be called a 'threshold of the second order' as contrasted with the thresholds of the first order at the energies of the lowest intermediate states. With the reaction of figure 14(a) the threshold of the second order occurs in the physical region. By means of processes in which more side-reactions take place successively, thresholds of the third and higher orders can be produced.

Thresholds of the second order were first noticed by Karplus, Sommerfield and Wichmann (1958, 1959), Oehme (1958) and Nambu (1958) in perturbation theory. They considered processes such as lambda–pion scattering rather than pion–two-nucleon scattering, and the origin of the new singularities appeared mysterious. They came to be known as 'anomalous thresholds'. By considering processes with deuterons, Oehme (1959) and Blankenbecler and Cook (1960) showed that the anomalous thresholds had to exist if the deuteron was to behave in any way realistically, and Mandelstam (1960 a) showed that they would appear of their own accord in a calculation of a relevant process from dispersion relations and unitarity. By pointing out that these thresholds often occurred in the physical region and that their position could be seen by inspection of the unitarity condition, Cutkosky finally cleared up the whole subject and gave them a simple physical interpretation.

In a two-body scattering process with thresholds of the second order, the double spectral function is non-zero in a region such as is shown in figure 15. The first-order threshold is represented by t_t, the second-order threshold by t_s. Certain processes such as sigma–sigma scattering or deuteron–deuteron scattering have third-order thresholds (Karplus, Sommerfield and Wichmann 1959, Mandelstam

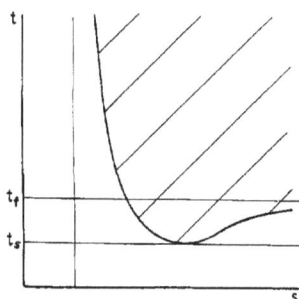

Figure 15. Double spectral function with a second-order threshold.

1959 a, Tarski 1960). Singularities then occur in the scattering amplitude in the complex (s, t) space and the double-dispersion representation no longer holds in its simple form.

§ 11. Unstable Elementary Particles

The statement which we have made several times, that our system of equations can be solved to give unique scattering amplitudes, needs modification. Castillejo, Dalitz and Dyson (1956) showed that an infinite number of solutions exist in the charged scalar theory without recoil, and this has since been found to be a very general feature of the equations in dispersion theory. The physical postulates of the non-Lagrangian approach to field theory have to be supplemented by an additional postulate if they are to yield a unique solution.

The origin of the Castillejo–Dalitz–Dyson ambiguity can best be seen by going back to the Lagrangian formalism. We remarked that, given the types and masses of the particles and the coupling constants, the Lagrangian is determined fully. Besides the stable particles, one also has to specify the unstable elementary particles. For instance, one could envisage a theory which contained, in addition to pions and nucleons, a baryon of zero strangeness whose mass was greater than $M + \mu$. If this baryon interacted strongly with the other particles it would not be seen, as it would decay into a pion and a nucleon with a lifetime of the order of magnitude of 10^{-23} sec. Nevertheless, the Lagrangian of the theory would contain operators for the creation and annihilation of the heavy baryon. It would therefore be quite different from the ordinary Lagrangian in pion–nucleon physics and would lead to different predictions. One therefore has to find some way of introducing the information that the system contains a given number of elementary particles. The assumption is made that, if the systems were to be placed in a box, the energies of the high-level states would not be shifted appreciably by the interaction. This is equivalent to the demand that the scattering phase shifts tend to zero or, more generally, to a multiple of π, at high energies. In the approximations which have

Dispersion Relations in Strong-coupling Physics

been solved, the phase shifts do so, albeit only logarithmically. One does not know the high-energy behaviour of the exact theory, and the methods we shall propose will only be rigorous if the phase shifts tend to zero. They are certainly adequate in the problems solved to date.

If a system of uncoupled particles is placed in a box, one can calculate the number of states below a certain energy, and this number will, of course, depend on the number of elementary particles. With the coupled system we *define* the number of elementary particles—stable or unstable—according to the number of states below a given energy. This definition is reasonable from the physical point of view. In theories with finite Lagrangians (which cannot be Lorentz-invariant!) and in which phase shifts tend to zero at infinity, the number of elementary particles defined in this manner is the same as the number which must be introduced into the Lagrangian.

The connection between the number of unstable particles and the number of states can be re-expressed as an equation between the number of unstable particles and the phase shift. The relation between the density of states and the phase shift is familiar in statistical mechanics and leads to the result

$$N_U - N_B = (1/\pi)\{\delta(\infty) - \delta(0)\}, \qquad \ldots\ldots(11.1)$$

where N_U is the number of unstable particles of a given quantum number, N_B the number of bound states, and $\delta(\infty)$ and $\delta(0)$ the phase shifts at infinity and at threshold. As long as $\delta(\infty)$ is an integral multiple of π, (11.1) is a perfectly rigorous consequence of the definition of the number of unstable particles.

In practice, above the threshold for production, scattering can take place through an infinite number of channels. The quantity δ must then be defined by summing the logarithms of the eigenvalues of the S-matrix with the appropriate quantum number (Ruderman and Sommerfield, private communication).

Equation (11.1) was first proved by Levinsohn (1949) for scattering by suitably restricted potentials.

The character of the Castillejo–Dalitz–Dyson solutions just fits in with these considerations. In their simplest solution (when there is no bound state) the phase shift at infinity is the same as it is at zero, so this corresponds to the absence of unstable particles. In their next solution the phase shift changes by π between zero and infinity. This solution contains two parameters in addition to the usual constants. It evidently corresponds to the presence of one unstable particle, and the two parameters represent its mass and the strength of its coupling with the other particles. The higher solutions show a corresponding behaviour.

The Castillejo–Dalitz–Dyson ambiguity was investigated for a soluble model by Dyson (1957). In this model the correspondence between the solutions and the number of unstable particles—or, in his terminology, the number of excited states of the scatterer—is as expected.

When the coupling is very weak, the phase shifts of the second Castillejo–Dalitz–Dyson solution would behave as in figure 16. This is a typical behaviour which one would expect if the pion and nucleon could combine to form a particle at an energy E_1.

Equation (11.1) provides in principle a distinction between an elementary particle and a bound state. If there existed two particles with certain quantum numbers one of which was elementary, the equation would not distinguish between them and, further, it would allow one to call an elementary particle a bound state provided one assumed an unstable elementary particle to exist. In other words, it only gives a meaning to the total number of elementary particles with a given quantum number. In this respect it is no different from a Lagrangian theory.

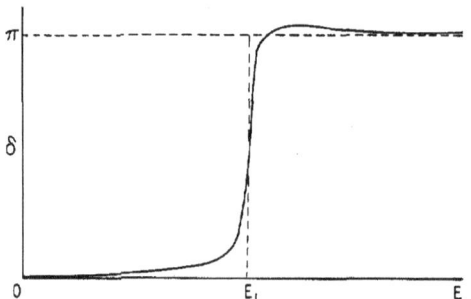

Figure 16. The second Castillejo–Dalitz–Dyson solution with weak coupling.

When solving our system of equations, the number of masses and coupling constants which one has to introduce as free parameters turn out to correspond to the number of elementary particles. With nucleon–nucleon scattering, for instance, the mass of the deuteron does not have to be introduced. The equations only have a solution provided a pole of a certain strength is present at a certain point; this is the pole due to the single-deuteron intermediate state.

In practice it would be impossible to find directly from experiment whether or not a given particle was elementary, as a measurement of $\delta(\infty)$ is not feasible. The use of (11.1) is to resolve the Castillejo–Dalitz–Dyson ambiguity when solving scattering problems. Outside strange-particle physics there is usually no doubt whether a particular particle is elementary. Both the deuteron and the 3–3 pion-pion resonance appear naturally in calculations or phenomenological investigations and it is presumed that they are not elementary. The suggestion of Fermi and Yang that the pion is not elementary and that there exists an unstable elementary vector particle is much harder to verify or disprove—except that, in current approximation schemes, systems with elementary vector particles do not appear to have solutions. This is in line with the fact that corresponding Lagrangians are not renormalizable.

When the number of unstable particles in a theory is specified, the Castillejo–Dalitz–Dyson ambiguity no longer exists. We do not know whether unstable elementary particles occur in nature; there is no physical reason against them. Calculations so far performed make the simplest assumption, that there are no unstable particles, and try to compare its consequences with experiment. The necessity of having to make such an assumption is of course not confined to dispersion-relation calculations. A feature which may be taken as experimental evidence on the subject is that such particles would be expected to exhibit themselves as resonances, and, in current approximation schemes, they would have to be S-state resonances for the equations to be soluble. The resonances that have been seen and analysed do not appear to be in an S-state, except when there are additional channels which could produce resonances that are not elementary particles.

Dispersion Relations in Strong-coupling Physics

§ 12. APPLICATIONS OF DISPERSION RELATIONS TO DYNAMICAL CALCULATIONS

12.1. *Pion–Pion Scattering*

Calculations of pion–pion scattering have been performed by Chew and Mandelstam (1960, 1961), Chew, Mandelstam and Noyes (1960) and Desai (1961). There were several reasons for adopting this as the first problem to be treated. The fact that the two crossed reactions are the same as the direct reaction means that we do not get coupling of the equations with those for other processes, and

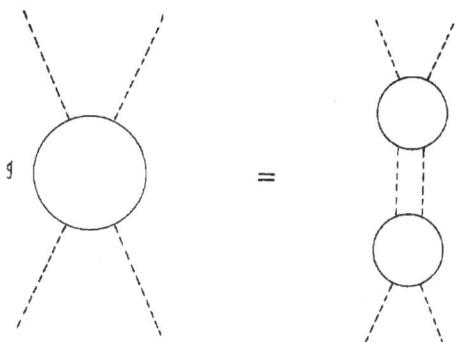

Figure 17. Unitarity approximation for π–π scattering.

the absence of spin also simplifies the problem greatly. Moreover, a knowledge of the pion–pion scattering amplitude is needed for a calculation of all other processes.

In the unitarity equation (5.1), the lowest intermediate state is the two-pion state itself. We may therefore write the unitarity equation

$$\mathscr{I}A_{\pi\pi\to\pi\pi} = kA^*_{\pi\pi\to\pi\pi} A_{\pi\pi\to\pi\pi}. \quad\quad\ldots\ldots(12.1)$$

If we are dealing with states of given isotopic spin and angular momentum, there is only one possible intermediate state and the summation sign in (5.1) disappears. Equation (12.1) may be represented diagrammatically as in figure 17. The reactions on the right are also pion–pion scattering. Thus the unitarity condition, the dispersion relations and the crossing relations involve pion–pion scattering only, and we get a system of equations whose solution is not very difficult. For every even value of the angular momentum there are two states involved, with $T = 0$ and $T = 2$; for an odd angular momentum there is only the $T = 1$ state.

Chew, Mandelstam and Noyes (1960) found one class of solutions which depended on a parameter λ.† The nature of this parameter was discussed in § 10; it is a fundamental constant of the theory corresponding to the term $\lambda\phi^4$ in the Lagrangian formalism. Scattering in all but the S-states was extremely small. The scattering in both S-states was repulsive for $\lambda > 0$, attractive for $\lambda < 0$, that in the $T = 0$ state being the larger. Solutions were obtained for a range of values

---

† There are three isotopic spin states involved so, according to § 10, three parameters might be expected. However, we can reduce the number to one by defining λ as the scattering amplitude (divided by 16π) at the unphysical point $s = -4\mu^2/3$, $u = -4\mu^2/3$. At this point $s = t = u$ according to (8.2) so that relations between the three amplitudes can be obtained from the crossing relation. It turns out that $\lambda_0 : \lambda_1 : \lambda_2 = 5 : 0 : 2$.

of λ between $-0{\cdot}5$ and $0{\cdot}3$. Below this range there was a bound $T = 0$ state, above it consistent solutions could not be obtained. The solutions were rather featureless.

Experimental evidence seems to indicate that there is a fairly narrow resonance in P-wave pion–pion scattering at a centre-of-mass energy, including both rest-masses, of about $4{\cdot}5\mu$. The original evidence came from the dispersion-relation calculation of nucleon electromagnetic structure, which will be treated below. The calculations disagreed totally with experiment unless such a resonance was postulated. Calculations of pion–nucleon scattering also seem to indicate a pion–pion resonance, but here the evidence is much weaker. There has recently been direct evidence of such a resonance. Pickup, Ayer and Salant (1960) and Rushbrooke and Radojičić (1960) have collected the experimental information on the reaction $\pi^- + p \to \pi^- + \pi^0 + p$ and have plotted a graph of the distribution of events against the Q-value of the outgoing pions; they find quite a marked peak just at the predicted value of $4{\cdot}5\mu$.

If one accepts the existence of a resonance, the solutions found by Chew, Mandelstam and Noyes are not those that occur in nature. Chew and Mandelstam (1961) have investigated the possibility of other solutions of their equations, and they found that solutions with a P-wave resonance can indeed exist. In this case their equations cannot be solved unless one introduces a high-energy cut-off. The necessity for this cut-off is probably due to the approximations made, which are bad at high energies, and a cut-off should appear naturally in a more complete calculation. At the moment, however, the position of the cut-off has to be introduced as an empirical constant. It is one of the phenomenological parameters mentioned in § 7 which compensates for our lack of knowledge of the high-energy region.

The position of the cut-off was arranged to give the resonance in its experimental position. The width of the calculated resonance, which did not depend sensitively on λ, was of the right order of magnitude, but rather too large. Approximations were made in solving the equations, and a better solution would probably yield a resonance which was narrower but still not narrow enough. It would be interesting to see whether the inclusion of more processes in the equations has the effect of narrowing the resonance.

If there is an appreciable amount of P-wave scattering the solutions for the S-wave also become modified, as the crossing relations connect the two partial waves. Desai (1961) has taken the experimental parameters for the P-wave scattering and has obtained solutions for the S-wave phase shifts. They differ somewhat from the case of Chew, Mandelstam and Noyes where there is no P-wave scattering. Again the range of λ for which solutions with no bound state could be obtained was $-0{\cdot}5 < \lambda < 0{\cdot}3$. For negative λ the $T = 0$ state is quite strongly attractive, the $T = 2$ state is weakly attractive and may even become repulsive at higher energies. For positive λ both states are repulsive.

Recently there has been some experimental evidence on S-state pion–pion scattering. Abashian, Booth and Crowe (1960) have investigated the spectrum of He_3 nuclei produced in the reaction $p + d \to He_3 + 2\pi^+$ or $p + d \to He_3 + \pi^+ + \pi^-$. They found slight but quite definite peaking of events towards the high-energy end of the He_3 spectrum, or, in other words, at low relative energies of the pions.

Dispersion Relations in Strong-coupling Physics

This would occur if there was an attractive force between the pions in an S-state. In the reaction $p + d \to H_3 + \pi^0 + \pi^+$ no peaking is observed, so that the attraction is probably in the state of isotopic spin zero. Desai found that values of λ between -0.15 and -0.2 would give the correct enhancement. This corresponds to a scattering length of $2\mu^{-1}$–$3\mu^{-1}$ in the $T = 0$ state. One cannot be certain that the entire effect is due to a final-state pion–pion interaction, as is assumed in the calculations, but it is at least reasonable that this accounts for a substantial portion. As the pions are not observed directly in the Abashian–Booth–Crowe experiment, an alternative explanation of the peak would be that a new neutral particle is being produced. There are difficulties in this explanation which, in any case, now appears unnecessary.

Only brief mention can be made here of an ambitious programme proposed by Chew and Frautschi (1960, 1961), K. Wilson (1961)† and McCauley (unpublished). Their scheme takes much more into account than any previous calculations, and one might hope to avoid having to introduce a cut-off altogether and to obtain the solutions in terms of one parameter λ only. Quite apart from the amount of computation involved, their approach still contains difficulties in principle, especially with regard to asymptotic behaviour of the scattering amplitude in the unphysical region. There is no reason to suppose that these difficulties cannot be overcome, and we would then have a powerful method of attack on scattering problems by means of dispersion relations.

Some people have raised objections to the fact that there exist two sets of solutions of the pion–pion integral equations without any means of preferring one to the other *a priori*. We do not feel that this is a serious objection, however, since in any case the solutions depend on a parameter which can assume a range of values and which is not fixed *a priori*. An additional doubling of the number of solutions should not alter things in principle. If one were able to obtain both sets of solutions in terms of one parameter, it would surely be possible to redefine the parameter so that it assumes a different range of values in each set. The value of the parameter would then determine the solution uniquely. To avoid risk of confusion we may remark that this ambiguity has nothing to do with the Castillejo–Dalitz–Dyson ambiguity discussed in the previous section.

To summarize the results obtained for pion–nucleon scattering: The integral equations predict a P-wave resonance, but they cannot yet determine its position nor do they give a good value for its width. Given the P-wave scattering the S-wave equations can be solved in terms of one parameter λ. One can then fix λ from the Abashian–Booth–Crowe experiment. There are at the moment no further experiments, direct or indirect, with which to compare the S-wave solutions.

12.2. Pion–Nucleon Scattering

The two reactions obtained from pion–nucleon scattering by crossing are pion–nucleon scattering again and the reaction $N + \bar{N} \to \pi + \pi$. The crossing relation couples the two reactions, and they have to be treated simultaneously.

The unitarity equation obtained by neglecting higher intermediate states is given in diagrammatic form in figure 18. The processes on the right of figure 18(*a*) are the pion–nucleon vertex, which is just the coupling constant g, and pion–nucleon

† Harvard preprint.

scattering again. On the right of figure 18(b) the processes are the pair-annihilation reaction again and pion–pion scattering. It is at this point that a knowledge of pion–pion scattering is necessary for the calculation of pion–nucleon scattering.

The equations do not enable one to calculate the transition amplitude for the reaction $N + \bar{N} \to \pi + \pi$ as a physical process. In the unitarity condition we have included the two-pion intermediate state but have neglected four-pion and higher states. The calculations will therefore be accurate only at energies below 4μ, the four-pion threshold. The reaction is not a physical process until the energy reaches the value $2M$. Nevertheless, the value of the transition amplitude in the unphysical region is needed as it occurs in the crossing relation for pion–nucleon scattering. Strictly speaking one is not permitted to use a unitarity equation in an unphysical region, but in this case it can be justified, at any rate within the framework of our approximations (Mandelstam 1960 a).

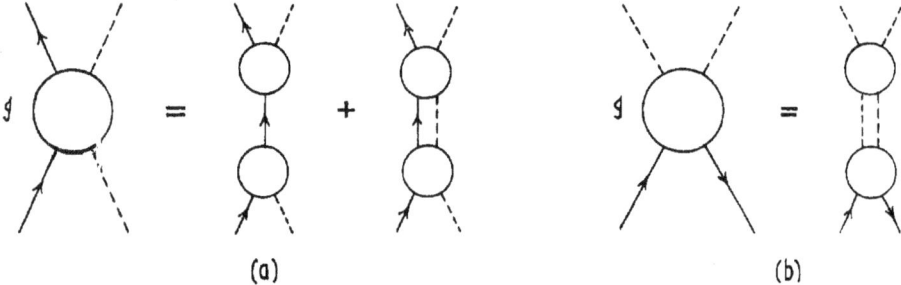

Figure 18. Unitarity approximation for π–N scattering.

The pion–nucleon integral equations have been treated by Frautschi and Walecka (1960) and, somewhat less ambitiously, by Bowcock, Cottingham and Lurié (1960 a, b, 1961). Kinematic relations have also been worked out by Frazer and Fulco (1960 c). Frautschi and Walecka found the expected resonance in the 3–3 state, but at rather too low an energy—just above threshold, in fact. The resonance could in principle have appeared below threshold as a bound state and, if we somewhat arbitrarily place its lower limit at the one-nucleon pole, we conclude that the calculations underestimate its energy by a factor of one half. By putting in an extra phenomenological constant one can arrange it to be in the correct position. The width then comes out correctly but, as the relation between the position and the width is given by the effective-range formula (§ 6.2), this is not significant. The shape does not agree particularly well with experiment, as the calculated scattering amplitude does not fall off sufficiently fast beyond the resonance. According to (6.2) a plot of $q^3 \cot \delta_{33}/\omega$ against ω should give a straight line. The experiments give a slightly falling curve whereas the calculations produce a slightly rising curve. However, Frautschi and Walecka showed that the inclusion of production processes would give an effect in the right direction and that the amount of production needed to give agreement with experiment was reasonable.

Bowcock, Cottingham and Lurié took the experimental parameters for the pion–nucleon resonance and calculated the small P-wave phase shifts. In order to avoid having to know the S-wave pion–pion scattering they treated only the difference $\delta_{11} - \delta_{31}$. The experimental results give a few degrees for this quantity at about 200 MeV, whereas previous calculations all gave about $-10°$. It was shown that the P-wave pion–pion resonance brought the results into agreement with experiment. A similar calculation by Frautschi, using a pion–pion resonance at the rather lower energy suggested by the calculations on the electromagnetic structure of nucleons, gave too large an effect. As a result the parameters used by Bowcock et al., with the pion–pion resonance at $4\cdot 5\mu$, are to be preferred. The nucleon-structure calculations are consistent with these parameters and, in fact, recent experimental data on nucleon structure support them. The direct evidence of Ayer, Pickup and Salant and of Rushbrooke and Radojičić also indicates a pion–pion resonance at 4–$4\cdot 5\mu$.

The value of the S-wave scattering lengths at threshold has not yet been reproduced in a reliable calculation.

12.3. Electromagnetic Structure of Nucleons

Much experimental effort has been put into the determination of the electromagnetic structure of nucleons by Hofstadter and his collaborators and, more recently, by R. R. Wilson and his collaborators. By scattering electrons off

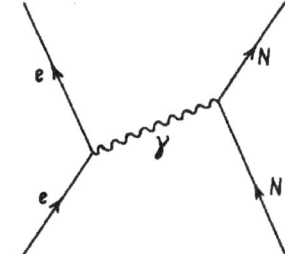

Figure 19. Exchange of one photon in electron–nucleon scattering.

protons and deuterons at various energies and measuring the differential cross section as a function of momentum transfer, one can obtain two 'form factors', the 'charge' and 'magnetic-moment' form factors, both of which are functions of the momentum transfer. They are the Fourier transforms of the charge and magnetic moment distributions of the nucleon. Instead of considering the form factors of the proton and neutron, it is more convenient for theoretical purposes to take the difference and the sum of the form factors, which are respectively a vector and a scalar in isotopic spin space. For reasons which will become evident presently, practically all the calculations have dealt with the isotopic vector form factor.

Both the charge and magnetic-moment isotopic vector form factors are observed to fall off as the momentum transfer increases, and recently the magnetic-moment form factor has been observed to change sign at high momentum transfers. At zero momentum transfer the charge form factor is just one half the electric charge and is not calculable, but the magnetic-moment form factor is the anomalous magnetic moment which should be obtainable by our methods.

In the approximation that the electromagnetic coupling is weak—and it is perhaps refreshing that we can be reasonably certain beforehand that here at least we have a good approximation—the process that takes place is given in figure 19.

An electron emits a γ-ray which is then absorbed by the nucleon, or the nucleon emits the γ-ray and it is absorbed by the electron. The electron–γ vertex is just the electric charge or the electron's magnetic moment, so that the process in which we are interested is

$$N + \gamma \to N.$$

The photon is 'virtual'; as it occurs only in an intermediate state the square of its momentum need not be zero, and is just the square of the momentum transfer in

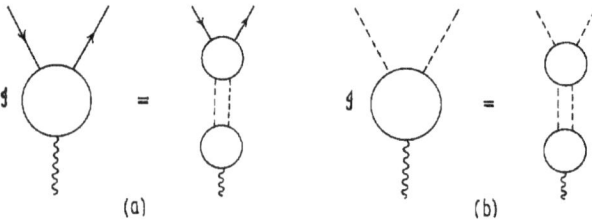

Figure 20. Unitarity approximation for the process $\gamma \to N + \bar{N}$.

the electron–nucleon scattering. We must thus investigate the process $N + \gamma \to N$ as a function of the square of the momentum of the photon. By crossing symmetry, we may consider instead the process

$$\gamma \to N + \bar{N}.$$

We have generally considered dispersion relations for scattering, but the amplitude for this process will satisfy a dispersion relation in the square of the momentum of the photon. Like most of the dispersion relations we have used, it has not been proved rigorously, but it is satisfied in any order of perturbation theory (Nambu 1957 b).

The unitarity approximation is given in figure 20(a). Only the intermediate two-pion state is taken into account. As it may easily be shown that intermediate states with an even number of pions occur in the isotopic vector form factor and states with an odd number of pions in the isotopic scalar form factor, this approximation can tell us nothing about the latter. One of the reactions in figure 20(a) is the process of $\gamma \to \pi + \pi$, which is the form factor of the pion. One also has to treat this process, therefore, and the unitarity approximation is given in figure 20(b). Now all the reactions on the right have either appeared in previous calculations or represent the quantities we are trying to calculate, and integral equations can be set up. The amplitude for pion–pion scattering is required, as it occurs directly in figure 20(b), and it is also necessary for the calculation of the process $\pi + \pi \to N + \bar{N}$, which occurs in figure 20(a). The amplitude for this last process is needed in the same unphysical region for which it had to be calculated in the previous problem.

When nucleon structure was first treated by means of dispersion relations nothing was known about pion–pion scattering. Chew, Karplus, Gasiorowicz and Zachariasen (1958) and also Federbush, Goldberger and Treiman (1958) performed calculations on the assumption that pion–pion scattering did not occur. The magnitude of the anomalous magnetic moment which they predicted was somewhat uncertain and there was no definite discrepancy with experiment. The shape of the form factors, however, disagreed badly with the data. As the momentum

Dispersion Relations in Strong-coupling Physics

transfer increased from zero, they did not fall off nearly fast enough. In other words, the theory predicted much too small a value for the charge and magnetic moment radii of the nucleon. The situation at the time was summarized by Drell (1958) who showed that the contribution from the low-energy intermediate states was underestimated by a factor of about five.

According to the approximation scheme, it is just the contribution from low-energy intermediate states which one should be able to calculate reliably, yet the calculations disagreed violently with experiment. Chew therefore suggested that the assumption of no pion–pion scattering was not correct and that there was probably a resonance in the P-wave, which was the only relevant state. The integral equations for the electromagnetic form factors were solved under this assumption by Frazer and Fulco (1959, 1960 a, b ; see also the lectures by Frazer in Screaton 1961). It looks as though we have quite a complicated system of equations, but, if the amplitudes for pion–pion scattering and pion–nucleon scattering are assumed known, a solution may be obtained analytically. Frazer and Fulco took the pion–nucleon scattering amplitude from experiment and showed that the calculations on nucleon structure would agree with the data provided that there was a fairly narrow resonance in P-wave pion–pion scattering at an energy of about $2 \cdot 5\mu - 3\mu$. We have observed that calculations on pion–pion scattering indicate a rather higher resonance energy—about $4 \cdot 5\mu$—and this still gave results not inconsistent with experiment. The fact that the magnetic form factor changes sign at high energies actually lends support to the higher value of the resonance energy.

The form factor of the nucleon was recalculated by Ball and Wong (1961) who assumed a resonance shape more like that predicted by Chew and Mandelstam, but their results did not differ significantly from those of Frazer and Fulco.

Not much theoretical work has been done on the isotopic scalar form factors of the nucleon. The lowest intermediate state involved is the three-pion state, but calculations involving three-particle intermediate states are difficult and have not yet been performed. Approximations will very probably be necessary. Experimentally the isotopic-scalar magnetic-moment form factor is small, but the isotopic-scalar charge form factor is fairly large. In fact, low-energy neutron–electron scattering indicates that the charge radius of the neutron is extremely small, so that the isotopic vector and isotopic scalar form factors balance one another. It thus appears that the three-pion intermediate state ($J = 1, T = 0$) is contributing appreciably. Nambu (1957 a) proposed that a new particle with these quantum numbers might exist. Chew (1960) suggested that it would be sufficient if there were a three-pion resonance, and pointed out that three pions could be in a $J = 1$, $T = 0$ state with each pair in the resonant $J = 1, T = 1$ state. The existence of a resonance was thus very feasible. Further, in a state with more than three pions, each pair could not be in a $J = 1, T = 1$ state, so that one could reasonably expect the multi-pion states to be less important than the two- and three-pion states.

Note added in proof.—

Convincing evidence has now been obtained for resonances in both the $J = 1$, $T = 1$ state, at $5 \cdot 3\mu$, and the $J = 1, T = 0$ state, at $5 \cdot 6\mu$ (Erwin *et al.* 1961, Pickup, Robinson and Salant 1961, Maglić *et al.* 1961, Pevsner *et al.*, to be published). Suspicions of further resonances are appearing.

12.4. Nucleon–Nucleon Scattering

The two reactions obtained by crossing in nucleon–nucleon scattering are both nucleon–anti-nucleon scattering, so that this process has to be treated together with nucleon–nucleon scattering. The unitarity approximations will be given by figure 21.

The intermediate processes on the right of figure 21(b) are $N+\bar{N}\to\pi+\pi$, the amplitude for which had to be calculated in treating pion–nucleon scattering. A

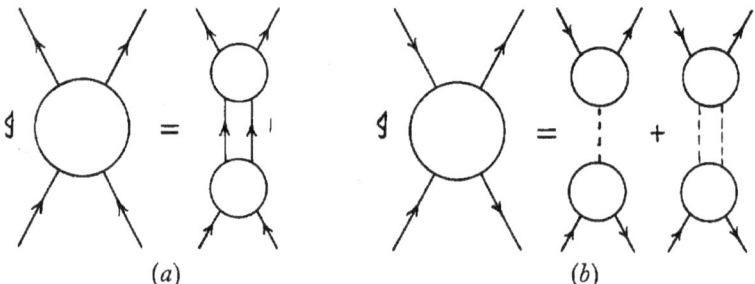

Figure 21. Unitarity approximation for the reactions $N+N\to N+N$ and $N+\bar{N}\to N+\bar{N}$.

knowledge of pion–pion scattering is thus necessary. Again, we only calculate the amplitude for nucleon–anti-nucleon scattering at unphysical energies below 4μ.

A preliminary calculation of the nucleon–nucleon phase shifts was made by Noyes and Wong (1959) who only took the first diagram in figure 21(b). The vertices are again given by the pion–nucleon coupling constant g, and a knowledge of pion–pion scattering is unnecessary. The approximation corresponds to allowing only one-pion exchange in the conventional formalism. As it is known that the one-pion exchange term does not give sufficient attraction, it is not surprising that a calculation with no phenomenological parameters fails to agree with experiment. If one parameter is inserted to give the correct scattering length the effective range is underestimated by about 60%. This again is to be expected, since the calculation essentially replaces the neglected forces by a zero-range force. If two phenomenological parameters are inserted to give the correct effective range and scattering length, the shape-dependent parameter obtained has the opposite sign to that which occurs in most potential models, $k\cot\delta$ decreasing below the straight line instead of increasing. Noyes claimed that preliminary experimental evidence supports his conclusion; this has been disputed by other physicists, who point out that it is very difficult to get a shape-dependent parameter of the type suggested by Noyes from a potential, even a velocity-dependent potential, with a repulsive core.

Wong (1959) has also calculated the ratio of D- to S-state in the deuteron and finds quite good agreement with the semi-phenomenological values. Noyes (1960) has recently proposed a method for using knowledge of the one-pion exchange term in a phenomenological analysis of phase shifts.

Equations with both terms of figure 21(b) included have been formulated by Goldberger, Grisaru, MacDowell and Wong (1960) and by Amati, Leader and Vitale (1960 a, b, c). In this case, unlike that with pion–pion and pion–nucleon scattering, one cannot use a partial-wave expansion in the crossing

Dispersion Relations in Strong-coupling Physics

relation. The reason is that nucleon–nucleon scattering can take place with the exchange of one pion, and the range of the force is μ^{-1} instead of $\tfrac{1}{2}\mu^{-1}$. However, if the diagram shown in figure 12 is computed explicitly, it can be shown that the use of a partial-wave expansion for the remainder of the amplitude is as legitimate here as it is in the other two processes. The expression for the scattering amplitude corresponding to figure 12 has been considered by P. Cziffra (1960),[†] who has worked out the double spectral function.

Taking two-pion intermediate states into account in the nucleon–anti-nucleon scattering is roughly equivalent to considering two-pion exchange in the conventional formalism, but the details of the approximations differ considerably. It is unlikely that one will obtain the repulsive core from the calculation, as Sakurai has pointed out that a core due to an even-pion intermediate state would have a different sign for different states. One will probably have to put in the core phenomenologically—for instance, one might require that the S-wave phase shifts change sign at the correct point. One would then hope to get other features of the low-energy scattering from the equations, but no calculations have yet been completed to date.

12.5. Other Applications

Applications of dispersion relations to photo-production of pions on nucleons have been treated by Ball (1960) and by Wong (1960). Photo-production of pions on pions, which occurs at an intermediate stage, has been considered by Gourdin and Martin (1960 a), who have also examined pion–photon scattering (1960 b).

As yet dispersion relations have not been applied in strange-particle physics to any great extent, mainly because of the numerous possible intermediate states. Scattering of pions from K-mesons—which will have to be treated before other calculations are performed—has been studied by Lee (1960), Lee and Cho (1961) and Gourdin, Noirot and Salin (1960). K-nucleon scattering has been examined by Ferrari, Frye and Pusterla (1960 a, b, c) and by Lee (1961). An effective-range formula has been derived, but otherwise many concrete results have not thus far been obtained.

The effects of strong interactions on weak interactions have also been studied using dispersion relations. Goldberger and Treiman (1958 a) derived a formula for the decay rate of the pion on the assumption that it proceeded through a universal Fermi interaction. Although they made severe approximations, they obtained remarkable agreement with experiment, and many physicists feel that their fairly simple formula is a consequence of some symmetry principle. Goldberger and Treiman have also studied the problem of form factors in β-decay (1958 b).

§ 13. Concluding Remarks

The results that have so far been obtained by using dispersion relations in dynamical calculations are fairly few in number but, if further experimental and theoretical work supports them, we feel they are significant. Perhaps the most important is the prediction that the P-wave pion–pion phase shift passes through a resonance at an energy of about $4\cdot5\mu$. This prediction was based on calculations of nucleon structure and, to a lesser extent, of pion–nucleon scattering from field

† *Thesis*, Lawrence Radiation Laboratory, preprint U.C.R.L. 9249.

theory, and was the first quantitative prediction that field theory has been able to make about a hitherto unmeasured quantity. We cannot be sure at this stage that the direct experimental evidence presented by Pickup, Ayer and Salant (1960) and Rushbrooke and Radojičić (1960) will not be nullified by further work but, unless it is, it establishes one success of the theory. Alternatively, if one starts from the direct evidence of the pion–pion resonance, one can make calculations on nucleon structure which agree with experiment. Of course, until one can reproduce the position and width of the resonance theoretically, they will not be purely fundamental calculations.

The prediction of the 3–3 resonance in pion–nucleon scattering is also worth mentioning, although it appears at too low an energy. Calculations taking into account some of the effects neglected are feasible and, if the position of the resonance agrees better with experiment, even if the agreement is still not quantitative, one will have to take them seriously. If inclusion of more effects worsens the agreement all the calculations will be suspect.

The small pion–nucleon phase shifts and the ratio of the asymptotic D- to S-state in the deuteron are other calculations of the theory which have been verified. Also, the fact that the present equations for pion–pion scattering do predict a resonance is not quite insignificant, even though they fail to give correctly the quantitative features.

Calculations are being done, or certainly will be done in the near future, to determine those quantities which the present equations do not give correctly, to improve the agreements with experiment that have been obtained, and to obtain further results. As was mentioned earlier in the article, there will probably be some quantities which could in principle be calculated but which will now have to be taken from experiment as our approximations are inadequate. The sum of the S-wave pion–nucleon scattering lengths may be such a quantity. Nevertheless, we would hope to be able to calculate sufficient quantities to verify that the fundamentals of the theory are correct in a limited region. At present we cannot really assert that we have done so. It may also turn out that the calculations will definitely disprove the theory, though the conclusions in such a case would always be more doubtful as one would not know whether the theory or the approximations were wrong.

The approach presented in this article has been from the point of view of local field theory. One of the main advantages of calculating with dispersion relations is that one can obtain equations between scattering amplitudes, which are the quantities measured in actual experiments. Nevertheless, the theory involves concepts other than scattering amplitudes, concepts which were essential in deriving the dispersion relations, whether one used perturbation theory or confined oneself to more rigorous methods. The very idea of locality was formulated in terms of the field operators, not in terms of the scattering amplitudes. We emphasized that causality applied to the scattering amplitude would not yield dispersion relations. Some physicists have suggested that one dispense with the field-theoretic substructure on which the dispersion relations were based and retain only the S-matrix, taking the analytic properties as postulates. In spite of the attractiveness of such an approach we feel that it has several drawbacks. Though quantum field theory may contain operators which do not correspond to measurements, the S-matrix

Dispersion Relations in Strong-coupling Physics

theory seems to go too far in the opposite direction and to contain less than could actually be measured; in particular, it is somewhat doubtful whether our ordinary macroscopic observations are limiting cases of scattering measurements. Another objection is that assumptions about the possibility of analytically continuing our functions are rather artificial unless they are deduced as mathematical consequences of some other postulates. Our present quantum field theory, in spite of its inadequacies, is based on well-worn concepts of classical field theory and quantum mechanics and would appear to be more natural. This, needless to say, is a matter of opinion about which it is difficult to argue and, in any case, one may maintain that such considerations should play no part in formulating a theory. One must keep an eye on all possibilities with regard to making a radical extension of the theory which all agree is necessary.

From a more practical point of view, an argument against the S-matrix formulation is that the principle of maximum analyticity does not give unambiguous results if it is applied to the scattering amplitude instead of to the Green's function. For instance, in problems with thresholds of the second order, it is possible either to include or to ignore the threshold if one confines oneself to the scattering amplitude. It is only when one extends the examination to the Green's function that one obtains a prescription for determining whether the singularity is there. Further, one has to use the unitarity condition in an unphysical region when considering the scattering amplitude, and one can only find whether or not it is justifiable by considering the Green's function. As it is known that all the content of a quantum field theory is contained in the Green's functions, it appears that one cannot eliminate some of the concepts by using the principle of maximum analyticity.

We may turn finally to the question of the general validity of the whole approach. Broadly speaking, one has to accept three sets of premises in order to perform calculations with dispersion relations; the truth of the conjectured analytic properties, the reliability of the approximation scheme and the principles underlying the derivation of the dispersion relations. The reasons for believing that the principle of maximum analyticity does follow from the principles of field theory have been given in the text and need not be repeated here; we feel that the assumption of their validity is the least doubtful of the three premises. The reliability of the approximation scheme is harder to judge. Ideally, one would like to be able to give a well-founded estimate of the error, but that seems to be very difficult. In this respect the situation is not peculiar to elementary-particle physics. More realistically one might observe whether including more terms in the equations made a substantial change in the result, and whether it improved the agreement with experiment. If this were done one should be able to get some idea of the validity of the approximations.

The truth or otherwise of the principles of quantum field theory on which everything in this article is based can, needless to say, only be established by comparing their consequences with experiment. The dynamical calculations should be of help, although the approximation scheme would not be expected to give quantitatively accurate results, and there would always be an uncertainty whether disagreements were due to the failure of the approximations or of the dispersion relations themselves. Forward dispersion relations for pion–nucleon scattering may be compared directly with experiment without any uncertain calculations, but we have seen that there are experimental difficulties in this. Perhaps the most promising experiments are those now being planned at Stanford for testing the validity of quantum electrodynamics at very high energies, as one would not suppose that one field theory was valid up to much higher energies than others. As quantum field theory almost certainly provides no explanation of the types of elementary particles that exist, one would expect it eventually to break down.

REFERENCES

ABASHIAN, A., BOOTH, N. E., and CROWE, K. M., 1960, Phys. Rev. Letters, **5**, 258.
AMATI, D., LEADER, E., and VITALE, B., 1960 a, Nuovo Cim., **17**, 68; 1960 b, Nuovo Cim., **18**, 409; 1960 c, Nuovo Cim., **18**, 458.
ANDERSON, H. L., DAVIDON, W. C., and KRUSE, U. E., 1955, Phys. Rev., **100**, 339.
ANDERSON, J. A., BANG, V. X., BURKE, P. G., CARMONY, D. D., and SCHMITZ, N., 1961, Phys. Rev. Letters, **6**, 365.
ASCOLI, R., 1960, Nuovo Cim., **18**, 754.
BALL, J. S., 1960, Phys. Rev. Letters, **5**, 73, and Thesis (University of California).
BALL, J. S., and WONG, D. Y., 1961, Phys. Rev. Letters, **6**, 29.
BJORKÉN, J. D., 1959, Bull. Amer. Phys. Soc., **4**, 448.
BLANKENBECLER, R., and COOK, L. F., 1960, Phys. Rev., **119**, 1475.
BLANKENBECLER, R., GOLDBERGER, M. L., KHURI, N. N., and TREIMAN, S. B., 1960, Ann. Phys., Lpz., **10**, 62.
BOGOLIUBOV, N. N., and SHIRKOV, D. V., 1957, Introduction to the Theory of Quantized Fields (New York: Interscience), ch. IX and Appendix.
BOWCOCK, J., COTTINGHAM, N., and LURIÉ, D., 1960 a, Nuovo Cim., **16**, 918; 1960 b, Phys. Rev. Letters, **5**, 386; 1961, Nuovo Cim., **19**, 142.
BREMERMANN, H. J., OEHME, R., and TAYLOR, J. G., 1958, Phys. Rev., **109**, 2178.
CAPPS, R. H., and TAKEDA, G., 1956, Phys. Rev., **103**, 1877.
CASTILLEJO, L., DALITZ, R. H., and DYSON, F. J., 1956, Phys. Rev., **101**, 453.
CHEW, G. F., 1958, Phys. Rev., **112**, 1380; 1959, Ann. Rev. Nucl. Sci., **9**, 29; 1960, Phys. Rev. Letters, **4**, 142; 1961, S-Matrix Theory of Strong Interactions (New York: Benjamin).
CHEW, G. F., and FRAUTSCHI, S. C., 1960, Phys. Rev. Letters, **5**, 580; 1961, Phys. Rev., **123**, 1478.
CHEW, G. F., GOLDBERGER, M. L., LOW, F. E., and NAMBU, Y., 1957 a, Phys. Rev., **106**, 1337; 1957 b, Phys. Rev., **106**, 1345.
CHEW, G. F., KARPLUS, R., GASIOROWICZ, S., and ZACHARIASEN, F., 1958, Phys. Rev., **110**, 265.
CHEW, G. F., and LOW, F. E., 1956, Phys. Rev., **101**, 1570; 1959, Phys. Rev., **113**, 1640.
CHEW, G. F., and MANDELSTAM, S., 1960, Phys. Rev., **119**, 467; 1961, Nuovo Cim., **19**, 752.
CHEW, G. F., MANDELSTAM, S., and NOYES, H. P., 1960, Phys. Rev., **119**, 478.
CHIU, H. Y., 1958, Phys. Rev., **110**, 1140.
CINI, M., and FUBINI, S., 1960, Ann. Phys., Lpz., **10**, 352.
CUTKOSKY, R. E., 1960 a, Phys. Rev. Letters, **4**, 624; 1960 b, J. Math. Phys., **1**, 429; 1961, Rev. Mod. Phys., **33**, 448.
CZIFFRA, P., MACGREGOR, M. H., MORAVCSIK, M. J., and STAPP, H. P., 1959, Phys. Rev., **114**, 880.
CZIFFRA, P., and MORAVCSIK, M. J., 1959, Phys. Rev., **116**, 226.
DAVIDON, W. C., and GOLDBERGER, M. L., 1956, Phys. Rev., **104**, 1119.
DESAI, B. R., 1961, Phys. Rev. Letters, **6**, 497.
DOWKER, J. S., 1961, Nuovo Cim., **20**, 182.
DRELL, S. D., 1958, 1958 Annual International Conference on High-Energy Physics (Geneva: CERN), p. 27.
DRELL, S. D., and ZACHARIASEN, F., 1961, Electromagnetic Structure of Nucleons (Oxford: University Press).
DYSON, F. J., 1957, Phys. Rev., **106**, 157; 1958, Phys. Rev., **110**, 1460.

Dispersion Relations in Strong-coupling Physics 161

EDEN, R. J., 1960 a, *Phys. Rev.*, **119**, 1763 ; 1960 b, *Phys. Rev.*, **120**, 1514 ; 1961 a, *Phys. Rev.*, **121**, 1567 ; 1961 b, *Lectures on the Use of Perturbation Methods in Dispersion Theory* (Physics Department : University of Maryland).
EDEN, R. J., LANDSHOFF, P. V., POLKINGHORNE, J. C., and TAYLOR, J. C., 1961 a, *Phys. Rev.*, **122**, 307 ; 1961 b, *J. Math. Phys.*, **2**, 656.
EFREMOV, A. V., MESHCHERYAKOV, V. A., SHIRKOV, D. V., and TZU, H. Y., 1960, *Nuclear Phys.*, **22**, 202.
ERWIN, A. R., MARCH, R., WALKER, W. D., and WEST, E., 1961, *Phys. Rev. Letters*, **6**, 627.
FEDERBUSH, P., GOLDBERGER, M. L., and TREIMAN, S. B., 1958, *Phys. Rev.*, **112**, 642.
FERRARI, F., FRYE, G., and PUSTERLA, M., 1960 a, *Phys. Rev. Letters*, **4**, 615 ; 1960 b, *Phys. Rev.*, **123**, 308 ; 1960 c, *Phys. Rev.*, **123**, 315.
FRAUTSCHI, S. C., 1960, *Phys. Rev. Letters*, **5**, 159.
FRAUTSCHI, S. C., and WALECKA, J. D., 1960, *Phys. Rev.*, **120**, 1486.
FRAZER, W. R., 1959, *Phys. Rev.*, **115**, 1763.
FRAZER, W. R., and FULCO, J. R., 1959, *Phys. Rev. Letters*, **2**, 365 ; 1960 a, *Phys. Rev.*, **117**, 1603 ; 1960 b, *Phys. Rev.*, **117**, 1609 ; 1960 c, *Phys. Rev.*, **119**, 1420.
GASIOROWICZ, S., 1960, *The Application of Dispersion Relations in Quantum Field Theory.* Lecture Notes (Munich: Max-Planck Institute).
GELL-MANN, M., 1956, *Proceedings of the Sixth Annual Rochester Conference on High-Energy Physics*, 1956, Sec. III (New York: Interscience), p. 30.
GELL-MANN, M., GOLDBERGER, M. L., and THIRRING, W., 1954, *Phys. Rev.*, **95**, 1612.
GOLDBERGER, M. L., 1955, *Phys. Rev.*, **99**, 979.
GOLDBERGER, M. L., GRISARU, M. T., MacDOWELL, S. W., and WONG, D. Y., 1960, *Phys. Rev.*, **120**, 2250.
GOLDBERGER, M. L., MIYAZAWA, H., and OEHME, R., 1955, *Phys. Rev.*, **99**, 986.
GOLDBERGER, M. L., NAMBU, Y., and OEHME, R., 1957, *Ann. Phys., Lpz.*, **2**, 226.
GOLDBERGER, M. L., and OEHME, R., 1960, *Ann. Phys., Lpz.*, **10**, 153.
GOLDBERGER, M. L., and TREIMAN, S. B., 1958 a, *Phys. Rev.*, **110**, 1178 ; 1958 b, *Phys. Rev.*, **111**, 354.
GOLDBERGER, M. L., WIGHTMAN, A. S., OMNÈS, R., KÄLLÉN, G., CHEW, G. F., TREIMAN, S. B., and YAMAGUCHI, Y., 1960, *Relations de Dispersion et Particules Elémentaires*. Ed. C. DeWitt and R. Omnès (Paris: Hermann).
GOURDIN, M., and MARTIN, A., 1960 a, *Nuovo Cim.*, **16**, 78 ; 1960 b, *Nuovo Cim.*, **17**, 224.
GOURDIN, M., NOIROT, Y., and SALIN, P., 1960, *Nuovo Cim.*, **18**, 651.
HAGEDORN, R., 1961, *Introduction to Field Theory and Dispersion Relations*. Lecture Notes (Geneva: CERN).
HAMILTON, J., 1958, *Phys. Rev.*, **110**, 1134.
JOST, R., and LEHMANN, H., 1957, *Nuovo Cim.*, **5**, 1598.
VAN KAMPEN, N., 1953 a, *Phys. Rev.*, **89**, 1072 ; 1953 b, *Phys. Rev.*, **91**, 1267.
KARPLUS, R., and RUDERMAN, M., 1955, *Phys. Rev.*, **98**, 771.
KARPLUS, R., SOMMERFIELD, C. M., and WICHMANN, E. H., 1958, *Phys. Rev.*, **111**, 1187 ; 1959, *Phys. Rev.*, **114**, 376.
KHURI, N. N., 1957, *Phys. Rev.*, **107**, 1148.
KHURI, N. N., and TREIMAN, S. B., 1958, *Phys. Rev.*, **109**, 198.
KLEIN, A., 1960, *J. Math. Phys.*, **1**, 41.
KLEIN, A., and ZEMACH, A. C., 1959, *Ann. Phys., Lpz.*, **7**, 440.
KRAMERS, H. A., 1927, *Atti Congr. Intern. Fisica, Como*, **2**, 545. See *Collected Scientific Papers* (Amsterdam: North-Holland).
KRONIG, R., 1926, *J. Opt. Soc. Amer.*, **12**, 547 ; 1946, *Physica*, **12**, 543.
LANDAU, L. D., 1959, *Nucl. Phys.*, **13**, 181.
LANDSHOFF, P. V., POLKINGHORNE, J. C., and TAYLOR, J. C., 1961, *Nuovo Cim.*, **19**, 939.
LANDSHOFF, P. V., and TREIMAN, S. B., 1961, *Nuovo Cim.*, **19**, 1249.
LEE, B. W., 1960, *Phys. Rev.*, **120**, 325 ; 1961, *Phys. Rev.*, **121**, 1550.
LEE, B. W., and CHO, K. S., 1961, *Nuovo Cim.*, **20**, 553.
LEHMANN, H., 1958, *Nuovo Cim.*, **10**, 579 ; 1959, *Nuovo Cim. Suppl.*, **14**, 153.
LEHMANN, H., SYMANZIK, K., and ZIMMERMANN, W., 1955, *Nuovo Cim.*, **1**, 205.

LEPORE, J. V., and WATSON, K. M., 1949, *Phys. Rev.*, **76**, 1157.
LEVINSOHN, N., 1949, *K. Danske Vidensk. Selsk., Mat.-Fys. Medd.*, **25**, No. 9.
LINDENBAUM, S. J., and STERNHEIMER, R. M., 1958, *Phys. Rev.*, **110**, 1174.
LOGUNOV, A. A., 1959, *Nucl. Phys.*, **10**, 71.
LOGUNOV, A. A., BELENKII, S. M., and TAVKHELIDZE, A. M., 1958, *Nuovo Cim.*, **10**, 953.
LOGUNOV, A. A., and ISAEV, P. S., 1958, *Nuovo Cim.*, **10**, 917.
LOW, F. E., 1955, *Phys. Rev.*, **97**, 1392.
MACDOWELL, S. W., 1959, *Phys. Rev.*, **116**, 774.
MACGREGOR, M. H., and MORAVCSIK, M. J., 1960, *Phys. Rev. Letters*, **4**, 524.
MACGREGOR, M. H., MORAVCSIK, M. J., and STAPP, H. P., 1959, *Phys. Rev.*, **116**, 1248.
MAGLIĆ, B., ALVAREZ, L., ROSENFELD, A., and STEVENSON, M. L., 1961, *Phys. Rev. Letters*, **7**, 178.
MANDELSTAM, S., 1958, *Phys. Rev.*, **112**, 1344; 1959 a, *Phys. Rev.*, **115**, 1741; 1959 b, *Phys. Rev.*, **115**, 1752; 1960 a, *Phys. Rev. Letters*, **4**, 84; 1960 b, *Nuovo Cim.*, **15**, 658.
MURASKIN, M., and NISHIJIMA, K., 1961, *Phys. Rev.*, **112**, 331.
NAMBU, Y., 1955, *Phys. Rev.*, **100**, 459; 1957 a, *Phys. Rev.*, **106**, 1366; 1957 b, *Nuovo Cim.*, **6**, 1064; 1958 *Nuovo Cim.*, **9**, 610.
NISHIJIMA, K., 1960, *Phys. Rev.*, **119**, 485.
NOYES, H. P., 1960, *Phys. Rev.*, **119**, 1736.
NOYES, H. P., and WONG, D. Y., 1959, *Phys. Rev. Letters*, **3**, 191.
OEHME, R., 1958, *Phys. Rev.*, **111**, 1430; 1959, *Nuovo Cim.*, **13**, 778.
OEHME, R., and TAYLOR, J. G., 1959, *Phys. Rev.*, **113**, 371.
PICKUP, E., AYER, F., and SALANT, E. O., 1960, *Phys. Rev. Letters*, **5**, 161.
PICKUP, E., ROBINSON, D. K., and SALANT, E. O., 1961, *Phys. Rev. Letters*, **7**, 192.
POLKINGHORNE, J. C., and SCREATON, G. R., 1960, *Nuovo Cim.*, **15**, 925.
PUPPI, G., and STANGHELLINI, A., 1957, *Nuovo Cim.*, **5**, 1305.
REGGE, T., 1959, *Nuovo Cim.*, **14**, 951; 1960, *Nuovo Cim.*, **18**, 947.
RUSHBROOKE, J. G., and RADOJIČIĆ, D., 1960, *Phys. Rev. Letters*, **5**, 567.
SALAM, A., 1956, *Nuovo Cim.*, **3**, 424.
SCREATON, G. R., 1961 (Ed.), *Dispersion Relations*. Scottish Universities' Summer School, 1960. Lectures by J. D. Jackson, J. C. Polkinghorne, W. Thirring, M. J. Moravcsik, G. F. Chew, W. R. Frazer, S. Fubini, and J. M. Jauch (Edinburgh: Oliver and Boyd).
SYMANZIK, K., 1957, *Phys. Rev.*, **105**, 743; 1958, *Progr. Theor. Phys., Japan*, **20**, 690; 1959, *Dispersion Relations*, Lecture Notes. University of Rome; 1960, *J. Math. Phys.*, **1**, 249.
TARSKI, J., 1960, *J. Math. Phys.*, **1**, 154.
WONG, D. Y., 1957, *Phys. Rev.*, **107**, 302; 1959, *Phys. Rev. Letters*, **2**, 406.
WONG, H. S., 1960, *Phys. Rev. Letters*, **5**, 70.

ANNALS OF PHYSICS: **19,** 1-24 (1962)

Quantum Electrodynamics Without Potentials

STANLEY MANDELSTAM

Department of Mathematical Physics, University of Birmingham, Birmingham, England

> A scheme is proposed for quantizing electrodynamics in terms of the electromagnetic fields without the introduction of potentials. The equations are relativistically covariant and do not require the introduction of unphysical states and an indefinite metric. Calculations carried out according to current quantization methods in the Coulomb or Lorentz gauges are justified in the new formalism. The theory exhibits an analogy between phases of operators and electromagnetic fields on the one hand, and coordinate systems and space curvature on the other. It is suggested that this analogy may be useful in quantizing the gravitational field.

I. INTRODUCTION

Current formulations of quantum electrodynamics all make use of potentials rather than fields; in other words, they work within particular gauges. The disadvantages of such gauges are sufficiently familiar merely to require mentioning here. Quantization in the Lorentz gauge, which is the most convenient for calculational purposes, has the ugly feature that one has to introduce unphysical states normalized with respect to an indefinite metric. While this procedure can probably be made mathematically sound, it is preferable to avoid introducing such states into the formulation if at all possible.

An alternative procedure is to quantize the field using the Coulomb gauge, in which no unphysical states appear. The Lagrangian is then, in appearance, not invariant with respect to Lorentz transformations; nevertheless, one can pass from one Lorentz frame to another by means of a canonical transformation (which is actually a gauge transformation), so that the physical predictions will be independent of the frame used for quantization. Further, it can be proved that the predictions of the theory quantized in the Lorentz gauge are the same as those of the theory quantized in the Coulomb gauge, so that calculations based on the former gauge can be justified without introducing an indefinite metric. The interconnection between different gauges has been considered in a recent paper by Zumino (*1*), who starts from the Coulomb gauge.

Use of the Coulomb gauge is thus mathematically sound and acceptable in the absence of a better alternative. As with the Lorentz gauge, however, it has drawbacks which one would prefer to avoid in that neither Lorentz-invariance

---

*Reprinted from *Annals of Physics* **19** (1962) 1–24. © 1962 Elsevier.

nor gauge-invariance are *manifestly* maintained. In fact, the fundamental operators which one introduces are not Lorentz- and gauge-covariant. Since it is fairly easy to obtain manifest covariance when quantizing any other field, it would seem highly desirable to bring the one exception into line.

The electromagnetic field has been quantized without making use of a particular gauge by Goldberg (2), who introduces the potentials into his theory but gives commutation rules for gauge-invariant quantities only. The procedure used is based on fundamental principles of quantum mechanics and is certainly correct, but his commutation rules do not appear to be sufficient to determine the theory. They could be supplemented to give a theory equivalent to that proposed here.

The approach to be described aims at quantizing the electromagnetic field without introducing the potentials at all. At first sight it does not appear to be possible to do so, since the potentials themselves enter into the equations of interacting electromagnetic and matter fields as they are usually formulated. This, however, is associated with the arbitrariness of fixing the phase factors in the operators of charged fields. We shall show in the following sections that it is possible to avoid introducing the potentials in the formulation of the theory. As in classical electrodynamics, the potentials will remain useful mathematical aids for calculation.

Current schemes for quantizing the electromagnetic field can be derived from the gauge-independent formalism. The method of obtaining the Coulomb gauge is very close to the corresponding method in classical electrodynamics. The operator formalism for the Lorentz gauge cannot be obtained without introducing unphysical states and an indefinite metric. We will show, however, that if one attempts to find a perturbation expansion for the gauge-independent theory, one is led fairly directly to the calculational rules of the Lorentz gauge.

The new formalism is not without complications of its own. It will be our object to show that such complications are inherent in nature and are not a feature of the formalism. They can be demonstrated experimentally in an experiment recently proposed by Aharonov and Bohm. It will appear that the effect manifested in this experiment bears a relation to the phase of the wave function or operator analogous to the relation between space curvature and the coordinate system. Space in which an electromagnetic field is present may be said to have a "phase" or "gauge" curvature with respect to charged particles. We shall explore this analogy in the following paper, where an attempt will be made to quantize the gravitational field in a manner which exhibits its relation to space curvature from the outset.

II. GAUGE-INVARIANT FIELD QUANTITIES

Our discussion will be confined to an electromagnetic field interacting with a scalar field, though the method can readily be extended to other fields. In the conventional method of quantization using the Lorentz gauge, we define field quantities ϕ and A_μ satisfying the commutation relations, if $x - y$ is spacelike,

$$[\phi(x), \phi(y)] = [\phi(x), \phi^*(y)] = [\phi^*(x), \phi^*(y)] = 0, \tag{2.1a}$$

$$[\dot\phi(x), \dot\phi(y)] = [\dot\phi(x), \dot\phi^*(y)] = [\dot\phi^*(x), \dot\phi^*(y)] = 0, \tag{2.1b}$$

$$[\phi(x), \dot\phi(y)] = [\phi^*(x), \dot\phi^*(y)] = 0, \tag{2.1c}$$

$$[\phi^*(x), \phi(y)] = [\phi(x), \phi^*(y)] = -i\delta^3(x - y), \tag{2.1d}$$

$$[A_\mu(x), A_\nu(y)] = [\dot A_\mu(x), \dot A_\nu(y)] = 0, \tag{2.2a}$$

$$[\dot A_\mu(x), A_\nu(y)] = -i\delta_{\mu\nu}\delta^3(x - y). \tag{2.2b}$$

The dots denote differentiation with respect to an arbitrarily chosen time direction. From the equations of motion it is easy to prove that the commutation rules are consistent, no matter what direction is chosen. The operators ϕ and ϕ^* will satisfy the equations of motion:

$$\left\{\left(\frac{\partial}{\partial x_\mu} - ieA_\mu\right)^2 - m^2\right\}\phi = 0, \tag{2.3a}$$

$$\left\{\left(\frac{\partial}{\partial x_\mu} + ieA_\mu\right)^2 - m^2\right\}\phi^* = 0, \tag{2.3b}$$

We shall not write down the equations of motion of the electromagnetic potentials, as they are peculiar to the Lorentz gauge and will not be preserved in our formulation—which will replace them with the Maxwell equations.

The transformation

$$\phi \to \phi e^{is}, \tag{2.4a}$$

$$\phi^* \to \phi^* e^{-is}, \tag{2.4b}$$

$$A_\mu \to A_\mu + \frac{\partial s}{\partial x_\mu}, \tag{2.4c}$$

is known as a gauge transformation, and it leaves—or, rather, it should leave—the equations of motion unaltered. Our aim is to formulate the theory in terms of operators which are themselves unaltered by a gauge transformation.

As far as the electromagnetic variables are concerned, it is well known that the appropriate gauge-invariant variables are the electromagnetic field strengths, defined by

$$F_{\mu\nu} = \frac{\partial A_\nu}{\partial x_\mu} - \frac{\partial A_\mu}{\partial x_\nu}. \tag{2.5}$$

The gauge-invariant matter field variables, though not quite so familiar, are also not difficult to write down. They are

$$\Phi(x, P) = \phi(x) \exp\left\{-ie \int_{-\infty}^{x} d\xi_\mu A_\mu(\xi)\right\}, \qquad (2.6a)$$

$$\Phi^*(x, P) = \phi^*(x) \exp\left\{ie \int_{-\infty}^{x} d\xi_\mu A_\mu(\xi)\right\}. \qquad (2.6b)$$

The integral on the right is to be taken over the spacelike path P.[1] The field quantities Φ thus depend not only on the point x but also on the whole of the path P. We could have written them $\Phi(P)$, so as to specify the end-point of a path is redundant once we have specified the path itself. Since, however, there is only one independent variable for each field point x, we have written in the x-dependence explicitly.

If the path in the definitions (2.6) is moved through an infinitesimal area $\sigma_{\mu\nu}$ at a point z, while the end-point x remains constant, it follows from Stokes' theorem that

$$\delta \int_{-\infty}^{x} d\xi_\mu A(\xi) = \{\partial_\mu A_\nu(z) - \partial_\nu A_\mu(z)\} \cdot \sigma_{\mu\nu}$$
$$= F_{\mu\nu}(z) \sigma_{\mu\nu}. \qquad (2.7)$$

Hence the change in the operators $\Phi(x, P)$ and $\Phi^*(x, P)$ caused by such a change in the path will be

$$\delta_z \Phi(x, P) = -ie\Phi(x, P) \cdot F_{\mu\nu}(z) \sigma_{\mu\nu}, \qquad (2.8a)$$

$$\delta_z \Phi^*(x, P) = ie\Phi^*(x, P) \cdot F_{\mu\nu}(z) \sigma_{\mu\nu}. \qquad (2.8b)$$

The dot indicates a symmetrized product.

The variables $\Phi(x, P)$, $\Phi^*(x, P)$, and $F_{\mu\nu}(x)$ are taken as our field variables in the gauge-independent formulation of quantum electrodynamics. The dependence of Φ and Φ^* on the path used to define them is given by (2.8). As this path-dependence is known, there is only one independent variable Φ for each field point x. In the final section we shall discuss the physical principles underlying the necessity for making the field variables of a gauge-independent theory path dependent. Let us first see how we can use such variables to construct a complete theory.

When two paths P_1 and P_2 differ by a finite area, we can find the change of Φ by repeated applications of (2.8). In order to ensure that the result be unique, two conditions must be fulfilled. First; if we choose a surface bounded by P_1 and P_2, and gradually move the path from P_1 to P_2, the result must not depend

---

[1] We shall allow the paths to have infinitesimal timelike portions in order to be able to perform variations or differentiations in the time direction.

QUANTUM ELECTRODYNAMICS

on the order in which two elements σ_1 and σ_2 of this surface are crossed. This will be the case if the F's at the two points commute.[2] Secondly, the result must not depend on which surface bounded by P_1 and P_2 we take within our four dimensions. If we have two such surfaces, then the difference in the integral $\int F_{\mu\nu}\sigma_{\mu\nu}$ over them will be given by the Gauss' theorem

$$\Delta \int F_{\mu\nu}\,\sigma_{\mu\nu} = \int \epsilon_{\kappa\lambda\mu\nu} \frac{F_{\mu\nu}}{\partial x_\lambda} \, d\tau, \qquad (2.9)$$

where τ is the volume between the surfaces. The consistency condition is thus

$$\epsilon_{\kappa\lambda\mu\nu} \frac{\partial F_{\mu\nu}}{\partial x_\lambda} = 0. \qquad (2.10)$$

These are the homogeneous Maxwell equations, which are fulfilled automatically if $F_{\mu\nu}$ is defined in terms of A_μ by (2.5), but have to be introduced as a consistency condition if the variables $F_{\mu\nu}$ are taken as fundamental.

The gauge-invariant derivatives of Φ are defined by the equations

$$\partial_\mu \Phi(x, P) = \lim_{dx_\lambda \to 0} \frac{\Phi(x + dx_\lambda, P') - \Phi(x, P)}{dx_\lambda}, \qquad (2.11)$$

where the path P' is obtained from P simply by giving it an extension of magnitude dx_μ in the μ-direction. In other words, the new path P' passes through the original point x. In terms of the conventional variables:

$$\begin{aligned}
\partial_\mu \Phi(x, P) &= \partial_\mu \left\{ \phi(x) \exp\left[-ie \int_{-\infty}^{x} A_\nu(\xi)\, d\xi_\nu \right] \right\} \\
&= \frac{\partial \phi(x)}{\partial x_\mu} \exp\left[-ie \int_{-\infty}^{x} A_\nu(\xi)\, d\xi_\nu \right] \\
&\quad - ieA_\mu(x)\phi(x) \exp\left[-ie \int_{-\infty}^{x} A_\nu(\xi)\, d\xi_\nu \right] \\
&= \left\{ \left(\frac{\partial}{\partial x_\mu} - ieA_\mu(x) \right) \phi(x) \exp\left[-ie \int_{-\infty}^{x} A_\nu(\xi)\, d\xi_\nu \right] \right\}.
\end{aligned} \qquad (2.12)$$

The necessity of adding a term $-ieA_\mu$ to the operator $\partial/\partial x_\mu$ to obtain a gauge-invariant quantity is, of course, well known.

It should be noted that the operators ∂_μ and ∂_ν do not commute if there is an electromagnetic field present. For, if the points (x_μ, x_ν), $(x_\mu + dx_\mu, x_\nu)$, $(x_\mu,$

---

[2] As we shall see in the next section, the magnetic and electric fields will not commute if σ_1 and σ_2 are adjacent. However, their commutator will be a c-number and, as the product in (2.8) is symmetrized, it still will not matter in which order we take the areas.

$x_\nu + dx_\nu)$, and $(x_\mu + dx_\mu, x_\nu + dx_\nu)$ are denoted by A, B, C and D,

$$\partial_\mu \partial_\nu \Phi(x, p) = \frac{\partial_\nu \Phi(x + dx_\mu, P + AB) - \partial_\nu \Phi(x, P)}{dx_\mu}$$

$$= \frac{\{\Phi(x + dx_\mu + dx_\nu, P + ABD) - \Phi(x + dx_\mu, P + AB)\} - \{\Phi(x + dx_\nu, P + AC) - \Phi(x, P)\}}{dx_\mu dx_\nu}.$$

The notation should be self-evident; $P + ABD$ indicates that the new path goes to A as before, then along the lines AB, BD. The other paths are similarly defined. In the expression for $\partial_\nu \partial_\mu \Phi(x, P)$, the first term in the numerator will be

$$\Phi(x + dx_\mu + dx_\nu, P + ACD).$$

Hence

$$(\partial_\mu \partial_\nu - \partial_\nu \partial_\mu)\Phi(x, P)$$

$$= \frac{\Phi(x + dx_\mu + dx_\nu, P + ABD) - \Phi(x + dx_\mu + dx_\nu, P + ACD)}{dx_\mu dx_\nu}$$

The numerator is just $-ie\Phi(x, P) \cdot F_{\mu\nu} dx_\mu dx_\nu$ by our fundamental formula (2.8), so that

$$(\partial_\mu\partial_\nu - \partial_\nu\partial_\mu)\Phi(x, P) = -ie\Phi(x, P) \cdot F_{\mu\nu}(x) \quad (2.13)$$

Before leaving this section we may note that the two formulas (2.8a) and (2.8b) can be unified into a single formula which, indeed, holds for the F's as well. For any operator X, the formula is

$$\delta_z X(P, x) = ie[J, X(x, P)] \cdot F_{\mu\nu}(z) \sigma_{\mu\nu}, \quad (2.14)$$

where J is the operator giving the total charge of the system.

III. COMMUTATION RULES AND EQUATIONS OF MOTION

We shall take as our Lagrangian density

$$L = -\{\partial_\mu \Phi^*(x)\}\{\partial_\mu \Phi(x)\} - m^2 \Phi^*(x)\Phi(x) - \tfrac{1}{4}\{F_{\mu\nu}(x)\}^2. \quad (3.1)$$

In appearance this is just the sum of the free-field Lagrangians for the two fields; the interaction comes about through the dependence of Φ on the F's in Eq. (2.8). As Φ and Φ^* always occur multiplied by one another, L will not depend on the path—this is immediately obvious from (2.14). Expressed in terms of the ϕ- and A-fields, (3.1) is the usual Lagrangian for interacting electromagnetic and charged scalar fields, except that the extra terms associated with the Lorentz gauge are absent.

In obtaining equations of motion from this Lagrangian, we must be careful about the meaning of gauge-invariant derivatives. We could proceed directly;

in order to make use of familiar techniques, however, we shall introduce potentials. The potentials now appear as a mathematical aid and are not present in the formulation of the theory; they will not appear in the equations of motion or the commutation rules.

From (2.10), we know that the electromagnetic field can be expressed in terms of potentials according to (2.5). This equation does not define the potentials uniquely and we shall not specify them further. We can then invert the procedure of the last section and define the ϕ's in terms of the Φ's by (2.6); the ϕ's, like the A's, are not defined uniquely. It follows from (2.8) that the ϕ's are independent of the path.

In terms of these path-independent but gauge-dependent variables, the Lagrangian (3.1) becomes the usual Lagrangian

$$L = -\left\{\frac{\partial \phi^*}{\partial x_\mu} + ieA_\mu \phi^*\right\}\left\{\frac{\partial \phi}{\partial x_\mu} - ieA_\mu \phi\right\} - m^2\phi^*\phi - \frac{1}{4}\{\partial_\mu A_\nu - \partial_\nu A_\mu\}^2. \quad (3.2)$$

As the variables are path-independent, it is now a straightforward matter to find the equations of motion and they turn out to be (2.3) together with the Maxwell equations. They can now be rewritten in terms of the gauge-independent variables in the form

$$(\partial_\mu^2 - m^2)\Phi = 0, \quad (3.3a)$$

$$(\partial_\mu^2 - m^2)\Phi^* = 0, \quad (3.3b)$$

$$\frac{\partial F_{\mu\nu}}{\partial x_\mu} + j_\nu = 0. \quad (3.4)$$

where j_μ is the current density, given by

$$j_\mu = -ie(\Phi^*\partial_\mu\Phi - \Phi\partial_\mu\Phi^*). \quad (3.5)$$

To obtain the commutation rules, we cannot apply the normal canonical method to the Lagrangian (3.2). The difficulty is due to the fact that one of the Maxwell equations contains no time derivative, so it is an equation of constraint rather than an equation of motion. The appropriate modification has been given by Bergmann and Goldberg (3) and by Goldberg (2). We shall apply instead the covariant formalism given by Peierls (4), which applies equally to Lagrangians with or without constraints. We can only obtain commutation rules between gauge-invariant quantities. To find the commutator between $\Phi(x_1)$ and another dynamical variable, we add to the Lagrangian (3.2) a term $\epsilon\delta^4(x - x_1)\Phi(x_1)$. On expressing the addition in terms of path-independent quantities, it becomes

$$\epsilon\delta^4(x - x_1)\phi(x_1) \exp\left\{-ie\int_\infty^{x_1} d\xi_\mu A_\mu(\xi)\right\}. \quad (3.6)$$

The new equations of motion are

$$\left\{\left(\frac{\partial}{\partial x_\mu} - ieA_\mu\right)^2 - m^2\right\}\phi = 0, \tag{3.7a}$$

$$\left\{\left(\frac{\partial}{\partial x_\mu} + ieA_\mu\right)^2 - m^2\phi^* + \epsilon\delta^4(x - x_1)\right\}\exp\left\{-ie\int_{-\infty}^{x_1} d\xi_\mu A_\mu\right\} = 0, \tag{3.7b}$$

$$\frac{\partial F_{\mu\nu}(x)}{x_\nu} + j_\nu(x) - \epsilon ie\left\{\int_{-\infty}^{x_1} d\xi_\nu \delta^4(x - \xi)\right\}\exp\left\{-ie\int_{-\infty}^{x_1} d\xi_\mu A_\mu(\xi)\right\} = 0 \tag{3.7c}$$

The last term in (3.7c) is obtained by taking the functional derivative of the exponential with respect to $A_\mu(x)$. We now assume that all variables are unchanged before the time corresponding to x_1. Immediately afterwards, the changes D_Φ in the variables due to the extra term in the Lagrangian will then be

$$D_\Phi\left\{\left(\frac{\partial}{\partial t} - ieA_\mu\right)\phi^*(x)\right\} = \epsilon\delta^3(x - x_1)\exp\left\{-ie\int_{-\infty}^{x_1} d\xi_\mu A_\mu\right\} \tag{3.8a}$$

$$D_\Phi\{F_{0i}(x)\} = \epsilon ie\int_{-\infty}^{x_1} d\xi_i \delta^3(x - \xi)\phi(x)\exp\left\{-ie\int_{-\infty}^{x_1} d\xi_\mu A_\mu(\xi)\right\} \tag{3.9b}$$

where the subscript i takes the values 1, 2, 3 only. The changes in the other variables are zero. Multiplying both sides of (3.8a) by $\exp\{-ie\int_{-\infty}^{x_1} d\xi_\mu A_\mu\}$, applying the Peierls rule

$$\epsilon[\Phi, X] = iD_\Phi X,$$

and repeating the procedure with Φ^*, Φ and Φ^*, instead of Φ, we derive the relations:

$$[\Phi(x, P), \Phi(y, P)] = [\Phi(x, P), \Phi^*(y, P)]$$
$$= [\Phi^*(x, P), \Phi^*(y, P)] = 0, \tag{3.10a}$$

$$[\dot\Phi(x, P), \dot\Phi(y, P)] = [\dot\Phi(x, P)\dot\Phi^*(y, P)]$$
$$= [\dot\Phi^*(x, P), \dot\Phi^*(y, P)] = 0, \tag{3.10b}$$

$$[\Phi(x, P), \dot\Phi(y, P)] = [\Phi^*(xP), \dot\Phi^*(y, P)] = 0, \tag{3.10c}$$

$$[\dot\Phi^*(x, P), \Phi(y, P)] = [\dot\Phi(x, P), \Phi^*(y, P)] = -i\delta^3(x - y), \tag{3.10d}$$

$$[\overset{(\cdot)}{\Phi}(x, P), F_{ij}(y)] = [\overset{(\cdot)}{\Phi}(x, P), F_{ij}(y)] = 0, \tag{3.11a}$$

$$[\overset{(\cdot)}{\Phi}(x, P), F_{0i}(y)] = -e\int_{-\infty}^{x} d\xi_i \delta^3(y - \xi)\overset{(\cdot)}{\Phi}(x, P), \tag{3.11b}$$

$$[\overset{(\cdot)}{\Phi}{}^*(x, P), F_{0i}(y)] = e\int_{-\infty}^{x} d\xi_i \delta^3(y - \xi)\overset{(\cdot)}{\Phi}(x, P). \tag{3.11c}$$

The dot denotes gauge-invariant differentiation with respect to t; if the dot is enclosed by a bracket the equation holds whether or not it is present. The integration on the right of (3.11) is to be performed over the path defining Φ. As usual, the preferred direction can be taken to be perpendicular to any spacelike surface, and the commutation relations are easily proved to be consistent.

One can obtain the commutation rules between the electromagnetic field operators by adding a term $\epsilon F_{\mu\nu}(x_1)\delta(x - x_1)$ to the Lagrangian. They are the same as in the free field case:

$$[F_{ij}(x), F_{i'j'}(y)] = [F_{0i}(x), F_{0i'}(y)] = 0, \qquad (3.12\text{a})$$

$$[F_{0i}(x), F_{jk}(y)] = -i\left\{\delta_{ij}\frac{\partial}{\partial y_k} - \delta_{ik}\frac{\partial}{\partial y_j}\right\}\delta^3(x - y). \qquad (3.12\text{b})$$

We may note that it is not necessary to state the complete commutation relations (3.11) between $\overset{(\cdot)}{\Phi}(x)$ or $\overset{(\cdot)}{\Phi}{}^*(x)$ and $F_{\mu\nu}(y)$. One could simply require that $\overset{(\cdot)}{\Phi}(x)$ and $\overset{(\cdot)}{\Phi}{}^*(x)$ commute with $F_{\mu\nu}(y)$ as long as F is not measured on a path used to define Φ. Equation (3.11) then follows from (2.8).

When interpreting the equations of motion for Φ and Φ^* one must bear in mind the meaning of the gauge-invariant derivative. Differentiation with respect to time adds an element in the time direction to the end of the path. One can then move the whole path in the time direction using (2.8). The equations of motion and the equations (2.8) enable one in principle to find the field operators for any spacelike path once they are known for spacelike paths on a given spacelike surface.

The equations of motion (3.3) and (3.4), the commutation rules (3.10)–(3.12), the equations (2.8) expressing the dependence of the operators on the paths, and the consistency conditions (2.10) thus give us a theory of interacting electromagnetic and charged scalar fields which involve only gauge-independent quantities.

IV. COULOMB GAUGE

Once we have a gauge-independent and relativistically covariant theory, we can introduce potentials and restrict their arbitrariness by a convenient rule. Unless we are prepared to introduce an indefinite metric such a rule cannot be covariant. Though this is often a disadvantage in practice, such special choices of the potentials, or such "special gauges" are useful. They are, of course, universally employed in nonrelativistic quantum mechanics.

The treatment is almost identical to the corresponding classical treatment. Because of Eqs. (2.10), we know that we can define the field strengths in terms of the potentials by (2.5). The potentials are arbitrary to within an operator gauge transformation

$$A_\mu \to A_\mu + \frac{\partial s}{\partial x_\mu},$$

and, by suitably choosing s, we can construct new potentials which still define the fields by (2.5) and which in addition satisfy the relation

$$\frac{\partial}{\partial x_i} A_i(x) = 0. \tag{4.1}$$

Before imposing (4.1) it is of course necessary to choose a particular time direction. As in the unquantized theory, we then find that

$$A_0(\mathbf{x}) = \frac{e}{4\pi} \int d^3x' \frac{j_0(\mathbf{x}')}{|\mathbf{x}' - \mathbf{x}|}, \tag{4.2}$$

so that A_0 is not an independent dynamical variable. In the integration on the right of (4.2) and subsequent similar integrations, the time is held fixed. A_i is given in terms of F_{ij} by the formula

$$A_i(\mathbf{x}) = \frac{1}{4\pi} \frac{\partial}{\partial x_j} \int d^3x' \frac{F_{ji}(\mathbf{x}')}{|\mathbf{x}' - \mathbf{x}|}. \tag{4.3}$$

The Coulomb-gauge ϕ and ϕ^* are now defined in terms of Φ, Φ^* and the potentials by (2.6) and, from (2.8), they are proved to be path-independent. The commutation rules (2.1) and the commutativity of $\overset{(\cdot)}{\phi}(x)$, $\overset{(\cdot)}{\phi}{}^*(x)$, and $A_i(x)$ with $A_j(y)$ are easily derived. From (4.3) and the formula

$$\dot{A}_i(\mathbf{x}) = -F_{0i}(\mathbf{x}) - \frac{e}{4\pi} \frac{\partial}{\partial x_i} \int d^3x' \frac{j_0(\mathbf{x}')}{|\mathbf{x}' - \mathbf{x}|}, \tag{4.4}$$

one can derive the commutation rules

$$[\dot{A}_i(\mathbf{x}), \dot{A}_j(\mathbf{y})] = 0, \tag{4.5a}$$

$$[\dot{A}_i(\mathbf{x}), A_j(\mathbf{y})] = -i\left\{\delta_{ij}\delta^3(\mathbf{x} - \mathbf{y}) + \frac{1}{4\pi}\frac{\partial^2}{\partial x_i \partial x_j} \frac{1}{|\mathbf{x} - \mathbf{y}|}\right\}. \tag{4.5b}$$

We have finally to show that $\overset{(\cdot)}{\phi}(\mathbf{x})$ and $\overset{(\cdot)}{\phi}{}^*(\mathbf{x})$ commute with $A_i(\mathbf{y})$. From (2.6a),

$$[\overset{(\cdot)}{\phi}(\mathbf{x}), \dot{A}_i(\mathbf{y})] = [\Phi(\mathbf{x}, P), \dot{A}_i(\mathbf{y})] \exp\left\{ie \int_{-\infty}^{x} d\xi_j A_j(\xi)\right\} \\
+ \Phi(\mathbf{x}, P)\left[\exp\left\{ie \int_{-\infty}^{x} d\xi_j A_j(\xi), \dot{A}_i(\mathbf{y})\right\}\right]. \tag{4.6}$$

Now

$$[\Phi(\mathbf{x}, P), \dot{A}_i(\mathbf{y})] = \left[\Phi(\mathbf{x}, P), -F_{0i}(\mathbf{y}) - \frac{e}{4\pi}\frac{\partial}{\partial y_i}\int d^3y \frac{j_0(\mathbf{y}')}{|\mathbf{y}' - \mathbf{y}|}\right],$$

$$= e \int_{-\infty}^{x} d\xi_i \delta^3(y - \xi)\Phi(\mathbf{x}, P) - \frac{e}{4\pi}\frac{\partial}{\partial y_i}\frac{1}{|\mathbf{x} - \mathbf{y}|},$$

from (3.11b) and (3.10). Also, from (4.5b),

$$\left[\exp\left\{ie\int_{-\infty}^{x}d\xi_j A_j(\xi)\right\}, \dot{A}_i(\mathbf{y})\right] = -e\int_{-\infty}^{x}d\xi_\mu \delta^3(\mathbf{y}-\xi)\exp\left\{ie\int_{-\infty}^{x}d\eta_j A_j(\mathbf{n})\right\}$$

$$-\frac{e}{4\pi}\frac{\partial}{\partial y_i \partial y_j}\int_{-\infty}^{x}d\xi_j \frac{1}{|\mathbf{y}-\xi|}\exp\left\{ie\int_{-\infty}^{x}d\xi_j A_j(\xi)\right\}$$

$$= -e\int_{-\infty}^{x}d\xi_\mu \delta^3(\mathbf{y}-\xi)\exp\left\{ie\int_{-\infty}^{x}d\eta_j A_j(\mathbf{n})\right\}$$

$$+\frac{e}{4\pi}\frac{\partial}{\partial y_i}\frac{1}{|\mathbf{x}-\mathbf{y}|}\exp\left\{ie\int_{-\infty}^{x}d\eta_j A_j(\mathbf{n})\right\}$$

Equations (4.7) and (4.8) together with (4.6) show that $\overset{(\cdot)}{\phi}(\mathbf{x})$ does commute with $A_i(\mathbf{y})$, and the proof that $\overset{(\cdot)}{\phi}^*(\mathbf{x})$ commutes with $A_i(\mathbf{y})$ follows in the same way.

The equations of motion (3.3) for Φ and Φ^* give the equations of motion (2.3) for ϕ and ϕ^* on using the formula (2.12) for the covariant derivative, while the equations of motion for A_i are obtained by putting (2.5) into (3.4) with $\mu = 1$, 2, and 3.

The Coulomb gauge is thus easily proved to yield the correct solutions of the gauge-invariant equations from which we start, just as in classical electromagnetic theory. The gauge-dependent quantities ϕ, ϕ^*, and A_i are regarded as auxiliary variable to calculate the gauge-independent quantities Φ, Φ^*, and $F_{\mu\nu}$.

V. PERTURBATION THEORY AND LORENTZ GAUGE

The method of the last section cannot be used to reproduce the Lorentz gauge. The reason is that the supplementary condition corresponding to (4.2) now involves the time, so that one cannot use it to eliminate one of the variables and then to obtain equal-time commutation relations for the others. One could attempt to show that the gauge-invariant operators, constructed according to (2.6) from the Lorentz-gauge operators with an indefinite metric, satisfied the correct commutation relations and equations of motion. This would provide a justification of the theory quantized in the Lorentz gauge.

We shall avoid referring to an indefinite metric and shall try to expand our gauge-independent theory in a perturbation series. It will be found that the terms in the perturbation series for our Green's functions can easily be obtained from the corresponding perturbation series in the Lorentz gauge. The use of the Lorentz gauge for calculations is then justified without the use of an indefinite metric. Actually the last section has already achieved just this, as we know that calculations of gauge-independent quantities in the Lorentz gauge give the same results as calculations in the Coulomb gauge. It is worth while, however, to obtain the Lorentz gauge without the intermediary of the Coulomb gauge. The procedure will also illustrate a method of working with path-dependent variables and will be useful for the corresponding problem in the quantized gravitational field, where no coordinate system corresponding to the Coulomb gauge has yet been found.

We shall obtain our perturbation terms by solving the differential equations for the Green's functions in a power series in e. Our attention will be confined to the lowest terms, but the method can readily be extended.

The lowest Green's functions are defined by the equations

$$G(xP_1\,;\,yP_2) = T\langle \Phi^*(x, P_1), \Phi(y, P_2) \rangle, \tag{5.1a}$$

$$G_{\mu\nu}(xP_1\,;\,P_2\,;\,z,w) = T\langle \Phi^*(x, P_1), \Phi(y, P_2), F_{\mu\nu}(z) \rangle, \tag{5.1b}$$

$$G_{\mu\nu,\lambda\sigma}(xP_1\,;\,yP_2\,;\,z,w) = T\langle \Phi^*(x, P_1)\Phi(y, P_2), F_{\mu\nu}(z), F_{\lambda\sigma}(w). \tag{5.1c}$$

The symbol T indicates as usual that the operators are to be ordered with the earlier ones to the right of the later ones.[3] When, however, the commutation rules contain derivatives of δ-functions, one must correct the definition of the T-product for equal times in order to obtain a covariant function. We thus define

$$T\{F_{0i}(x), F_{0j}(y)\} = F_{0i}(x)\cdot F_{0j}(y) + i\delta^4(x-y) \tag{5.1d}$$

for $x = y_0$. The T-products are then covariant, as may be verified from the commutation rules. In the conventional formalism, the definition is equivalent to

$$T\{F_{\mu\nu}(x), F_{\lambda\sigma}(y)\} = \left(\frac{\partial}{\partial x_\mu}\delta_{\nu\rho} - \frac{\partial}{\partial x_\nu}\delta_{\mu\rho}\right)\left(\frac{\partial}{\partial x_\lambda}\delta_{\sigma\tau} - \frac{\partial}{\partial x_\sigma}\delta_{\lambda\tau}\right)T\{A_\rho(x), A_\tau(y)\},$$

with the differential operators outside the time-ordered product. The extra term in (5.1d) arises from differentiating the T-operator.

The Green's functions satisfy the differential equations:

$$(-\partial^2_{x,\mu} + m^2)G(xP_1, yP_2) = (-\partial^2_{y,\mu} + m^2)G(xP_1, yP_2)$$
$$= -i\delta^4(x-y), \tag{5.2}$$

$$(-\partial^2_{x,\lambda} + m^2)G_{\mu\nu}(xP_1, yP_2, z)$$
$$= (-\partial^2_{y,\lambda} + m^2)G_{\mu\nu}(xP_1 yP_2, z) = 0, \tag{5.3a}$$

$$\frac{\partial}{\partial z_\mu}G_{\mu\nu}(xP_1, yP_2, z) + T\langle \Phi^*(x, P_1), \Phi(y, P_2), j_\nu(z) \rangle = 0. \tag{5.3b}$$

---

[3] In performing the time-ordering, we associate with an operator all the points along the path defining it. We restrict the paths so as to avoid one operator having points which are timelike earlier and points which are timelike later than those of another operator.

The δ-function on the right of (5.2) comes from the equal-time commutator. The equation for the electron-photon Green's function will only be required in lowest order, so that the term containing j may be neglected:

$$\frac{\partial}{\partial z_\mu} G^{(0)}_{\mu\nu,\lambda\sigma}(xP_1; yP_2; z, w)$$
$$= -i\left\{\frac{\partial}{\partial w_\lambda}\delta_{\nu\sigma} - \frac{\partial}{\partial w_\sigma}\delta_{\nu\lambda}\right\}\delta^4(z-w)G(xP_1, yP_2), \quad (5.4a)$$

$$\frac{\partial}{\partial w_\lambda} G^{(0)}_{\mu\nu,\lambda\sigma}(xP_1; yP_2; z, w)$$
$$= -i\left\{\frac{\partial}{\partial z_\mu}\delta_{\sigma\nu} - \frac{\partial}{\partial z_\nu}\delta_{\sigma\mu}\right\}\delta^4(z-w)G(xP_1, yP_2), \quad (5.4b)$$

$$(-\partial^2_{x,\rho} + m^2)G^{(0)}_{\mu\nu,\lambda\sigma}(xP_1; yP_2; z, w) = (-\partial^2_{y,\rho} + m^2)G^{(0)}_{\mu\nu,\lambda\sigma}(xP_1; yP_2; z, w) \quad (5.4c)$$
$$= -i\delta^4(x-y)T\langle F_{\mu\nu}(z), F_{\lambda\sigma}(w)\rangle.$$

Though the electromagnetic interaction does not occur explicitly in (5.2) and (5.3a), it enters via the dependence of the operators on the path, so that the equations cannot be solved exactly. In lowest-order perturbation theory the path dependence may be neglected, so that the solutions of (5.2), (5.3), and (5.4) give the familiar results:

$$G^{(0)}(x; y) = \frac{-i}{(2\pi)^4}\int dp\, e^{ip(x-y)}\frac{1}{p^2 + m^2}, \quad (5.5)$$

$$G^{(0)}_{\mu\nu}(x; y; z) = 0, \quad (5.6)$$

$$G^{(0)}_{\mu\nu,\lambda\sigma}(x; y; z; w) = -\frac{1}{(2\pi)^8}\int dp\, dq\, e^{ip(x-y)+iq(z-w)}$$
$$\cdot \frac{q_\mu q_\lambda \delta_{\nu\sigma} + q_\nu q_\sigma \delta_{\mu\lambda} - q_\mu q_\sigma \delta_{\nu\lambda} - q_\nu q_\lambda \sigma_{\mu\sigma}}{(p^2 + m^2)q^2}. \quad (5.7)$$

To find the three-point Green's function in first-order theory, it is necessary to take the path dependence into account to lowest order. Equation (2.8a) transformed into an equation for the Green's function is

$$-\delta_{P_1,w}G^{(1)}_{\mu\nu}(xP_1, yP_2, z) = \delta_{P_2,w}G^{(1)}(xP_1, yP_2, z)$$
$$= -ieG^{(0)}_{\mu\nu,\sigma\tau}(xP_1; yP_2; z; w)\sigma_{\mu\nu}. \quad (5.8)$$

The operation of changing the path may alter the order of the operators and give rise to an equal-time commutator from (3.11b) and (3.11c), but this is accounted for by the δ-function on the right of (5.8) coming from the definition (5.1d).

Our procedure is to find a path integral which satisfies (5.8). The difference between $G^{(1)}_{\mu\nu}(xP_1, yP_2, z)$ and this path integral will be independent of the path. To do this construct *any* function $g^{(0)}_{\mu\nu,\lambda}(x; y; z; w)$ which satisfies the equation

$$\frac{\partial}{\partial w_\lambda} g^{(0)}_{\mu\nu,\sigma}(x; y; z; w) - \frac{\partial}{\partial w_\sigma} g^{(0)}_{\mu\nu,\lambda}(x; y; z; w) = G^{(0)}_{\mu\nu,\lambda\sigma}(x; y; z; w). \quad (5.9)$$

The functions designated by lower-case g's will later be identified with Green's functions in the Lorentz gauge. The subscripts are associated with the variables to the right of the second semicolon, two subscripts denoting an electromagnetic field variable and one a potential. Thus,

$$g^{(0)}_{\mu\nu,\lambda}(x; y; z; w) \approx T\langle \phi^*(x), \phi(y), F_{\mu\nu}(z), A_\lambda(w) \rangle, \quad (5.10)$$

as is evident from (5.9). (In lowest order we do not distinguish between ϕ and Φ). Equation (5.10) has been written to indicate the connection between the present calculation and the Lorentz gauge. It is not part of our calculation, which is gauge-independent.

If we construct a path integral of g over the variable w, an application of Stokes' theorem to the definition (5.9) shows that the integral depends on the path by the equation

$$\delta_w \int dw_\sigma g^{(0)}_{\mu\nu,\sigma}(x; y; z; w) = G^{(0)}_{\mu\nu,\lambda\sigma}(x; y; z; w)\sigma_{\mu\nu}. \quad (5.11)$$

The function $g^{(1)}_{\mu\nu}(x; y; z)$ is now defined by the equation

$$G^{(1)}_{\mu\nu}(xP_1, yP_2, z) = g^{(1)}(x; y; z) - ie \int_x^y dw_\lambda g^{(0)}_{\mu\nu,\lambda}(x; y; z; w). \quad (5.12)$$

The integral $\int_x^y$ is taken from x to $-\infty$ along P_1 and then from $-\infty$ to y along P_2. It follows from (5.8) and (5.12) that $g^{(1)}(x; y; z)$ is path-independent.

As usual, an equation of the type (5.9) does not define $g^{(0)}(x; y; z; w)$ uniquely. A possible solution is

$$g^{(0)}_{\mu\nu,\lambda}(x; y; z; w) = \frac{-i}{(2\pi)^8} dp\, dq e^{ip(y-x)+iq(z-w)} \frac{q_\mu \delta_{\nu\lambda} - q_\nu \delta_{\mu\lambda}}{(p^2 + m^2)q^2}, \quad (5.13)$$

which corresponds to the Feynman form of the Lorentz gauge.

From (5.3a), (5.4c), and (5.12) we obtain the differential equation for $g^{(1)}_{\mu\nu}(x; y; z)$:

$$\left(-\frac{\partial^2}{\partial x^2} + m^2\right) g^{(1)}_{\mu\nu}(x; y; z) = ie\left(2\frac{\partial}{\partial x_\lambda} + \frac{\partial}{\partial w_\lambda}\right) g^{(0)}_{\mu\nu,\lambda}(x; y; z; w)_{w=x}. \quad (5.14)$$

On substituting (5.13) into the right-hand side and dividing by the operator on the left, we finally obtain

$$g^{(1)}_{\mu\nu}(x;y;z) = \frac{-ie}{(2\pi)^8} \int dp\,dq\, e^{-i(p+q)x+ipy+iqz} \frac{q_\mu(2p+q)_\nu - q_\nu(2p+q)_\mu}{\{(p+q)^2+m^2\}(p^2+m^2)q^2}. \quad (5.15)$$

Both terms on the right of (5.15) are now known, so that the Green's function $G^{(1)}_{\nu\mu}(xP_1, yP_2, z)$ has been calculated to first order. The expression (5.15) is just the usual formula for the Green's function $T\langle \phi^*(x), \phi(y), F_{\mu\nu}(z)\rangle$ in the Lorentz gauge. In fact, if the terms $g^{(1)}_{\mu\nu}(x;y;z)$ and $g^{(0)}_{\nu\mu,\lambda}(x;y;z;w)$ are identified with their respective Lorentz-gauge Green's function by the result just obtained and (5.10), it is observed that the right-hand side of (5.12) is the lowest term in the perturbation series for

$$T\left\langle \phi^*(x) \exp\left\{ie \int_{-\infty}^x dw_\lambda \cdot A_\lambda(w)\right\}, \phi(y) \exp\left\{-ie \int_{-\infty}^y dw_\lambda A_\lambda(w)\right\}, F_{\mu\nu}(z) \right\rangle.$$

The Lorentz-gauge Green's functions are now regarded as being auxiliaries for the calculation of the gauge-independent Green's function $G_{\mu\nu}(xP_1, yP_2, z)$.

To illustrate the calculation of a radiative correction we shall calculate

$$G(xP_1, yP_2)$$

in second order. The path dependence is again given by (2.8):

$$\begin{aligned}-\delta_{P_1,z}G^{(2)}(xP_1, yP_2) &= \delta_{P_2,z}G^{(2)}(xP_1, yP_2) \\ &= -ieG^{(1)}_{\mu\nu}(xP_1, yP_2, z)\sigma_\mu.\end{aligned} \quad (5.16)$$

To obtain a function independent of the path from which $G^{(2)}(xP_1, yP_2)$ may be calculated, we define two functions $g^{(1)}_\mu(x;y;z)$ and $g^{(0)}_{\mu,\lambda}(x;y;z;w)$ by the equations

$$\frac{\partial}{\partial z_\mu} g^{(1)}_\nu(x;y;z) - \frac{\partial}{\partial z_\nu} g^{(1)}_\mu(x;y;z) = g^{(1)}_{\mu\nu}(x;y;z), \quad (5.17)$$

$$\frac{\partial}{\partial z_\mu} g^{(0)}_{\nu,\lambda}(x;y;z,w) - \frac{\partial}{\partial z_\nu} g^{(0)}_\mu(x;y;z;w) = g^{(0)}_{\mu\nu,\lambda}(x;y;z;w). \quad (5.18)$$

They correspond respectively to $T\langle\phi^*(x), \phi(y), A_\mu(z)\rangle$ and

$$T\langle\phi^*(x), \phi(y), A_\mu(z), A_\lambda(w)\rangle$$

in the conventional theory. It then follows from (5.12) that the function $g^{(2)}(x;y)$ defined by

$$G^{(2)}(xP_1, yP_2) = g^{(2)}(x;y) - ie \int_{-\infty}^y dz_\mu\, g^{(1)}_\mu(x;y;z) \\ - \frac{e^2}{2} \int_{-\infty}^y dz_\mu \int_{-\infty}^y dw_\lambda(x;y;z,w) \quad (5.19)$$

is independent of the path. Possible solutions of (5.18) are:

$$g_\mu^{(1)}(x; y; z) = -\frac{e}{(2\pi)^8} \int dp\, dq e^{-i(p+q)x+ipy+iqz} \frac{(2p+q)_\mu}{\{(p+q)^2 + m^2\}(p^2+m^2)q^2}, \quad (5.20)$$

$$g_{\mu,\lambda}^{(0)}(x; y; z; w) = -\frac{1}{(2\pi)^8} \int dp\, dq e^{ip(y-x)+iq(z-w)} \frac{\delta_{\mu\lambda}}{(p^2+m^2)q^2}. \quad (5.21)$$

The differential equation (5.2) applied to (5.19) gives the equation

$$\left(-\frac{\partial^2}{\partial x_\mu^2} + m^2\right) g^{(2)}(x; y) = ie\left(-2\frac{\partial}{\partial x_\mu} - \frac{\partial}{\partial z_\mu}\right) g_\mu^{(1)}(x; y; z)_{z=x}$$
$$+ ie \int_x^y dz_\mu \left(-\frac{\partial^2}{\partial x_\mu^2} + m^2\right) g_\mu^{(1)}(x; y; z) - e^2 g_{\mu,\mu}^{(0)}(x; y; x; x)$$
$$+ e^2 \int_x^y dz_\mu \left(2\frac{\partial}{\partial x_\lambda} + \frac{\partial}{\partial w_\mu}\right) g_{\mu,\lambda}^{(0)}(x; y; z; w)_{w=x}$$
$$+ \frac{e^2}{2} \int_x^y dz_\mu \int_x^y dw_\lambda \left(-\partial_{x,\mu}^2 + m^2\right) g_{\mu\lambda}^{(0)}(x; y; z; w).$$

(5.2)

The last term vanishes by (5.4c), and the second and fourth terms cancel according to (5.20) and (5.21). Inserting the formulas (5.2) and (5.2) for

$$g_\mu^{(1)}(x; y; z) \quad \text{and} \quad g_{\mu,\mu}^{(0)}(x; y; x; x)$$

and dividing by the operator $[(\partial^2/\partial x_\mu^2) + m^2]$, we obtain as our formula for $g^{(2)}(x; y)$

$$g^{(2)}(x; y) = \frac{e^2}{(2\pi)^8} \int dp\, e^{ip(y-x)} \frac{1}{(p^2+m^2)^2} \int dq \left\{\frac{(2p+q)^2}{\{(p+q)^2+m^2\}q^2} + \frac{4}{q^2}\right\}. \quad (5.2)$$

This is the conventional formula for the charged particle Green's function in the second order. The second term in the curlybracket represents the process where a photon is emitted and absorbed at the same point due to the term $-e^2\phi^*(x)\phi(x)\{A_\mu(x)\}^2$ in the Lagrangian. From our point of view, $g^{(2)}(x; y)$ is an auxiliary quantity for the calculation of $G^{(2)}(xP_1; yP_2)$ by the formula (5.19), all the terms on the right of which are now known. We also observe that the right-hand side of (5.19) is the expansion up to second order of the Green's function

$$T\left\langle \phi^*(x) \exp\left\{ie \int_{-\infty}^x dz_\mu\, A_\mu(z)\right\}, \phi(y) \exp\left\{-ie \int_{-\infty}^y dz_\mu\, A_\mu(z)\right\}\right\rangle$$

in the Lorentz gauge.

It thus appears that a prescription for calculating the gauge-independent Green's functions in any order of perturbation theory is first to calculate the

Green's functions in the Lorentz gauge and then to use (2.5) and (2.6), the latter expanded in a perturbation series, to obtain the gauge-independent Green's functions. In this way the Lorentz gauge appears naturally in the process of obtaining a perturbation series for gauge-independent quantum electrodynamics. The derivation is completely covariant.

VI. REDUCTION FORMULAS

It is the purpose of this section to point out that the path dependence of operators affects neither the classification of the states of a system, e.g., according to the number and momenta of incoming or outgoing particles, nor the procedure of calculating matrix elements of operators between such states. In the usual formalism, one would determine the possible masses of the particles by finding the singularities of the Green's function in momentum space. From the coefficients of these singularities for more complicated Green's functions, one can calculate the S-matrix or the matrix element of an operator. For instance, if one wanted to find the matrix elements of the operator $\phi(x)$ between a state containing two incoming charged particles and a single-particle state, one would consider the integral

$$\int dy dz dw \, e^{-ipy-iqz+irw} T \langle \phi^*(y)\phi^*(z)\phi(w)\phi(x) \rangle. \tag{6.1}$$

This function would have singularities when $p^2 = m^2$, $q^2 = m^2$, $r^2 = m^2$. By finding the coefficient of the singularity when p, q, and r have the appropriate orientations relative to one another, one can find the required matrix element. The method of doing so is given by the reduction formulas of Lehmann et al. (5).

In electrodynamics, the procedure would be to associate with each point y a spacelike path $P(y)$. We further demand that the prescription for specifying the path be independent of the point—for instance, we could take all the paths to be straight lines in a given direction. It would be possible to be more general and only to require that the prescription for specifying the path be independent of the point at sufficiently large distances from a given point. We could then repeat the procedure of the path-independent case, using instead of (6.1) the integral

$$\int dy dz d \, e^{-ipy-iqz+iz} T \langle \Phi^*\{y, P(y)\} \Phi^*\{z, P(z)\} \Phi\{w, P(w)\} \Phi\{x, P'\} \rangle. \tag{6.2}$$

The paths associated with the first three operators, which correspond to the states in which we are interested, are fixed by our prescription. The path associated with the last operator, whose matrix element we wish to find, must be specified when defining the operator and will not, in general, be the path fixed by our prescription.

That it is justified to use the Lehman-Symanzik-Zimmermann method for path-dependent operators, and that the result will be independent of the paths used in our prescription, is an immediate consequence of the assumptions which go into their formalism. The operators $\phi^*(x)$, $\phi^*(y)$, and $\phi(w)$ in (6.1) can be any operators with the correct quantum numbers and which transform under the momentum operation according to

$$[p, \phi(x)] = i \frac{\partial \phi(x)}{\partial x}. \qquad (6.3)$$

The result obtained will be independent of the operators chosen. Further, we need only require that (6.3) be satisfied at sufficiently large distances. It is not necessary to demand that ϕ transforms covariantly under the angular-momentum operation, and the procedure could be used for problems without spherical symmetry.

Our operators defined with the prescribed paths satisfy the requirements just mentioned, so that the Lehmann-Symanzik-Zimmermann formalism can be applied to them. The operators over which we integrate must first be correctly renormalized, and the renormalization constants will in general depend on the paths chosen. For a given choice of path they will depend on the direction of motion and on the spin orientation, as the paths cannot be specified in a spherically symmetrical way. The matrix elements, however, will be independent of the method of specifying the paths associated with the operators over which we integrate. They will, of course, depend on the path used to define the operator whose matrix element we wish to find.

The reasoning just given also holds if there are bound states, on applying the formalism developed by Nishijima (6), Mandelstam (7), or Zimmermann (8).

The essential features of our argument is that the newer covariant formalisms do not associate a particular operator with a particular state, except in so far as conservation laws are concerned. The arbitrariness in specifying the paths does not therefore propagate itself into the specification of the states.

Instead of using definite paths as has just been described, we could work within a definite gauge. In particular we could take Coulomb gauge operators, so that calculations of scattering cross-sections or of matrix elements (of gauge-independent quantities!) in this gauge would give the same results as gauge-independent calculations. It should also not be difficult to prove that the coefficients of the singularities of the Lorentz-gauge Green's functions are the same as those of the gauge-independent Green's functions—always after allowing for differences in renormalization constants. This would verify that the conventional S-matrix calculations are correct.

VII. CONCLUDING REMARKS

The theory outlined in this paper does not involve the electromagnetic potentials except as calculational aids. It is manifestly Lorentz-invariant and introduces no unphysical states or indefinite metric. The dependence of the field operators on the paths may seem a heavy price to pay for these features, desirable as they are. We feel, however, that this path dependence is a fundamental physical property. Failure to take this into account leads to the difficulties or inelegance of current methods of quantizing the electromagnetic field.

The path dependence is intimately connected with the arbitrariness in the choice of the phase factors in the operators of charged fields. One can choose the phase factors arbitrarily at one point, which we have taken to be at infinity. Once this has been done, however, the phase factors at a neighboring point will be fixed. If one chooses different phase factors at two neighboring points one has to add extra terms to the equations of motion. Such terms are unphysical and can be removed by a gauge transformation, which re-establishes the correct choice of phase at the second point.

The correct choice of the phase factors at a field point a finite distance away from the reference point cannot be determined directly. One can, however, construct a path between the two points and pass continuously from the reference point to the field point, keeping the phase factors of all the operators the same along the path. By doing so one obtains the path-dependent operators discussed in the foregoing sections. If one obtained the same result independently of the path chosen, the particles in the theory would have no interaction with the electromagnetic field.

In nature, however, the operator may depend on the path, and it does so when there is an electromagnetic field present. The dependence is given by Eqs. (2, 8). In the presence of an unquantized electromagnetic field, two operators at the same point but defined by different paths simply differ by a phase factor. If the electromagnetic field is quantized, the two operators will differ by a "q-number phase factor."

The dependence of operators on the path used to define them can be directly measured in the experiment proposed by Aharonov and Bohm. They take the simplest situation where the wave functions of charged particles are not subjected to second quantization, and the experiment exhibits the path dependence of the wave functions in the presence of an unquantized electromagnetic field. The proposal is to separate and recombine a beam of electrons and to examine the interference pattern produced. The path of neither beam passes through an electromagnetic field, but the region between them does contain a field. Each of the beams would then obey the equations of motion in the absence of a field; nevertheless, their phases would be displaced relative to one another and the interference fringes would be altered. No experiment performed at any point in the electron beams would reveal a difference between this situation and the field-free situation, where the interference fringes are unaffected.

The path dependence of the operators of charged fields thus appears to be a deep-seated phenomenon, and there is no point in writing the fundamental equations of electromagnetism in such a way as to obscure it. In the unquantized theory of an electromagnetic field interacting with a charged field, it would not really be necessary to write the equations in a form free from potentials. As long as one kept the equations gauge-covariant (as distinct from gauge-independent) throughout, one would have a theory which essentially contained all the features of a correct theory. By performing a gauge transformation so as to remove the potentials along a certain path, one would obtain the path-dependent operators. In a quantized theory, however, one has to start with gauge-independent operators, as it is only between such operators that the commutation rules are given. Once the commutation rules between gauge-independent operators are known, potentials corresponding to a special gauge may be introduced.

The electromagnetic field viewed in this light bears a striking analogy to the Riemann tensor of general relativity. The coordinate system instead of the phase is now the arbitrary quantity. Once the system has been specified at one point, we must obtain it at a neighboring point by parallel displacement if we do not wish to add extra terms to the equations of motion. As long as the coordinate systems obtained at a point by parallel displacement from a reference point do not depend on the path, there is no permanent gravitational field. They may, however, depend on the path, and the difference between the field variables for two neighboring paths with the same end-point can be expressed in terms of the Riemann tensor.

When the result of a parallel displacement depends on the path, we say that the space is curved. Correspondingly, we might introduce a notion of "phase" or "gauge" curvature, which implies a dependence of the process of fixing the phase on the path chosen to do so. Gauge curvature is less universal than ordinary curvature, as there is an arbitrariness in phase associated with each conservation law, though only the charge conservation law appears to give rise to a path dependence. In the following paper we shall make use of the analogy between gauge and ordinary curvature to quantize the gravitational field.

APPENDIX. NON-SPACELIKE PATHS

In the main text we obtained a complete theory of quantum electrodynamics without introducing paths which were not space like. It may be of interest, however, to show that this restriction can be removed at the expense of extra complication. When quantizing the gravitational field it will be necessary to consider these more general paths, as we cannot be certain whether or not a particular path is spacelike.

QUANTUM ELECTRODYNAMICS

We shall again begin from the conventional formalism. To define a path-dependent operator it is now necessary to specify not only the shape of the path, but also the order in which the various elements are to be taken. The ordering of the operator itself with respect to the elements of the path must also be specified. We then write equations corresponding to (2.6).

$$\Phi(x, P) = O\phi(x) \exp\left\{-ie \int_{-\infty}^{x} d\xi_\mu A_\mu(\xi)\right\}, \quad (\text{A.1a})$$

$$\Phi^*(x, P) = O\phi(x) \exp\left\{ie \int_{-\infty}^{x} d\xi_\mu A_\mu(\xi)\right\}, \quad (\text{A.1b})$$

where the symbol O indicates that the exponential is to be regarded as a product of exponentials for the various elements of the path, ordered in the appropriate manner with respect to one another and to $\phi(x)$.

When writing down a product of two operators it is necessary to specify the ordering of each element of one path with respect to all the elements of both paths and with respect to both operators. In addition, the ordering of the two operators must be specified, needless to say. We can then write down definitions corresponding to (A.1). A product of two operators $X(x, P_1)$ and $Y(y, P_2)$ will be denoted by the symbol $\overline{XY}(x, y, P_1, P_2)$. Only if the operator X and all the elements of its path are ordered before (or after) the operator Y and all the elements of its path will it be a true product

$$X(x, P_1)Y(y, P_2)\{\text{or } Y(yP_2)X(xP_1)\}.$$

We may point out that this complication occurs as long as one of X and Y is path-dependent.

The equations corresponding to (2.8) will now be

$$\delta_z\Phi(x, P) = -ie\overline{\Phi F_{\mu\nu}}(x, z, P)\sigma_{\mu\nu}, \quad (\text{A.2a})$$

$$\delta_z\Phi^*(x, P) = ie\overline{\Phi^* F_{\mu\nu}}(x, z, P)\sigma_{\mu\nu}. \quad (\text{A.2b})$$

The ordering of the path elements on the right of (A.2) is the same as on the left, while F is to be ordered in the position corresponding to the path element which is being varied. It is easily checked that these formulas follow from (A.1). Similar formulas may be obtained for the path dependence of products of operators.

In writing down commutation relations, one must now consider the commutation of two operators with one another, of an operator with an element of a path, and of two path elements. As usual, one can only write down equal-time commutation rules in simple terms. The commutation relations between two operators will be the same as in (3.10)–(3.12), except that (3.11b) and (3.11c) become simply

$$[\overline{\overset{(\cdot)}{\Phi}, F_{0i}}(x, y, P)]_{\Phi,F} = [\overline{\overset{(\cdot)}{\Phi}, F_{0i}}(x, y, P)_{\Phi,F}} = 0. \tag{A.3}$$

The subscripts Φ, F indicate that the orders of Φ and F are to be interchanged, but that the ordering of both operators with the path elements remains unchanged. The terms on the right of (3.11) are due to the commutation of $F_{0i}(y)$ with the path elements of $\overset{(\cdot)}{\Phi}(x, P)$—in conventional terminology, they are due to the commutation of $F_{0i}(y)$ with the potentials in the path integral—and thus they do not appear in (A.3). The commutators in this formulation of the theory thus take a simpler form than they did in the main text.

From (3.11b) and (3.11c) we can also obtain the commutation relations between the operator F_{0i} and an element of a path defining Φ. They become

$$[\overline{\overset{(\cdot)}{\Phi}, F_{0i}}(x, y, P)]_{p,F} = -e \int d\xi_i \, \delta^3(y - \xi) \Phi(x, P), \tag{A.4a}$$

$$[\overline{\overset{(\cdot)}{\Phi}{}^*, F_{0i}}(x, y, P)]_{p,F} = e \int d\xi_i \, \delta^3(y - \xi) \Phi^*(x, P). \tag{A.4b}$$

The subscripts p, F mean that the ordering of F and an element p of the path defining Φ is to be altered; all other orderings are to be unchanged. The integral on the right is over the path p. All other equal-time commutators between operators and path elements are zero, and so are all equal-time commutators between path elements.

The equations (A.4) follow from the formula (A.3) and the condition that F_{0i} commutes with a path element separated from it by a spacelike distance.

It may be checked that the left-hand side of (3.3), and of (3.4) with $\nu = 1, 2,$ or 3, commute with all operators or elements. These equations are therefore also valid if Φ, Φ^*, F, and j form parts of generalized products with the path elements arranged arbitrarily. The left-hand side of (3.4), which we denote by $\tilde{j}_0$, has the following commutation relations:

$$[\overline{\overset{(\cdot)}{\Phi}, \tilde{j}_0}(x, y, P)]_{\Phi,j} = e\delta(x - y) \overset{(\cdot)}{\Phi}(x, P), \tag{A.5a}$$

$$[\overline{\overset{(\cdot)}{\Phi}, \tilde{j}_0}(x, y, P)]_{p,j} = -ed\xi_i \frac{\partial \delta^3(x - y)}{\partial \xi_i} \overset{(\cdot)}{\Phi}(x, P). \tag{A.5b}$$

In the commutators between $\overset{(\cdot)}{\Phi}{}^*$ and $\tilde{j}_0$, the sign is reversed. Equations (A.5) show that the result of commuting an operator or path element with $\tilde{j}_0$ is to effect a gauge transformation. If $\tilde{j}_0$ is commuted with an operator and all the elements of its path the result will be zero. This is of course necessary for the consistency of the equation $\tilde{j}_0 = 0$. .

Once the equal-time commutators between path elements and operators are known, we can obtain the generalized products introduced here, in terms of true products, for spacelike paths. We should then like to use the equations of motion to calculate them for arbitrary paths. Before doing so, however, we require one supplementary equation. The commutator between a path element and another element will have the same general form as (A.4):

$$[\overset{(i)}{\Phi}, B(x, y, P_1, P_2)]_{p_1,x} = d\xi_\mu G_{\mu,x}\{\xi, B(y, P_2)\}\Phi(x, P_j). \quad (A.6)$$

Such a formula is true whether the element p_1 of the path P_1 defining Φ is commuted with the operator $B(x, P_2)$ itself or with an element p_2 of its path (if it is path-dependent). This operator or element has been denoted by X. The function G on the right will depend on the operator B and on the nature of X, but it will not depend on x or P_1. The position of the element p_1 has been denoted by ξ, and the element itself by $d\xi_\mu$. All elements must of course be ordered in the same position on both sides of (A.6). G will not be known explicitly except for spacelike paths.

From the path-dependence equation (2.8) we can obtain one equation for G:

$$\frac{\partial}{\partial \xi_\mu} G_{\nu,x}\{\xi, B(y, P_2)\} - (\partial/\partial\xi_\nu)G_{\mu,x}\{\xi, B(y, P_2)\} = -ie[\overline{F_{\mu\nu}(\xi), B(y, P_2)}], \quad (A.7)$$

We also require an equation for $\partial G_\mu/\partial \xi_\mu$ before we can calculate it in principle. We shall find the required equation in the conventional formalism and shall introduce it as a further postulate into our formalism.

The function G_μ on the conventional formalism is just

$$G_{\mu,x}\{\xi, B(y, P_2)\} = -ie[A_\mu(\xi), B(y, P_2)]_{A,x}, \quad (A.8)$$

so that we require an equation for the commutator between $\partial A_\nu/\partial \xi_\nu$ and another operator. To find such an equation let us examine the commutator of $\tilde{j}_\nu$ and that operator where $\tilde{j}_\nu$ is defined as the left-hand side of (3.4). In the conventional formalism

$$\tilde{j}_\nu(\xi) = \frac{\partial^2 A_\nu}{\partial \xi_\mu^2} + j_\nu - \frac{\partial}{\partial \xi_\nu}\frac{\partial A_\mu}{\partial \xi_\mu}. \quad (A.9)$$

Now, if we are using the Gupta-Bleuler gauge, the sum of the first two terms is zero and it will therefore commute with all elements or operators. We may therefore assert that there exists an operator function $H_x\{\xi, B(y, P_2)\}$ defined by

$$[\tilde{j}_\nu(\xi), B(y, P_2)]_{j,x} = (\partial/\partial\xi_\nu)H_x\{\xi, B(y, P_2)\}. \quad (A.10)$$

From (A.8), (A.9), and (A.10) it follows that

$$\frac{\partial}{\partial \xi_\mu} G_{\mu,x}\{\xi, B(y, P_2)\} = H_x\{\xi, B(y, P_2)\}. \quad (A.11)$$

In the present formalism we shall therefore postulate the divergence condition (A.11), with H defined by (A.10). The equations of motion, supplemented by (A.11), now allow operators to be calculated for arbitrary paths once they are known for spacelike paths.

When making the correspondence between the present formalism and the conventional formalism for general paths, we had to take a particular gauge—the Gupta-Bleuler gauge. The distinguishing feature of this gauge is its causal commutation rules. A "gauge-independent" operator is defined not as one which is completely independent of the gauge used for the quantized potentials, but as one which remains unchanged under a c-number gauge transformation. Such an operator, when acting on a physical state, reproduces a physical state, and its matrix elements between physical states are independent of the components of scalar and longitudinal photons present. When formulating the theory in terms of such operators we need not therefore mention the unphysical states.

Acknowledgment

The author would like to acknowledge a valuable discussion with Dr. Irwin Goldberg.

Received: February 12, 1962

Note added in proof: Several people have raised the question whether the theory formulated here is local or not. The path dependence of operators may be regarded as a kind of nonlocality. However, the physical basis for this nonlocality is understood and has been disussed in the paper. Further, the nonlocality introduces no arbitrariness. The requirement that two operators commute when their paths are spacelike separated is just as restrictive as the spacelike commutativity in other field theories.

REFERENCES

1. B. Zumino, *J. Math. Phys.* **1**, 1 (1960).
2. I. Goldberg, *Phys. Rev.* **112**, 1361 (1958).
3. P. G. Bergmann and I. Goldberg, *Phys. Rev.* **98**, 531 (1955).
4. R. E. Peierls, *Proc. Roy. Soc. (London)* **213**, 143 (1952).
5. H. Lehmann, K. Symanzik, and W. Zimmermann, *Nuovo cimento* **1**, 425 (1955).
6. K. Nishijima, *Prog. Theoret. Phys. (Kyoto)* **10**, 549 (1953); *ibid.* **12**, 279 (1954).
7. S. Mandelstam, *Proc. Roy. Soc. (London)* **233**, 248 (1955).
8. W. Zimmermann, *Nuovo cimento* **10**, 597 (1958).

Feynman Rules for Electromagnetic and Yang–Mills Fields from the Gauge-Independent Field-Theoretic Formalism

S. Mandelstam

Phys. Rev. **175** (1968) 1580–1623

Abstract

The Feynman rules for the Yang–Mills field, originally derived by Feynman and DeWitt from S-matrix theory and the tree theorem, are here derived as a consequence of field theory. Our starting point is the gauge-independent, path-dependent formalism which we proposed earlier. The path-dependent Green's functions in this theory are expressed in terms of auxiliary, path-independent Green's functions in such a way that the path-dependence equation is automatically satisfied. The formula relating the path-dependent to the auxiliary Green's functions is similar to the classical formula relating the path-dependent field variables to the potentials. By using a notation similar but not identical to Schwinger's functional notation, the infinite set of equations satisfied by the Green's function can be replaced by a single equation. When the equation for the auxiliary Green's functions of electromagnetism is solved in a perturbation series, the usual Feynman rules result. For the Yang–Mills field, however, one obtains extra terms; such terms correspond precisely to the closed loops of fictitious scalar particles introduced by Feynman, DeWitt, and Faddeev and Popov.

Quantization of the Gravitational Field

Stanley Mandelstam

Department of Mathematical Physics, University of Birmingham, Birmingham, England

The scheme proposed in the preceding paper for the gauge-independent quantization of the electromagnetic field is here applied to the coordinate independent quantization of the gravitational field. Einstein's theory is first reformulated so as to avoid reference to a coordinate system. The quantization of the resulting theory is then carried through. No mention is made of unphysical variables such as the metric tensor except for the purpose of linking the present formalism with more conventional theories, the gravitational field is described by the Riemann tensor and its connection with space curvature is clear from the outset. First-order perturbation calculations are carried out and give results equivalent to those of "flat-space" theories.

I. INTRODUCTION

When one attempts to quantize Einstein's theory of gravitation one is immediately faced with the difficulty of interpreting a quantized metric. Our present quantum field theory assumes the existence of a c-number coordinate system as a fundamental postulate, and no one has yet succeeded in generalizing it to a q-number coordinate system.

Present quantum theories of the gravitational field generally use "flat space." They start with the Lagrangian of the classical theory and apply the normal methods of quantization to it, treating the variables $g_{\mu\nu}$ as ordinary variables which have no connection with the metric. Owing to ambiguities in the ordering of the factors such a procedure cannot be carried through exactly. However, it may be that one can write down a Lagrangian to any order in the gravitational constant and obtain consistent equations of motion up to that order. Such a program was originally suggested by Gupta (1) and carried out by him to first order. Feynman has recently made an extensive investigation on the problem of renormalization in this theory.

Quantization in flat space can only be regarded as a provisional solution of the problem for several reasons. One would like to be able to formulate the equations of a theory exactly, even though approximations have to be made in their solution. This cannot be done, or, at any rate, has not yet been done, in the flat-space gravitation theory. Also, one has to introduce an indefinite metric and unphysical states, just as in quantizing electrodynamics with the

Lorentz gauge. The coordinate condition used by Gupta is similar in appearance to the Lorentz condition in electrodynamics.

But the main objection to this method of quantization lies surely in the physical sacrifices it makes by going to flat space. The variables specifying the coordinates are numbers without physical significance which can be chosen in an infinite variety of ways. On the other hand, distances in space–time, which are physically significant entities, are related to the coordinates and the field variables in a manner which has not been elucidated when the metric is quantized. It may be possible to add to the theory a prescription for interpreting its results physically. If it could then be shown that the physical predictions of the theory were independent of the coordinate conditions used, and that they tended to the predictions of unquantized general relativity in the classical limit, we would have a satisfactory theory.

Some progress has actually been made in this direction by Thirring (2), who has shown essentially that, if one performed general c-number coordinate transformations, one could make appropriate changes in the variables $g_{\mu\nu}$ of the quantized theory, so as to leave physical quantities unaltered. This certainly indicates the connection of the Gupta variables to the metric, and is a step in the direction of putting the theory on a better physical footing. However, the ambiguity in specifying coordinate conditions in quantized general relativity is much larger than can be covered by a c-number coordinate transformation, and the basic difficulties of the "flat space" approach remain. Also, Thirring restricted his treatment to lowest-order perturbation theory.

The method to be proposed here avoids mentioning quantities such as $g_{\mu\nu}$ which are dependent on the coordinate system. The fundamental variables of the gravitational field will be the components of the Riemann tensor, and the connection between these variables and "space curvature" will be evident from the outset. There is thus no ambiguity associated with coordinate conditions, and all "distances" which appear in the theory are physical distances, which could in principle be measured, rather than the completely unphysical coordinate differences of the flat-space theory. We may therefore claim that the ideas which underlie the classical theory of gravitation are present in the foundations of the quantized theory.

The concepts on which the coordinate-independent theory of gravitation are based are closely analogous to the concepts underlying the gauge-independent theory of electrodynamics, which formed the subject of the preceding paper (3), hereafter referred to as I. We start with a reference point which we take to be at infinity as before. Our attention will be confined to systems where there is no matter at sufficiently large distances. We construct a coordinate basis, i.e., a set of orthogonal directions, at the reference point. If we then want to specify an operator at another point, we must first define the position of the second point, without using a coordinate system. This we do by giving a prescription for constructing a path from the reference point to the field point. As we construct the path, we take the coordinate basis with us by parallel displacement, and we use this basis to fix the direction of the next path element. To illustrate what we mean, a specimen prescription may run as follows: Move a distance d_1 from the reference point in the x-direction, to the point P_1 taking the coordinate basis along by parallel displacement. Now move a further distance d_2 in the y-direction (defining the y-direction with respect to the coordinate basis which has been taken to P_1). This takes us to a point P_2. Move the coordinate system by parallel displacement to P_2 as before. Finally, move a distance d_3 from P_2 in the x-direction—defined with respect to the coordinate basis at P_2. Perform the required measurement at the point just reached.

The theory of the gravitational field to be described operates in terms of these path-dependent variables. It thus avoids reference to a coordinate system. If it turned out that our measurements depended only on the total displacement in each of the four directions, and not on the other details of the path, we would have a "flat space" and would not require to introduce path-dependent variables. In the presence of a gravitational field a measurement will depend on the details of the path, and the difference between two operators with neighboring paths will be given by a familiar formula of Einstein's theory, or, strictly of Riemannian geometry. It resembles the corresponding path-dependence formula of electrodynamics, but the Riemann tensor replaces the electromagnetic field. The relation between this tensor and space curvature is thus evident at an early stage. Because of the path-dependence formula, the number of *independent* variables will be equal to the number of points, not the number of paths.

As is to be expected, the theory of gravitation is much more complicated than electrodynamics. One basic reason for the complication is that, while in electrodynamics the electromagnetic field variables are not path-dependent, all the variables are path-dependent in the presence of a gravitational field. Also, the path-dependence equation for $\Phi(x, P)$ in the electromagnetic case involved only $\Phi(x, P)$ and the electromagnetic field, whereas in the gravitational case it also involves derivatives of Φ. In the classical theory, this has the effect of displacing the end-points of two paths relative to one another. The manner in which the displacement comes about should become clear in the next section.

In classical gravitational theory, unlike electrodynamics, one cannot tell whether two paths lead to the same point from a knowledge of the specifications of the paths alone. However, the question has a meaning and can in principle be answered if we have sufficient knowledge of the Riemann tensor. For it is a consequence of the path-dependence formula that, in certain cases, *any* measurement performed at the end of two different paths will yield the same result. If an observer at the end of one path sends out a signal, the observer at the end of the other will perceive it. We may therefore say that the two paths lead to the same point in space-time, and the physical consequences of two observers being at the same point will be fulfilled. In the quantized gravitational theory the conception of two paths leading to the same point loses its meaning, as the Riemann tensor is now an operator. It was always expected that the uncertainty principle would affect the precise location of points in a quantized theory of gravitation, and this is the form the uncertainty takes.

The quantized theory of the gravitational field can be expanded in a perturbation series by a procedure very similar to that used in electrodynamics. We shall see that in lowest order it agrees formally with the flat-space theory, though, as in the case of the electromagnetic field, the functions must be interpreted differently.

In order to avoid complications about ordering of factors we shall begin with the classical theory of the gravitational field. We shall formulate it in terms of path-dependent variables, and shall avoid quantities such as $g_{\mu\nu}$ which depend on the coordinate system. In the following sections we shall attempt to quantize the theory.

II. COORDINATE-INDEPENDENT THEORY OF THE CLASSICAL GRAVITATIONAL FIELD

GENERAL FORMALISM

We shall take as our example the interaction between a scalar and a gravitational field. The equation expressing the path dependence of operators will first be required and here we shall treat a vector field as well, as a scalar field is rather a special case.

Suppose we have two paths P_1 and P_2, defined according to the same prescription except that, after a portion of P_3 common to both has been traversed, the two differ by a small area $\sigma_{\kappa\lambda}$. According to a fundamental formula of Riemannian geometry, two vectors taken by parallel displacement along the paths which were identical before reaching the small area will differ after passing it by an amount

$$da_\mu = \tfrac{1}{2} R_{\mu\nu\kappa\lambda} a_\nu \sigma_{\kappa\lambda} . \tag{2.1}$$

In other words, all vectors are rotated by an amount $-\tfrac{1}{2} R_{\mu\nu\kappa\lambda}\sigma_{\kappa\lambda}$ in the $\mu\nu$-plane. As our theory always measures quantities in terms of a local Euclidean coordinate system taken along the path by parallel displacement, there is no distinction between upper and lower indices.

The vectors defining our coordinate system are also subjected to the rotation (2.1). The two coordinate systems will therefore not be the same after the area $\sigma_{\kappa\lambda}$ has been passed, and it is this difference which is responsible for the path dependence of the field variables. The rotation will have two effects. First, all vector operators will appear to be rotated by an amount $\tfrac{1}{2} R_{\mu\nu\kappa\lambda}\sigma_{\kappa\lambda}$ relative to the coordinate system; spinor and tensor operators will be similarly affected. Secondly, the path itself will be altered, since directions are always defined relative to the local coordinate system taken along the path by parallel displacement. The direction of P_2 will therefore be rotated relative to that of P_1 by an amount $-\tfrac{1}{2} R_{\mu\nu\kappa\lambda}\sigma_{\kappa\lambda}$ in the $\mu\nu$ plane, and, after passing the area $\sigma_{\mu\lambda}$, the two paths will begin to diverge.

The path dependence of a vector field variable in a weak gravitational field will therefore be given by the equation

$$\delta_z A_\mu(x, P_1) = \tfrac{1}{2} R_{\mu\nu\kappa\lambda}(z, P_3) A_\nu(x, P_1)\sigma_{\kappa\lambda} \\ - \tfrac{1}{2} R_{\iota\nu\kappa\lambda}(z, P_3)(x - z)_\iota \frac{\partial A_\mu(x, P_1)}{\partial x_\nu} \sigma_{\kappa\lambda}. \quad (2.2)$$

z is the point on P_1 at which the path is being varied, and P_3 is that part of P_1 leading to z. The second term represents the effect of the change of the path.

The two terms in (2.2) can be comprised in a single formula

$$\delta_z A(x, P_1) = -\frac{i}{4} R_{\mu\nu\kappa\lambda}(z, P_3)[J_{\mu\nu}(z, P_3), A(x, P_1)]\sigma_{\kappa\lambda}. \quad (2.3)$$

$J(z, P_3)$ is the total angular momentum of the system *about* z, the end-point of P_2, and the symbol [ ] denotes i times the Poisson bracket. One can check directly from the fundamental Poisson-bracket relations that (2.3) is equivalent to (2.2). Alternatively, it is known quite generally that the effect of taking the Poisson bracket of the angular momentum and another variable is to rotate the system, so that (2.3) could have been obtained without the intermediary of (2.2). Equation (2.3) is thus true for any scalar, tensor, or spinor quantity A, and we shall henceforth restrict ourselves to scalar and gravitational fields.

In the general case of a strong gravitational field, (2.2) is no longer correct. We shall now *define* the symbol $i[J_{\mu\nu}(z, P_3), A(x, P_1)]\alpha_{\kappa\lambda}$ to mean the change in the variable $A(x, P_1)$ caused by rotating the local coordinate system at the end of P_3 by an amount $\sigma_{\kappa\lambda}$ in the $\mu\nu$ plane, so that (2.3) is still true. This definition is adopted with a view to quantizing the field, as it agrees with the usual definition of the angular-momentum operation in quantum mechanics. The rotation will affect both the components of tensors (or spinors) and the direction of the path after the end of P_3. We thus reach the conclusion mentioned in the introduction that one effect of separating the paths by the area $\sigma_{\mu\nu}$ is to displace their end-points.

Equation (2.3) resembles Eq. (2.14) of I in its form. The angular-momentum operator now replaces the charge operator, as the former bears the same relation to coordinate transformations as the latter to gauge transformations.

We have emphasized that one cannot determine whether or not two paths lead to the same point without knowledge of the Riemann tensor for intermediate paths. The coordinate x in an expression such as $A(x, P_1)$ simply refers to the total displacement along the path in the local coordinate system, and the fact that two points have the same value of x does not imply that they are the same point if the paths are different. We could dispense with the variable x altogether, but it will prove convenient to keep this notation.

Covariant derivatives can be defined in the same way as gauge-invariant derivatives in electrodynamics. They will just be the ordinary derivatives in the local coordinate system, and are therefore the equivalent of covariant derivatives as they are usually defined in Riemannian geometry. We shall denote them by the symbol $\partial_\kappa(x)A(x, P)$. One can also define a covariant derivative of a path-dependent operator taken at a point z along the path. This simply corresponds to adding the element dx_κ, not at the end of the path, but at the point z. Such derivatives will be written $\partial_\kappa(z)A(x, P)$.[1] The formula for the commutator of two differential operators which replaces (2.13) of I is

$$\{\partial_\kappa(z)\partial_\lambda(z) - \partial_\lambda(z)\partial_\kappa(z)\} A(x, P)$$
$$= -\frac{i}{2} R_{\mu\nu\kappa\lambda}(z, P')[J_{\mu\nu}(z, P'), A(x, P)], \quad (2.4)$$

a well-known result. P' is the portion of P leading to z. Differential operators $\partial_\kappa(z)$ and $\partial_\lambda(w)$, referring to two different points, commute. To avoid proliferation of brackets, we shall adopt the convention that a differential operator or a product of differential operators acts only on the factor immediately to the right. If a series of factors to the right of one or more differential operators is enclosed in brackets the operators act on the whole bracket, of course.

We must impose integrability conditions on the Riemann tensor, in order that the difference between variables defined according to two paths separated by a finite area be unique. The difference between the integral of (2.3) taken over two small surfaces with the same bounding curve will be

$$\frac{i}{4}\epsilon_{\rho\sigma\mu\nu} \int d\tau_\rho \partial_\sigma \{R_{\mu\nu\kappa\lambda}(z, P_3)[J_{\mu\nu}(z, P_3), A(x, P_1)]\}, \quad (2.5)$$

where the integral is over the volume between the surfaces, and ∂_σ denotes co-

---

[1] To avoid confusion we emphasize that the variable z in the expression $\partial(z)$ is not the variable with respect to which we are differentiating. As our dynamical variables depend only on a single path, there is no ambiguity regarding the variable with respect to which we are differentiating. We shall later adopt a special convention for δ-functions, which depend on two variables.

variant differentiation with respect to the end-point of P_3. The differential operator, besides operating on R, will also operate on J if σ is the same as κ or λ. For (2.5) to be zero, two conditions must therefore be fulfilled

$$\epsilon_{\rho\nu\mu\lambda} R_{\mu\nu\kappa\lambda} = 0, \tag{2.6}$$

$$\epsilon_{\rho\sigma\kappa\lambda} \partial_\sigma R_{\mu\nu\kappa\lambda} = 0. \tag{2.7}$$

The first of these equations represents the usual symmetry requirements on R. Taken in conjunction with the antisymmetry of $R_{\mu\nu\kappa\lambda}$ in the variables $\mu\nu$ and $\kappa\lambda$, they imply that R is symmetric under the interchange of the pair $\kappa\lambda$ with $\mu\nu$ and also that

$$\epsilon_{\mu\nu\kappa\lambda} R_{\mu\nu\kappa\lambda} = 0. \tag{2.8}$$

The set of equations (2.7) are the Bianchi identities.

PATH-DEPENDENT VARIABLES IN FIRST APPROXIMATION

Before proceeding further it will be useful to obtain expressions for our path-dependent variables in terms of the path-independent but coordinate dependent variables of the usual formalism. We shall work in terms of the variables

$$\gamma_{\mu\nu} = g_{\mu\nu} - \delta_{\mu\nu}$$

and shall restrict ourselves to the case where $\gamma_{\mu\nu}$ is small and of second order, while its derivatives are small and of first order. The formulas will therefore apply either in a small region in an arbitrary gravitational field, or in any region in a weak gravitational field. For vectors in the ordinary formalism we shall use neither the covariant nor contravariant components, but "neutral" components defined by

$$a_\mu = g_{\mu\nu}^{\frac{1}{2}} a^\nu = g^{(1/2)\mu\nu} a_{(\nu)} \tag{2.9}$$

where a^ν is the contravariant νth component and $a_{(\nu)}$ the covariant νth component of a. The variable a_μ is just the μth component of a in that Euclidean coordinate system which is obtained from the local nonEuclidean system by distortion without rotation. As the neutral components are path-independent but coordinate-dependent, they will be regarded as conventional variables. In fact, such variables have to be used for spinors.

We thus require a formula for a path-dependent variable $A(x, P)$ in terms of a coordinate-dependent variable $a(x)$. The difference between the two variables arises from two sources. The coordinate system which we taken along P by parallel displacement (and in which we measure $A(x, P)$) is rotated relative to the nonEuclidean coordinate system in which we measure $a(x)$; also, the coordinates of the end-point of P in the nonEuclidean system are slightly different from x, the total displacement in the local Euclidean system. To calculate the first effect, we make use of the formula giving the change of the components of a vector a, in the nonEuclidean system, on being subjected to a parallel displacement dz_μ:

$$da_\mu = -\frac{1}{2}\left\{\frac{\partial \gamma_{\mu\lambda}(z)}{\partial z_\nu} - \frac{\partial \gamma_{\nu\lambda}(z)}{\partial z_\mu}\right\} a_\nu\, dz_\lambda. \tag{2.10}$$

This formula applies to the neutral components defined by (2.9) and holds only when the γ's are small. The coordinate system thus undergoes a rotation of

$$\frac{1}{2}\left\{\frac{\partial \gamma_{\mu\lambda}(z)}{\partial z_\nu} - \frac{\partial \gamma_{\nu\lambda}(z)}{\partial z_\mu}\right\} dz_\lambda$$

in the $\mu\nu$ plane relative to the nonEuclidean system. The difference between $A(x, P)$ and $a(x)$ due to this rotation will be

$$\frac{i}{4}\left\{\frac{\partial \gamma_{\mu\lambda}(z)}{\partial z_\nu} - \frac{\partial \gamma_{\nu\lambda}(z)}{\partial z_\mu}\right\} [J_{\mu\nu}(z), a(x)]\, dz_\lambda,$$

where $J_{\kappa\lambda}(z)$ is the angular momentum about the point of rotation. Integrating over the path, we find for the total difference between $A(x, P)$ and $a(x)$ due to this rotation:

$$\{A(x, P) - a(x)\}_1 = \frac{i}{4}\int_P^x dz_\lambda \left\{\frac{\partial \gamma_{\mu\lambda}(z)}{\partial z_\nu} - \frac{\partial \gamma_{\nu\lambda}(z)}{\partial z_\mu}\right\} [J_{\mu\nu}(z), a(x)]. \tag{2.11}$$

This includes the effect of the rotation of the coordinate system on the remainder of the path.

For the other contribution to the difference between $A(x, P)$ and $a(x)$, we observe that a displacement dz in the Euclidean system will increase the local nonEuclidean coordinate by $dz_\mu - \tfrac{1}{2}\gamma_{\mu\lambda}\, dz_\lambda$. The μ coordinate of the end-point of the path in the nonEuclidean system is thus $x_\lambda - \tfrac{1}{2}\int_P^x dz_\lambda \gamma_{\mu\lambda}$, so that

$$\{A(x, P) - a(x)\}_2 = -\frac{1}{2}\int_P^x dz_\lambda\, \gamma_{\mu\lambda}\, \frac{\partial a(x)}{\partial x_\mu}. \tag{2.12}$$

On adding (2.11) and (2.12), we arrive at the formula

$$A(x, P) = a(x) + \frac{i}{4}\int_P^x dz_\lambda \left\{\frac{\partial \gamma_{\mu\lambda}(z)}{\partial z_\nu} - \frac{\partial \gamma_{\nu\lambda}(z)}{\partial z_\mu}\right\} [J_{\mu\nu}(z), a(x)]$$
$$-\frac{1}{2}\int_P^x dz_\lambda\, \gamma_{\mu\lambda}(z)\, \frac{\partial a(x)}{\partial x_\mu}. \tag{2.13}$$

Equation (2.13) is the analogue of Eq. (2.6) of I when the γ's are small. We could obtain formulas correct to any order in the γ's, but there do not appear to be exact closed formulas.

When $\gamma_{\mu\nu}$ is small, the Riemann tensor is given by the formula

$$R_{\mu\nu\kappa\lambda}(z) = \frac{1}{2}\left(\frac{\partial^2 \gamma_{\mu\kappa}}{\partial z_\nu \partial z_\lambda} - \frac{\partial^2 \gamma_{\nu\kappa}}{\partial z_\mu \partial z_\lambda} - \frac{\partial^2 \gamma_{\mu\lambda}}{\partial z_\nu \partial z_\kappa} - \frac{\partial^2 \gamma_{\nu\lambda}}{\partial z_\mu \partial z_\kappa}\right). \tag{2.14a}$$

The equation (2.3) for the path-dependence of the field variable then follows on applying Stokes' theorem to (2.13). There will be a contribution from differentiating $J_{\mu\nu}(z)$ in the first term of (2.13) with respect to z, but since

$$[J_{\mu\nu}(z), a(x)] = i\left\{(x-z)_\mu \frac{\partial a}{\partial x_\nu} - (x-z)_\nu \frac{\partial a}{\partial x_\mu}\right\} \tag{2.15}$$

$$+ \text{ terms independent of } z,$$

it can easily be shown that this contribution just cancels that from the second term of (2.13).

From (2.13) we can obtain formulas for covariant derivatives of path-dependent variables in terms of path-independent variables. If, for instance, the variable A is replaced by a scalar Φ, we may rewrite (2.13) as

$$\Phi(x, P) = \phi(x) + \int dz_\lambda\, I\{z, \phi(x)\}, \tag{2.16}$$

where I_λ is short for the integrand on the right of (2.13), with $a(x)$ replaced by $\phi(x)$. Differentiating (2.16), we obtain

$$\partial_\mu \Phi(x, P) = \frac{\partial \phi(x)}{\partial x_\mu} - \frac{1}{2}\gamma_{\lambda\mu}\frac{\partial \phi(x)}{\partial z_\lambda} + \int dz_\lambda\, I_\lambda\left\{z, \frac{\partial \phi(x)}{\partial z_\mu}\right\}. \tag{2.17}$$

The second term results from differentiating the second integral on the right of (2.13) with respect to its upper limit. (Differentiation of the first integral on the right of (2.13) to its upper limit gives no contribution, as $[J_{\mu\nu}(x), a(x)] = 0$]. We can now differentiate (2.17) again. This time there is a contribution from both integrals on differentiating them with respect to their upper limits, and we obtain

$$\partial_\nu \partial_\mu \Phi(x, P) = \frac{\partial^2 \phi(x)}{\partial x_\nu \partial x_\mu} - \frac{1}{2}\gamma_{\lambda\mu}\frac{\partial^2 \phi(x)}{\partial x_\lambda \partial x_\nu} - \frac{1}{2}\gamma_{\lambda\nu}\frac{\partial^2 \phi(x)}{\partial x_\lambda \partial x_\mu}$$
$$- \frac{1}{2}\left\{\frac{\partial \gamma_{\lambda\mu}(x)}{\partial x_\nu} + \frac{\partial \gamma_{\lambda\nu}(x)}{\partial x_\mu} - \frac{\partial \gamma_{\nu\mu}(x)}{\partial x_\lambda}\right\}\frac{\partial \phi(x)}{\partial x_\lambda} + \int dz_\lambda\, I_\lambda\left(z, \frac{\partial^2 \phi(x)}{\partial x_\nu \partial x_\mu}\right) \tag{2.18}$$

The second and third terms on the right of (2.18) arise because the derivatives on the left of the equation are with respect to the local Euclidean coordinates, whereas those on the right are with respect to the nonEuclidean coordinates.

The third term on the right is the familiar correction to the covariant derivative in the conventional formalism.

EQUATIONS OF MOTION

We shall assume that the scalar field satisfies the Klein-Gordon equation and the gravitational field the Einstein equations. They can be expressed in the new formalism precisely as in the old, so that

$$\{-\partial_\mu^2(x) + m^2\}\Phi(x, P) = 0, \quad (2.19)$$

$$R_{\mu\nu\mu\lambda}(x, P) - \tfrac{1}{2}\delta_{\lambda\nu}R_{\mu\rho\mu\rho}(x, P) = -\kappa T_{\lambda\nu}(x, P), \quad (2.20a)$$

where T is the symmetrical energy tensor for the matter field, given by

$$T_{\lambda\nu}(x, P) = \{\partial_\lambda(x)\Phi(x, P)\}\{\partial_\nu(x)\Phi(x, P)\}$$
$$- \tfrac{1}{2}\delta_{\lambda\nu}(\{\partial_\mu(x)\Phi(x, P)\}^2 + m^2\{\Phi(x, P)\}^2). \quad (2.20b)$$

The constant κ in (2.20a) is Einstein's gravitational constant.

DELTA FUNCTIONS AND KRONECKER DELTA'S

Before writing down the Poisson brackets we require definitions of δ-functions and Kronecker deltas. It would be most satisfactory if one could formulate a generalization of distribution theory, but we shall adopt the more straightforward method of regarding the theory as a limiting case of one with δ-functions replaced by finite peaks.

The δ-function will depend on two paths, P_1 and P_2, the total displacements of which, in their local Euclidean system, will be denoted by x and y. We shall suppose that the shorter of the two paths (P_1 say) coincides with the longer over its entire length, except possibly for an infinitesimal portion at the end. It will be sufficient to restrict our definition to such cases. We may further subdivide the pairs of paths into two cases.

1. If P_1 is shorter than P_2 by a finite amount, the function $\delta^3(x - y)$ is zero when the path between x and y is spacelike and is otherwise undefined. By a spacelike path we mean one between any two points of which signals cannot propagate. The determination whether a particular path is spacelike is a complicated process, like the process of determining whether two paths lead to the same point, but it can in principle be done. When dealing with the quantum theory we shall have to avoid using delta functions for paths which differ by a finite amount.

2. If P_1 and P_2 are almost the same length, so that they differ only by their final portions leading to x and y, we define the function $\delta^3(x - y)$ to be a peaked function of $(x - y)$, provided that the infinitesimal final portions of the paths are smooth. The argument $x - y$, according to our previous definitions of x and y,

will be the displacement between the end-points of the paths in the local Euclidean system. The arguments of the δ-function are really to be regarded as paths rather than points, since all functions occurring in our theory are path-dependent but, as we are restricting our definition to pairs of paths which do not diverge by a finite amount, the δ-function will not depend on the details of the paths other than x and y, at any rate for case 2. It is thus convenient to use the notation $\delta^3(x - y)$ and this is, in fact, the reason for writing our variables as $\Phi(x, P)$ rather than simply as $\Phi(P)$.

If the final portions of P_1 and P_2 are not smooth, we must alter the definition of the delta function to agree with the path-dependence equation. As the paths involved, and therefore the areas between paths, are infinitesimal, this alteration will have no effect unless we differentiate the δ-function. The process of differentiation necessarily produces curves which are not smooth, so that we must then take care. The modification of the definition of the delta function for curves with non-smooth final portions depends on the Kronecker deltas multiplying it. If the subscript α is associated with the point x, μ with the point y, the new definition is

$$\delta_{\alpha\mu}\delta(x - y) = \{\delta_{\alpha\mu}\delta(x - y)\}_0 + \tfrac{1}{2}R_{\alpha\mu\kappa\lambda}(x)\{\delta(x - y)\}_0 \sigma_{\kappa\lambda}, \quad (2.21)$$

where the subscript 0 indicates the ordinary peaked function of $(x - y)$, and $\sigma_{\kappa\lambda}$ is the area by which the final portions of our paths differ from smooth curves. This equation is true to first order in the "lack of smoothness." It is easy to check that the change in (2.21) caused by a variation in the end portions of the paths is given by the fundamental path-dependence equation (2.3).

From (2.21) it follows that

$$\partial_\kappa(x)\partial_\lambda(x)\{\delta_{\alpha\mu}\delta(x - y)\} = (\partial_\kappa(x)\partial_\lambda(x)\{\delta_{\alpha\mu}\delta(x - y)\})_0 + \tfrac{1}{2}R_{\alpha\mu\kappa\lambda}(x),$$

the second term arising from differentiating the area $\sigma_{\mu\nu}$. It is thus clear that the second term in (221) is important if we are to differentiate more than once, and also that it gives results consistent with (2.4). We cannot manipulate Kronecker deltas by the usual rules if they occur under two or more differential operators. If we were to differentiate more than three times we would have to expand (2.21) to higher order in the lack of smoothness.

When writing down differential operators acting on δ-functions, such as $\partial_\alpha(x)\delta(x - y)$, the coordinate x in the differential sign will refer both to the variable of differentiation (involving the plus-minus relationship in the flat space) and to the point at which the differentiation is being performed. The contraction

$$\partial_\alpha(x)\{\delta_{\alpha\kappa}\delta(x - y)\} = -\partial_\kappa(y)\delta(x - y) \quad (2.22)$$

is permissible, as we are contracting subscripts referring to the same point. With more complicated expressions, one can use this rule only if no differential operators are interposed between the subscripts. If one had an expression such as

$$\partial_\alpha(x)\partial_\lambda(y)\{\delta_{\alpha\kappa}\delta_{\beta\rho}\delta(x-y)\},$$

one would first have to interchange the orders of differentiation, which is permissible since x and y are different points. Thus,

$$\partial_\alpha(x)\partial_\lambda(y)\{\delta_{\alpha\kappa}\delta_{\beta\rho}\delta(x-y)\} = \partial_\lambda(y)\partial_\alpha(x)\{\delta_{\alpha\kappa}\delta_{\beta\rho}\delta(x-y)\}$$
$$= -\partial_\lambda(y)\partial_\kappa(y)\{\delta_{\beta\rho}\delta(x-y)\}.$$

By (2.4) this can be written

$$-\partial_\kappa(y)\partial_\lambda(y)\{\delta_{\beta\rho}\delta(x-y)\} + \frac{i}{2}R_{\mu\nu\kappa\lambda}(y)[J_{\mu\nu},\delta_{\beta\rho}\delta(x-y)].$$

If the expression of interest contains no further derivatives, we may treat the Kronecker delta in the second term by the usual rules and obtain

$$\partial_\alpha(x)\partial_\lambda(y)\{\delta_{\alpha\kappa}\delta_{\beta\rho}\delta(x-y)\} = -\partial_\kappa(y)\partial_\lambda(y)\{\delta_{\beta\rho}\delta(x-y)\} + R_{\beta\rho\kappa\lambda}(y)\delta(x-y).$$

The presence of the second term on the right of this equation indicates that direct contraction is not permissible.

Poisson Brackets

We now turn to the problem of finding the Poisson brackets between our path-dependent variables. We have not succeeded in writing the Lagrangian in terms of these variables. There is no difficulty in writing down a Lagrangian density, but it is not clear how to integrate it over volume. The procedure we shall adopt will therefore be to take the usual Einstein Lagrangian of the conventional formalism, and from it find the Poisson brackets between path-dependent variables. When quantizing the field we shall obtain commutation rules from these Poisson brackets by the correspondence principle, and shall then verify that all equations are consistent.

The Peierls formalism has been applied to the gravitational field by DeWitt (4), and we can adapt his results to our theory. It is first necessary to express the path-dependent variables in terms of the conventional variables. We shall limit ourselves to stating Poisson-bracket relation between two variables whose paths coincide for all but an infinitesimal portion of the length of the shorter. We can then use a coordinate system which, along the path of interest, is the local Euclidean system. As we shall show below, we then have to work to first order in the γ's and to second order in their derivatives. It is therefore adequate to use (2.13) for the relation between path-dependent and coordinate dependent variables. The coordinate-dependent elements of the Riemann tensor are given in terms of the γ's, to sufficient accuracy, by

$$r_{(\mu)(\nu)(\kappa)(\lambda)}(x) = \left\{ \frac{\partial^2 \gamma_{\mu\kappa}}{\partial x_\nu \partial x_\lambda} - \frac{\partial^2 \gamma_{\nu\kappa}}{\partial x_\mu \partial x_\lambda} - \frac{\partial^2 \gamma_{\mu\lambda}}{\partial x_\nu \partial x_\kappa} + \frac{\partial^2 \gamma_{\nu\lambda}}{\partial x_\mu \partial x_\kappa} \right\}$$

$$+ \frac{1}{4} \left\{ \frac{\partial \gamma_{\rho\mu}}{\partial x_\kappa} + \frac{\partial \gamma_{\rho\nu}}{\partial x_\mu} - \frac{\partial \gamma_{\kappa\mu}}{\partial x_\rho} \right\} \left\{ \frac{\partial \gamma_{\rho\mu}}{\partial x_\lambda} + \frac{\partial \gamma_{\rho\lambda}}{\partial x_\nu} - \frac{\partial \gamma_{\nu\lambda}}{\partial x_\rho} \right\} \quad (2.14\text{b})$$

$$- \frac{1}{4} \left\{ \frac{\partial \gamma_{\rho\mu}}{\partial x_\lambda} + \frac{\partial \gamma_{\rho\lambda}}{\partial x_\mu} - \frac{\partial \gamma_{\lambda\mu}}{\partial x_\rho} \right\} \left\{ \frac{\partial \gamma_{\rho\nu}}{\partial x_\kappa} + \frac{\partial \gamma_{\rho\kappa}}{\partial x_\nu} - \frac{\partial \gamma_{\nu\kappa}}{\partial x_\rho} \right\}.$$

We have written the subscripts of r in brackets to indicate that they are the covariant components, from which the neutral components may be obtained from (2.9), with $g^{\mu\nu} = 1 - {}^{1\cdot}2\gamma_{\mu\nu}$; we have used a small r as it is a coordinate-dependent variable, from which the path-dependent variable must be obtained using (2.13).

DeWitt's prescription is then to proceed as if the γ's had the simple equal-time Poisson-brackets:

$$\left[\frac{\partial \gamma_{\alpha\beta}(x)}{\partial x_0}, \gamma_{\mu\nu}(y) \right] = -2i\kappa \{ \delta_{\alpha\mu} \delta_{\beta\nu} + \delta_{\alpha\nu} \delta_{\beta\nu} - \delta_{\alpha\beta} \delta_{\mu\nu} \}. \quad (2.23)$$

The Poisson bracket between two γ's is zero. That between two time derivatives is given by the derivative of (2.23) with respect to y_0, which can only be set equal to zero if it is not subsequently to be differentiated more than once. It should be emphasized that (2.23) is an algorithm for calculating Poisson brackets between quantities which are independent of the coordinate system. Poisson brackets between quantities such as the γ's have no other meaning.

Expressing the path-dependent quantities in terms of coordinate-dependent quantities by (2.14b), (2.9), and (2.13), and using (2.23) for the Poisson-bracket relations between the γ's, one may calculate directly the Poisson brackets between path-dependent variables. One can only write down explicit results for spacelike paths. After the calculation has been performed, quantities such as $\gamma_{\alpha\beta}\delta(x-y)$ or $\{\partial \gamma_{\alpha\beta}(x)/\partial x_\gamma\}\delta(x-y)$ may be put equal to zero, as we are in the local Euclidean system. This is why it is sufficient to work to first order in the γ's and second order in their derivatives. If, however, we have a quantity such as

$$\frac{\partial \gamma_{\alpha\beta}(x)}{\partial x_\gamma} \frac{\partial \delta(x-y)}{\partial x_\delta},$$

it is not zero, even in the local Euclidean system. We must first express it as

$$\frac{\partial}{\partial x_\delta} \left\{ \frac{\partial \gamma_{\alpha\beta}(x)}{\partial x_\gamma} \delta(x-y) \right\} - \frac{\partial^2 \gamma_{\alpha\beta}(x)}{\partial x_\delta \partial x_\gamma} \delta(x-y)$$

and the second term will not be zero. In an expression such as

$$\partial^2 \{\delta_{\gamma\mu} \delta(x-y)\}/\partial x_\alpha \partial x_\beta,$$

it is thus necessary to take into account the difference between the ordinary and the covariant derivative in the usual formalism.

Before stating the results of our procedure when applied to path-dependent quantities we shall establish one or two points of notation. The Poisson brackets will be written

$$[A, B] = [A, B]_1 + [A, B]_2 + [A, B]_3 \tag{2.24}$$

where $[A, B]_1$ does not involve an integration, $[A, B]_2$ involves an integral over the path defining one variable, and $[A, B]_3$ an integral over the paths defining both variables. These integrals result from taking the Poisson brackets between the γ's in the integral of (2.13) and other variables and are analogous to the terms on the right of (3.11b) and (3.11c) in I. We shall refer to the three terms of (2.24) as the first, second, and third parts of the Poisson brackets. Another convenient point of notation is to write

$$\underset{\alpha\leftrightarrow\beta}{A}\, f_{\alpha\beta} = f_{\alpha\beta} - f_{\beta\alpha}\,. \tag{2.25a}$$

$$\underset{\alpha\leftrightarrow\beta}{S}\, f_{\alpha\beta} = f_{\alpha\beta} + f_{\beta\alpha} - \delta_{\alpha\beta}f_{\gamma\gamma}\,. \tag{2.25b}$$

We add the third term to the last definition owing to the form of the Poisson-bracket relations (2.23).

For a gravitational field interacting with a neutral scalar field, the Poisson brackets between operators defined with spacelike paths then become:

$$[\Phi(x, P_1), \Phi(y, P_2)]_1 = [\dot\Phi(x, P_1), \dot\Phi(y, P_2)]_1 = 0, \tag{2.26a}$$

$$[\dot\Phi(x, P_1), \Phi(y, P_2)]_1 = -i\delta^3(x-y), \tag{2.26b}$$

$$[R_{abcd}(x, P_1), \dot\Phi(y, P_2)]_1 = [R_{0abc}(x, P_1), \Phi(y, P_2)]_1 = 0, \tag{2.27a}$$

$$[R_{0abc}(x, P_1), \dot\Phi(y, P_2)]_1$$
$$= \frac{i\kappa}{2} \underset{b\leftrightarrow c}{A}\, \underset{a\leftrightarrow c}{S}\, \partial_b(x)\{\delta_{a0}\,\delta_{c\lambda}\,\delta^3(x-y)\}\partial_\lambda(y)\Phi(y, P_2), \tag{2.27b}$$

$$[R_{abcd}(x, P_1), R_{ijkl}(y, P_2)] = 0, \tag{2.28a}$$

$$[R_{0abc}(x, P_1), R_{ijkl}(y, P_2)]_1$$
$$= -\frac{i\kappa}{2} \underset{\substack{i\leftrightarrow j\\ k\leftrightarrow l\\ b\leftrightarrow c}}{A}\, \underset{a\leftrightarrow c}{S}\, \partial_b(x)\partial_k(y)\partial_i(y)\{\delta_{aj}\,\delta_{cl}\,\delta^3(x-y)\}$$
$$-\frac{i\kappa}{2} \underset{\substack{k\leftrightarrow l\\ b\leftrightarrow c}}{A}\, \underset{a\leftrightarrow c}{S}\, R_{ijkm}(y, P_2)\partial_b(x)\{\delta_{am}\,\delta_{cl}\,\delta^3(x-y)\}, \tag{2.28b}$$

$[R_{0abc}(x, P_1), R_{0ijk}(y, P_2)]_1$

$$= -\frac{i\kappa}{2} \underset{\substack{j \leftrightarrow k \\ b \leftrightarrow c}}{A} \partial_b(x) \partial_j(y) \Big(\underset{0 \leftrightarrow i}{A} \underset{i \leftrightarrow k}{S} \partial_0(y) \{\delta_{ai} \delta_{ck} \delta^3(x-y)\} $$

$$- \underset{0 \leftrightarrow a}{A} \underset{i \leftrightarrow k}{S} \partial_0(x) \{\delta_{ai} \delta_{ck} \delta^3(x-y)\} \quad (2.28c)$$

$$- \frac{i\kappa}{2} \underset{\substack{j \leftrightarrow k \\ b \leftrightarrow c}}{A} \underset{a \leftrightarrow c}{S} R_{0ij\mu}(y, P_2) \partial_b(x) \{\delta_{a\mu} \delta_{ck} \delta^3(x-y)\}$$

$$- \frac{i\kappa}{2} \underset{\substack{j \leftrightarrow k \\ b \leftrightarrow c}}{A} \underset{i \leftrightarrow k}{S} R_{0abc}(x, P_1) \partial_j(y) \{\delta_{\epsilon i} \delta_{ck} \delta^3(x-y)\}.$$

The first term in (2.28c), like the second and third, would vanish in a flat space; however, in a curved space, we must keep it if we are subsequently going to differentiate the result. The terms on the right of (2.28b) taken individually are not symmetric under the interchange $i, j \leftrightarrow k, l$ (the first term because of the noncommutativity of the operators $\partial_k(y)$ and $\partial_i(y)$). It is nevertheless easy to show from (2.4) that the sum is symmetric.

The elements of the Riemann tensor with two zero subscripts are given in terms of the other dynamical variables by the equations of motion, so that it is unnecessary to include them in our fundamental Poisson-bracket relations.

The second part of the Poisson-brackets between two operators A and B will consist of two terms, one of which depends only on A and the other only on B in its form. We therefore write

$$[A, B]_2 = [A, B]_{2A} + [A, B]_{2B}.$$

The terms on the right of this equation then assume the form[2]

$$[\Phi(x, P_1), A(y, P_2)]_{2\Phi} = 0, \quad (2.29a)$$

$[\dot{\Phi}(x, P_1), A(y, P_2)]_{2\dot{\Phi}}$

$$= -\frac{\kappa}{2} \int_{P_2} dz_\lambda \{ \underset{\lambda \leftrightarrow \mu}{S} \delta_{0\lambda} \delta_{a\mu}\} \delta^3(x-z) \partial_a(x) \Phi(x, P_1) [J_{\mu 0}(z), A(y, P_2)], \quad (2.29b)$$

$[R_{abcd}(x, P_1), A(y, P_2)]_{2R}$

$$= -\frac{\kappa}{2} \underset{\substack{a \leftrightarrow b \\ c \leftrightarrow d}}{A} \underset{b \leftrightarrow d}{S} \int_{P_2} dz_\lambda\, \partial_c(x) \partial_a(x) \{\delta_{b\lambda} \delta_{d\mu} \delta^3(x-z)\} [J_{\mu 0}(z), A(y, P_2)] \quad (2.30a)$$

$$- \frac{\kappa}{2} \underset{c \leftrightarrow d}{A} \int_{P_2} dz_\lambda\, R_{abc\epsilon}(x, P_1) \{ \underset{\epsilon \leftrightarrow d}{S} \delta_{\epsilon\lambda} \delta_{d\mu}\} \delta^3(x-z) [J_{\mu 0}(z), A(y, P_2)],$$

[2] As our paths are not necessarily perpendicular to the time axis, in the local Euclidean co-ordinates, but are only restricted to be space-like, we use Greek sub-scripts for the path elements.

$$[R_{0abc}(x, P_1), A(y, P_2)]_{2R}$$

$$= -\frac{\kappa}{2} A \underset{b \leftrightarrow c}{S} \underset{a \leftrightarrow c}{S} \int_{P_2} dz_\lambda (\partial_b(x) \partial_\mu(z) \{\delta_{a\lambda} \delta_{c\nu} \delta^3(x - z)\} [J_{\mu\nu}(z), A(y, P_2)]$$

$$- i \partial_b(x) \{\delta_{a\lambda} \delta_{c\nu} \delta^3(x - z) \partial_\nu(z) A(y, P_2)\}) \qquad (2.30b)$$

$$- \frac{\kappa}{2} A \underset{\substack{0 \leftrightarrow a \\ b \leftrightarrow c}}{S} \underset{a \leftrightarrow c}{S} \int_{P_2} dz_\lambda \, \partial_b(x) \partial_0(x) \{\delta_{a\lambda} \delta_{c\mu} \delta^3(x - z)\} [J_{\mu 0}(z), A(y, P_2)]$$

$$- \frac{\kappa}{2} A \int_{P_2} dz_\lambda \, R_{0abe}(x, P_1) \underset{e \leftrightarrow c}{S} \delta_{e\lambda} \delta_{c\mu} \delta^3(x - z) [J_{\mu 0}(z), A(y, P_2)].$$

In the integrals over z, it is taken for granted that the path associated with z is that part of P_2 leading to z. We shall not write the path explicitly in expressions such as $J_{\mu\nu}(z)$.

The third part of the Poisson bracket takes the same form for all variables; it is

$$[A(x, P_1), B(y, P_2)]_3 = \frac{i\kappa}{2} \delta_{\alpha\lambda} \delta_{\gamma\nu} \underset{\lambda \leftrightarrow \nu}{S} \int_{P_1} dz_\alpha \int_{P_2} dz_\lambda' \{\partial_\mu(z') \delta^3(z - z')$$

$$\times [J_{\gamma 0}(z), A(x, P_1)][J_{\mu\nu}(z'), B(y, P_2)] + \partial_\beta(z') \delta^3(z - z')[J_{\beta\gamma}(z), A(x, P_1)] \qquad (2.31)$$

$$\times [J_{\nu 0}(z'), B(y, P_2)] + i\delta^3(z - z')[J_{\gamma 0}(z), A(x, P_1)]\partial_\nu(z') B(y, P_2)$$

$$- i\delta^3(z - z')\partial_\gamma(z) A(x, P_1)[J_{\nu 0}(z'), B(y, P_2)].$$

When using these relations to find the Poisson brackets between derivatives of Φ and the Riemann tensor, additional contributions to the first part of the Poisson bracket will arise from the differentiation of the integral in (2.30) with respect to its endpoint. Thus, for instance,

$$[R_{0abc}(x, P_1), \partial_\mu \Phi(y, P_2)]_1 = \frac{i\kappa}{2} A \underset{\substack{b \leftrightarrow c \\ a \leftrightarrow c}}{S} \partial_b(x) \{\delta_{a\mu} \delta^3(x - y)\} \partial_c \Phi(y, P_2). \qquad (2.32)$$

The first term on the right of (2.30b) does not contribute since

$$[J_{\mu\nu}(y), \Phi(y, P_2)] = 0.$$

CONSISTENCY OF THE POISSON-BRACKET RELATIONS

As in the case of Eq. (3.11) of I, we could have obtained (2.30) from the path-dependence equations and the requirement that this part of the Poisson bracket be zero if y is not on P_1. This follows if we can verify that (2.30) is consistent with the path-dependence equations. We are at the moment only interested in spacelike variations of the paths, as the paths themselves are restricted to be spacelike. To take as our example (2.30b) with $A = \Phi$, the change of the left-hand side resulting from a change in the path P_2 at the point z by an area σ_{kl} is

$$\Delta_L[R_{0abc}(x, P_1), \Phi(y, P_2)]$$
$$= -\frac{i}{4}[R_{0abc}(x, P_1), R_{\mu\nu kl}(z)[J_{\mu\nu}(z), \Phi(y, P_2)]]\sigma_{kl}, \quad (2.33)$$

on applying (2.3) to $\Phi(y, P_2)$. As the point z is on P_2, the associated path is that part of P_2 leading to z, and we shall not write it explicitly. The Poisson bracket on the right of (2.33) is as usual the sum of three parts. According to (2.28b) and (2.28c), the first part is

$$\Delta_{L1}[R_{0abc}(x, P_1), \Phi(y, P_2)] = -\frac{\kappa}{2} \underset{b \leftrightarrow c}{A} \underset{a \leftrightarrow c}{S} \partial_b(x)\partial_k(z)\partial_\mu(z)\{\delta_{al}\delta_{c\nu}\delta^3(x-z)\}$$
$$\times [J_{\mu\nu}(z), \Phi(y, P_2)]\sigma_{kl}$$

$$-\frac{\kappa}{2} \underset{\substack{b \leftrightarrow c \\ 0 \leftrightarrow a}}{A} \underset{a \leftrightarrow c}{S} \partial_b(x)\partial_k(z)\partial_0(x)\{\delta_{al}\delta_{c\mu}\delta^3(x-z)\}[J_{0\nu}(z), \Phi(y, P_2)]\sigma_{kl} \quad (2.34)$$

$$+\frac{\kappa}{4} \underset{b \leftrightarrow c}{A} \underset{a \leftrightarrow c}{S} R_{\mu\nu k p}(z)\partial_b(z)\{\delta_{al}\delta_{cp}\delta^3(x-z)\}[J_{\mu\nu}(z), \Phi(y), P_2)]\sigma_{kl}$$

$$+\frac{\kappa}{4} \underset{b \leftrightarrow c}{A} R_{0abc}(x, P_1)\partial_k(z)\} \underset{c \leftrightarrow a}{S} \delta_{el}\delta_{c\mu}\delta^3(x-z)\}[J_{\mu 0}(z), \Phi(y, P_2)]\sigma_{kl}.$$

We shall leave the remaining parts of (2.33) for the moment. Looking at the right of (2.30b), we first calculate the variation in the path integral due to the change in P_2, neglecting the variation in the function $\Phi(y, P_2)$ itself. The application of Stokes' theorem is fairly straightforward; the first term in the first integral on the right of (2.30b) gives a contribution

$$\Delta_{K1a}[R_{0abc}(x, P_1), \Phi(y, P_2)]$$
$$= -\frac{\kappa}{2} \underset{b \leftrightarrow c}{A} \underset{a \leftrightarrow \nu}{S} (\partial_b(x)\partial_k(z)\partial_\mu(z)\{\delta_{al}\delta_{c\nu}\delta^3(x-z)\} \quad (2.35a)$$
$$\times [J_{\mu\nu}(z), \Phi(y, P_2)] + i\partial_b(x)\partial_k(z)\{\delta_{al}\delta_{c\nu}\delta^3(x-z)\}\partial_\nu(z)\Phi(y, P_2))\sigma_{kl}.$$

The second term in (2.35a) arises from differentiating $J_{\mu\nu}(z)$ in (2.30b) with respect to z, as

$$\partial_k(z)J_{\mu\nu}(z) = -i\delta_{\nu k}\partial_\mu(z) + i\delta_{\mu k}\partial_\nu(z).$$

The contribution to Δ_{Kl} from the second term in the first integral of (2.30b) is given by

$$\Delta_{K1b}[R_{0abc}(x, P_1), \Phi(y, P_2)]$$
$$= \frac{\kappa}{2} \underset{b \leftrightarrow c}{A} \underset{a \leftrightarrow c}{S} i\partial_b(x)\partial_k(z)\{\delta_{al}\delta_{c\nu}\delta^3(x-z)\partial_\nu(z)\}\Phi(y, P_2)\sigma_{kl}. \quad (2.35b)$$

We have written $\partial_\nu(z)$ inside the curly bracket, since this operator depends on z and will therefore have a derivative with respect to z. In fact, it is not difficult to see that

$$\partial_k(z)\{\delta_{al}\delta_{cp}\delta^3(x-z)\partial_p(z)\}\Phi(y,P_2) - \partial_k(z)\{\delta_{al}\delta_{cp}\delta^3(x-z)\}\partial_p(z)\Phi(y,P_2)$$
$$= -\frac{i}{2}R_{\mu\nu k p}(z)\delta_{al}\delta_{cp}\delta^3(x-z)[J_{\mu\nu}(z),\Phi(y,P_2)]. \quad (2.36)$$

The contribution to Δ_{R1} from the second and third integrals of (2.30b) is

$$\Delta R_{ic}[R_{0abc}(x,P_1),\Phi(y,P_2)] = -\frac{\kappa}{2}\underset{\substack{b\leftrightarrow c \\ 0\leftrightarrow a}}{A}\underset{a\leftrightarrow c}{S}\partial_k(z)\partial_b(x)\partial_0(x)\{\delta_{el}\delta_{c\mu}\delta^3(x-z)\}$$
$$\times [J_{\mu 0}(z),\Phi\}y,P_2)]\sigma_{kl} + \frac{\kappa}{2}\underset{b\leftrightarrow c}{A}R_{0abe}(x,P_1)\partial_k(z) \quad (2.35c)$$
$$\cdot\{\underset{e\leftrightarrow c}{S}\delta_{el}\delta_{c\mu}\delta^3(x-z)\}[J_{\mu 0}(z),\Phi(y,P_2)]\sigma_{kl}.$$

Adding (2.35a), (b), and (c) and applying (2.36), we reproduce (2.34). In the same spirit it can be shown that the second and third parts of the Poisson bracket on the right of (2.33) are together equal to the change of the right-hand side of (2.35a) due to the variation of P_2 in the argument of Φ, plus the change of the right-hand side of (2.31) with $A = R_{0abc}$, $B = \Phi$. The consistency of the Poisson-bracket relation with the path-dependence equation is thus established.

It remains to show that the Poisson-bracket relations are Lorentz-invariant and consistent with the equations of motion. One can write all the relations (2.26)–(2.31) in a formally covariant manner by setting $\delta^3(x-y) = -\partial_0(x)\cdot\Delta(x-y)$. This has been done in Appendix I. Even for spacelike paths such equations would not be entirely correct, as they would imply relations involving the Riemann tensor with two zero subscripts, R_{0a0b}. We have already emphasized that this quantity is not to be regarded as a fundamental dynamical variable when stating Poisson-bracket relations, but is given in terms of the other variables by the equations of motion. It is therefore necessary to add correction terms to the Poisson-bracket relations, so that the new relations will be valid to first order in the timelike portion of the path when applied to variables other than R_{0a0b}, and to zeroth order when applied to R_{0a0b}. If such correction terms can be found so as to ensure the vanishing of the Poisson bracket between any variable and a quantity which should be zero according to the equations of motion or the Bianchi identities, and if the extra terms are covariant, we shall have established both the covariance of the theory and the consistency of the Poisson-bracket relations with the equations of motion. It will finally be necessary to verify the consistency of the Poisson-bracket relations with the path-dependence equation for timelike variations of the path.

We shall illustrate the procedure by taking the Poisson bracket between $R_{\alpha\beta\gamma\delta}(x, P_1)$ and $\Phi(y, P_2)$. The covariant form of (2.27a) is simply

$$[R_{\alpha\beta\gamma\delta}(x, P_1), \Phi(y, P_2)]_0 = 0, \tag{2.36}$$

where the subscript 0 indicates that the correction term just discussed has not yet been included. The correction term will now be determined from the requirement that the Poisson bracket between $R_{\alpha\beta\gamma\delta}(x, P_1)$ and $(\partial_\mu^2 - m^2)\Phi(y, P_2)$ be zero. This last Poisson bracket will consist of two parts, one from differentiation of (A1.2b) with respect to y, and the other from setting $A(y, P_2)$ in (A1.5) equal to $\partial_\mu \Phi(y, P_2)$ and differentiating with respect to the end-point of the path, in analogy with the derivation of (2.32). On adding the two contributions we find that

$$[R_{\alpha\beta\gamma\delta}(x, P_1), \partial_\mu^2(y)\Phi(y, P_2)]_{0,1} = -i\kappa \underset{\substack{\alpha\leftrightarrow\beta\\\gamma\leftrightarrow\delta}}{A} \underset{\beta\leftrightarrow\delta}{S} \partial_\gamma(x)\partial_\alpha(x)\partial_\mu(y)\{\delta_{\beta\mu}\delta_{\delta\nu}$$

$$\times \Delta(x-y)\partial_\nu(y)\Phi(y, P_2)\} - i\kappa \underset{\substack{\alpha\leftrightarrow\beta\\\gamma\leftrightarrow\delta}}{A} \delta_{\beta\delta}\partial_\gamma(x)\partial_\alpha(x)\partial_\mu(y)\{\Delta(x-y)\partial_\mu(y)\Phi(y, P_2)\} \tag{2.37}$$

$$- i\kappa \underset{\gamma\leftrightarrow\delta}{A} R_{\alpha\beta\gamma\epsilon}(x, P_1) \underset{\epsilon\leftrightarrow\delta}{S} \partial_\mu(y)\{\delta_{\epsilon\mu}\delta_{\delta\nu}\Delta(x-y)\partial_\nu(y)\Phi(y, P_2)\}$$

$$- i\kappa R_{\alpha\beta\gamma\delta}(x, P_1)\partial_\mu(y)\{\Delta(x-y)\partial_\mu(y)\Phi(y, P_2)\}.$$

In order that the Poisson bracket on the left of (2.37) should vanish, one must therefore supplement (2.36) by a term resembling the right-hand side of (2.37), but with $\Delta(x-y)$ replaced by $-\Gamma(x-y)$, where

$$\Gamma(x-y) = \tfrac{1}{4}\epsilon(x_0 - y_0)\theta\{-(x-y)\}^2. \tag{2.38a}$$

The function Γ is chosen so that

$$\partial_\mu^2(x)\Gamma(x-y) = \partial_\mu^2(y)\Gamma(x-y) = \Delta(x-y). \tag{2.38b}$$

This term takes our relations to the accuracy we require and, for nearly space-like paths, it is only nonzero when $x = y$. We may therefore perform contractions as in (2.22) and replace $\Phi(y, P_2)$ by $\Phi(x, P_1)$. Accordingly, the correction term is

$$[R_{\alpha\beta\gamma\delta}(x, P_1), \Phi(y, P_2)]_{c,1} = -i\kappa \underset{\substack{\alpha\leftrightarrow\beta\\\gamma\leftrightarrow\delta}}{A} \underset{\beta\leftrightarrow\delta}{S} \partial_\gamma(x)\partial_\alpha(x)\{\partial_\beta(x)\Gamma(x-y)\partial_\delta(x)\Phi(x, P_1)\}$$

$$- i\kappa \underset{\substack{\alpha\leftrightarrow\beta\\\gamma\leftrightarrow\delta}}{A} \delta_{\beta\delta}\partial_\gamma(x)\partial_\alpha(x)\{\partial_\epsilon(x)\Gamma(x-y)\partial_\epsilon(x)\Phi(x, P_1)\} \tag{2.39}$$

$$- i\kappa \underset{\gamma\leftrightarrow\delta}{A} R_{\alpha\beta\gamma\epsilon}(x, P_1) \underset{\epsilon\leftrightarrow\delta}{S} \partial_\epsilon(x)\Gamma(x-y)\partial_\delta(x)\Phi(x, P_1)$$

$$- i\kappa R_{\alpha\beta\gamma\delta}(x, P_1)\partial_\epsilon(x)\Gamma(x-y)\partial_\epsilon(x)\Phi(x, P_1).$$

It is important to observe that, even when the sum of (2.39) and (2.36) is used for the Poisson bracket between $R_{\alpha\beta\gamma\delta}(x, P_1)$ and $\Phi(y, P_2)$, the Poisson bracket in (2.37) will not be zero if R has two zero subscripts. In that case, *both* of the factors in the Poisson bracket would be given by the equations of motion, and one would have to work to higher order in the timelike portions of the path to calculate it. Nevertheless, if we accept (2.39) when two subscripts are not zero, we must accept it in this last case as well. Not only is this necessary for covariance, but, unless we do so, the Poisson brackets between the Bianchi identities and Φ will not be zero. We can see at once, in fact, that the first two terms of (2.39) satisfy the Bianchi identities except for expressions involving elements of the Riemann tensor which arise from interchanging differential operators. These expressions, together with the last two terms of (2.39), will not themselves satisfy the Bianchi identities. The difference is made up by correction terms (A1.8) to (2.29a) (just as (2.39) was a correction term to (2.36)). When these terms are differentiated with respect to the end-point of the path, they provide further contributions to the Poisson brackets between the Bianchi identities and Φ, which cancel the unbalanced contributions from (2.39).

We next verify that the Poisson bracket between the Einstein equations and Φ is zero, with the same correction term (2.39) for the Poisson bracket between $R_{\alpha\beta\gamma\delta}$ and Φ. We shall work to first order in the timelike portion of the path, so that our results are adequate for $\dot{\Phi}$ as well as for Φ. In (2.39) we may then neglect terms which have fewer than two time derivatives acting on $\Gamma(x - y)$, and may therefore neglect the last two terms altogether. Thus, from this equation,

$$[R_{\gamma\alpha\gamma\beta}(x, P_1) - \tfrac{1}{2}\delta_{\alpha\beta}R_{\gamma\delta\gamma\delta}(x, P_1), \Phi(y, P_2)]_{c,1}$$
$$= i\kappa \underset{\alpha \leftrightarrow \beta}{S} \Delta(x - y)\partial_\alpha(x)\partial_\beta(x)\Phi(x, P_1). \quad (2.40)$$

We can also evaluate the second part of the Poisson bracket by contracting two of the subscripts on the Riemann tensor in (A1.5). It turns out that many of the terms cancel and most of the others can be formed into an over-all derivative. The relation (A1.5) must then be supplemented with a correction term given by (A1.9); as before, the justification of such a term is that it enables us to obtain consistency with this and all other relations. The final result is:

$$[R_{\gamma\alpha\gamma\beta}(x, P_1) - \tfrac{1}{2}\delta_{\alpha\beta}R_{\gamma\delta\gamma\delta}(x, P_1), \Phi(y, P_2)_{2R}]$$
$$= \kappa \underset{\alpha \leftrightarrow \beta}{S} \int_{P_2} \partial z_\lambda \, \partial_\lambda(z)\Big(\partial_\alpha(x)\partial_\mu(z)\{\delta_{\beta\nu}\Delta(x - z)\}[J_{\mu\alpha}(z), \Phi(y, P_2)]$$
$$- \frac{i}{2}\partial_\alpha(z)\{\delta_{\beta\nu}\Delta(x - z)\}\partial_\nu(z)\Phi(y, P_2)\Big) - \frac{\kappa}{2}\underset{\alpha \leftrightarrow \beta}{S}\int_{P_2} dz_\lambda R_{\gamma\alpha\gamma\epsilon}(x, P_1) \quad (2.41)$$
$$\times \Big((\partial_\mu(z)\{\underset{\lambda \leftrightarrow \nu}{S}\delta_{\epsilon\lambda}\delta_{\beta\nu}\Delta(x - z)\}[J_{\mu\nu}(z), \Phi(y, P_2)]$$
$$- i\{\underset{\lambda \leftrightarrow \nu}{S}\delta_{\epsilon\lambda}\delta_{\beta\nu}\Delta(x - z)\}\partial_\nu(z)\Phi(y, P_2)\Big)$$

or, by integrating the first term,
$[R_{\gamma\alpha\gamma\beta}(x, P_1) - \tfrac{1}{2} \delta_{\alpha\beta} R_{\gamma\delta\gamma\delta}(x, P_1), \Phi(y, P_2)]_{2R}$

$$= -i\kappa \underset{\alpha \leftrightarrow \beta}{S} \partial_\alpha(x) \{\Delta(x - z) \partial_\beta(x) \Phi(x, P_1)\} - \frac{\kappa}{2} \underset{\alpha \leftrightarrow \beta}{S} \int_{P_2} dz_\lambda R_{\gamma\alpha\gamma\epsilon}(x, P_1)$$

$$\times \left(\partial_\mu(z) \{ \underset{\lambda \leftrightarrow \nu}{S} \delta_{\epsilon\lambda} \delta_{\beta\gamma} \Delta(x - z) \} [J_{\mu\nu}(z), \Phi(y, P_2)] \right. \quad (2.42)$$

$$\left. - i\{ \underset{\lambda \leftrightarrow \nu}{S} \delta_{\epsilon\lambda} \delta_{\beta\nu} \Delta(x - z) \} \partial_\nu(z) \Phi(y, P_2) \right).$$

The Poisson bracket between $T_{\alpha\beta}(x, P_1)$ and $\Phi(y, P_2)$ is easily evaluated. The covariant equivalent of (2.26b) is, of course just given by (A1.1), so that, from the definition (2.20b) of T:

$$[T_{\alpha\beta}(x, P_1), \Phi(y, P_2)]_1 = i \underset{\alpha \leftrightarrow \beta}{S} \partial_\alpha(x) \Delta(x - y) \partial_\beta(x) \Phi(x, P_1). \quad (2.43)$$

The second part of the Poisson bracket between T and Φ can be found from (A1.4b) with $A = \Phi$; the correction terms do not contribute in the approximation to which we are working. Thus

$$[T_{\alpha\beta}(x, P_1), \Phi(y, P_2)]_{2T} = -\frac{\kappa}{2} \underset{\alpha \leftrightarrow \beta}{S} \int_{P_2} dz_\lambda \{T_{\alpha\epsilon}(x, P_1) - \tfrac{1}{2} \delta_{\alpha\epsilon} T_{\delta\delta}(x, P_1)\}$$

$$\times (\partial_\mu(z) \{ \underset{\lambda \leftrightarrow \nu}{S} \delta_{\epsilon\lambda} \delta_{\beta\nu} \Delta(z - z) \} [J_{\mu\nu}(z), \Phi(y, P_2)] \quad (2.44)$$

$$- i\{ \underset{\lambda \leftrightarrow \nu}{S} \delta_{\epsilon\lambda} \delta_{\beta\nu} \Delta(x - z) \} \partial_\nu(z) \Phi(y, P_2)).$$

It now follows that, in the Poisson bracket

$$[R_{\gamma\alpha\gamma\beta}(x, P_1) - \tfrac{1}{2} \delta_{\alpha\beta} R_{\gamma\delta\gamma\delta}(x, P_1) + \kappa T_{\alpha\beta}(x, P_1), \Phi(y, P_2)],$$

the term (2.43) cancels (2.40) and the first term of (2.42). Also, as (2.44) and the second term of (2.42) have the same form, they will cancel by virtue of the Einstein equations. The third parts of the Poisson brackets will similarly cancel, so that the compatibility of the Poisson-bracket relations with Einstein's equations is verified. We may observe that we could not have obtained consistency without the term $-\kappa T_{\lambda\mu}$ on the right of (2.20a).

Generally, it is possible to arrange the correction terms to the Poisson-bracket relations in such a form that each of the three parts of the Poisson bracket between an equation of motion and a dynamical variable is zero. The one exception is the set of Einstein equations which involve equations of constraint as well as equations of motion. In proving their consistency with the equations of motion we had to make use of a cancellation between the first and second parts of the Poisson bracket.

The consistency of all the Poisson-bracket relations and equations of motion can be proved in a similar way. Finally, by subjecting the correction terms (A1.7), (A1.8), and (A1.9) to precisely the same manipulations as those leading to (2.35) one can prove their consistency with the path-dependence equation, for spacelike or timelike variations of the path. In fact, it can be seen on inspection that the relation between (A1.8) and (A1.7) is the same as that between (A1.5) and (A1.3) or their noncovariant forms (2.30) and (2.28). There is little point in going through all the consistency proofs in detail. It is guaranteed that they will work, as our formalism is equivalent to that of Einstein, which we know to be consistent. For the purposes of quantization, however, it is important to demonstrate the manner in which consistency proofs may be carried through directly.

ENERGY TENSOR AND ENERGY-MOMENTUM VECTOR

The quantity

$$\mathfrak{J}_{\lambda\nu}(x, P) = T_{\lambda\nu}(x, P) + \frac{1}{\kappa} R_{\mu\nu\mu\lambda}(x, P) - \frac{1}{2\kappa} \delta_{\lambda\nu} R_{\mu\rho\mu\rho}(x, P) \qquad (2.45)$$

may be regarded as the energy tensor for the system. The first term is the contribution from the scalar field, the last two from the gravitational field. Equation (2.43), with $\alpha = 4$, is the familiar result that the effect of taking the Poisson bracket of $T_{4\nu}(x)$ with a dynamical variable is to displace the coordinate at the point x in the ν-direction. In a gravitational field the coordinate involved is the local Euclidean coordinate. The last two terms of (2.45) will have the same effect on the Riemann tensor as $T_{\lambda\nu}$ has on the scalar field. By examining the first term of Eq. (2.41) one observes that the Poisson bracket of $\mathfrak{J}_{4\nu}$ with a dynamical variable also displaces the local Euclidean coordinates used to fix the path, so that the net effect is to leave the variable unchanged. (The second term of (2.43) cancels with (2.44)). The path dependence of the variables is thus one way of understanding how in gravitational theory, unlike in other field theories, the energy tensor can vanish.

Quite separate from the operation of deforming the local Euclidean coordinates at a point is the operation of displacing all the paths rigidly by a fixed amount, or, in other words, of changing the coordinates of the reference point. The energy-momentum vector performs this operation. It is nonzero and cannot be expressed as a spare integral in a coordinate independent theory. In the conventional theory, as is well known, it can be expressed as the integral of a noncovariant quantity.

III. QUANTIZATION OF THE THEORY

SPACELIKE CHARACTER OF PATHS ASSUMED KNOWN

The theory of the gravitational field developed in the previous section operates only with coordinate-independent quantities and can therefore provide us with a basis for quantization. We cannot simply quantize it as it stands, however, as it abounds in equations where the ordering would be ambiguous. The fundamental path-dependence equation (2.3) is an example, as we would not know how to order the two factors on the right, which may not commute.

The ambiguities of ordering can be removed completely by means of a procedure similar to that used in the Appendix of I. There we only had to use it for nonspacelike paths, but here it will be necessary to use it in general. To define an operator we specify not only the shape of the path, but the ordering of all the elements of the path with respect to one another and to the operator itself. We shall then state "commutation relations" between the elements, giving the difference between two operators in which the ordering of a pair of elements is reversed. Similarly, the "commutation relations" between an element and an operator will have to be given. The commutation relations, like ordinary commutation relations, can only be explicitly stated for spacelike paths. We shall assume for the moment that we know whether or not a particular path is spacelike.

As in the Appendix to I we can also define a generalized product of two operators $\overline{A(x, P_1)B(y, P_2)}$ in which the relative orderings of all the elements of the two paths, and of the operators themselves, are given. Only if the elements of one path and its operators are all before (or after) the elements of the other path and its operator will this be a true product. Knowing the commutation relations between the elements and the operators for spacelike paths, we can in principle find the generalized products from the true products.

When forming a derivative $\partial_\mu(x)\Phi(x, P_1)$ of an operator, the result will depend on the ordering of the path element immediately adjacent to the operator with the operator itself and with the other elements. We may thus define the ordering of the differential operator with Φ and with the path elements according to the position of the element which is being added. One may similarly define the ordering of the differential operator in an expression such as $\partial_\mu(z)\Phi(x, P_1)$, where the added element is not the end of the path.

With these more general types of operators there is no longer any ambiguity in writing down the path-dependence equation (2.3). It becomes

$$\delta_z A(x, P_1) = -\frac{i}{4} \overline{R_{\mu\nu\kappa\lambda}(z, P_3)[J_{\mu\nu}(z, P_3), A(x, P_1)]} \sigma_{\kappa\lambda}. \qquad (3.1)$$

The elements of the path on the right of (3.1) are to be ordered relative to each other and to $[J, A]$ in the same way as on the left, while R is ordered in the same position as the element of the path which is being varied. The elements of the path surrounding the area of variation will be ordered symmetrically or, in other words, the expression will be averaged over the different possible orderings.

The consistency conditions (2.6) and (2.7), and the equations of motion (2.19) and (2.20) are the same as in the classical theory and need not be restated. In the Bianchi identities (2.7) and the equation of motion (2.19) the differential operators and the functions on which they operate will be ordered symmetrically. For the Bianchi identities this follows from their derivation as integrability conditions, and is a consequence of the symmetrical ordering of the elements surrounding the area of variation in (3.1). We shall see below why we must also adopt the symmetrical ordering in (2.19). In the definition of the energy tensor (2.20b) the products on the right will be generalized products, and the ordering of the elements in all the terms of (2.20a) must of course be the same. The differential operator $\partial_\lambda(x)$ commutes with $\Phi(x, P)$ so that there is no ambiguity of ordering them in (2.20b).

We can now write down the commutation relations directly from the classical theory. As in the Appendix to I, we define $[A(x, P_1), B(y, P_2)]_{A,B}$ as the difference between two generalized products in which the ordering of A and B is reversed. We can similarly define $[A(x, P_1)B(y, P_2)]_{A,p_2}$ or $[A(x, P_1)B(y, P_2)]_{p_1,p_2}$ where p_1 and p_2 are elements of P_1 and P_2 respectively, as the difference between generalized products in which the ordering of the operator or elements denoted by subscripts is reversed. Further, there will be commutators such as $[A(x, P_1)]_{A,p_1}$ or $[A(x, P_1)]_{p_{1a},p_{1b}}$, in which the ordering of a path element and its own operator, or another element of the same path, is interchanged.

The commutators of the first type will be identical to (2.26)–(2.28). We rewrite them for completeness:

$$\overline{[\Phi(x, P_1), \Phi(y, P_2)]}_{\Phi,\Phi} = [\dot{\Phi}(x, P_1), \dot{\Phi}(y, P_2)]_{\dot{\Phi},\dot{\Phi}} = 0, \quad (3.2a)$$

$$\overline{[\dot{\Phi}(x, P_1), \Phi(y, P_2)]}_{\dot{\Phi},\Phi} = -i\delta^3(x - y), \quad (3.2b)$$

$$\overline{[R_{abcd}(x, P_1), \dot{\Phi}(y, P_2)]}_{R,\dot{\Phi}} = [R_{0abc}(x, P_1), \Phi(y, P_2)]_{R,\Phi} = 0, \quad (3.3a)$$

$$\overline{[R_{0abc}(x, P_1), \dot{\Phi}(y, P_2)]}_{R,\dot{\Phi}}$$

$$= \frac{i\kappa}{2} \underset{b \leftrightarrow c}{A} \underset{a \leftrightarrow c}{S} \partial_b(x) \{\delta_{a0} \delta_{c\lambda} \delta^3(x - y)\} \partial_\lambda(y) \Phi(y, P_2), \quad (3.3b)$$

$$\overline{[R_{abcd}(x, P_1), R_{ijkl}(y, P_2)]}_{R,R} = 0, \quad (3.4a)$$

$$\overline{[R_{0abc}(x, P_1), R_{ijkl}(y, P_2)]}_{R,R}$$

$$= -\frac{i\kappa}{2} \underset{\substack{i \leftrightarrow j \\ k \leftrightarrow l \\ b \leftrightarrow c}}{A} \underset{a \leftrightarrow c}{S} \partial_b(x) \partial_k(y) \partial_i(y) \{\delta_{aj} \delta_{cl} \delta^3(x - y)\} \quad (3.4b)$$

$$+ \frac{i\kappa}{2} \underset{\substack{k \leftrightarrow l \\ b \leftrightarrow c}}{A} \underset{a \leftrightarrow c}{S} R_{ijkm}(y, P_2) \partial_b(x) \{\delta_{am} \delta_{cl} \delta^3(x - y)\},$$

$$\overline{[R_{0abc}(x, P_1), R_{0ijk}(y, P_2)]}_{R,R} = -\frac{i\kappa}{2} \underset{\substack{j \leftrightarrow k \\ b \leftrightarrow c}}{A} \partial_b(x) \partial_j(y) (\underset{0 \leftrightarrow i}{A} \underset{i \leftrightarrow k}{S} \partial_0(y)$$

$$\times \{\delta_{ai}\delta_{ck}\delta^3(x-y)\} - \underset{\substack{0 \leftrightarrow a \; i \leftrightarrow k}}{A \quad S} \partial_0(x)\{\delta_{ai}\delta_{ck}\delta^3(x-y)\})$$

$$+\frac{i\kappa}{2} \underset{\substack{j \leftrightarrow k \; a \leftrightarrow c \\ b \leftrightarrow c}}{A \quad S} R_{0ij\mu}(y,P_2)\partial_b(x)\{\delta_{a\mu}\delta_{ck}\delta^3(x-y)\} \qquad (3.4c)$$

$$-\frac{i\kappa}{2} \underset{\substack{j \leftrightarrow k \; i \leftrightarrow k \\ b \leftrightarrow c}}{A \quad S} R_{0abe}(x,P_1)\partial_j(y)\{\delta_{ei}\delta_{ck}\delta^3(x-y)\}.$$

In these equations it is of course taken for granted that the ordering of all path elements on the right with respect to one another and to R is the same as on the left.

The commutation relations between an operator and a path element can be written down from Eqs. (2.29) and (2.30). The integration will no longer be present, as we are now only concerned with single path elements. Thus

$$\overline{[\Phi(x,P_1)A(y,P_2)]}_{\Phi,P_2} = 0, \qquad (3.5a)$$

$$\overline{[\dot\Phi(x,P_1)A(y,P_2)]}_{\Phi,P_2}$$
$$= -\frac{\kappa}{2} dz_\lambda \{\underset{\lambda \leftrightarrow \mu}{S} \delta_{0\lambda}\delta_{a\mu}\{\delta^3(x-z)\{\partial_\alpha(x)\Phi(x,P_1)\}\overline{[J_{\mu 0}(z),A(y,P_2)]}, \qquad (3.5b)$$

$$\overline{[R_{abcd}(x,P_1)A(y,P_2)]}_{R,P_2}$$
$$= -\frac{\kappa}{2} \underset{\substack{a \leftrightarrow b \; b \leftrightarrow d \\ c \leftrightarrow d}}{A \quad S} dz_\lambda \overline{\partial_c(x)\partial_a(x)}\{\delta_{b\lambda}\delta_{d\mu}\delta^3(x-z)\}\overline{[J_{\mu 0}(z),A(y,P_2)]}$$
$$+\frac{\kappa}{2} \underset{c \leftrightarrow d}{A} dz_\lambda \overline{R_{abce}(x,P_1)}\{\underset{\epsilon \leftrightarrow d}{S} \delta_{\epsilon\lambda}\delta_{d\mu}\delta^3(x-z)\}\overline{[J_{\mu 0}(z),A(y,P_2)]}, \qquad (3.6a)$$

$$\overline{[R_{0abc}(x,P_1)A(y,P_2)]}_{R,P_2}$$
$$= -\frac{\kappa}{2} \underset{\substack{b \leftrightarrow c \; a \leftrightarrow c}}{A \quad S} dz_\lambda (\partial_b(x)\partial_\mu(z)\{\delta_{a\lambda}\delta_{c\nu}\delta^3(x-z)\}\overline{[J_{\mu\nu}(z),A(y,P_2)]}$$
$$- i\partial_b(x)\delta_{a\lambda}\delta_{c\nu}\delta^3(x-z)\partial_\nu(z)A(y,P_2)) \qquad (3.6b)$$
$$- \frac{\kappa}{2} \underset{b \leftrightarrow c}{A} dz_\lambda \underset{0 \leftrightarrow a}{A} \overline{\partial_b(x)\partial_0(x)}\{\underset{\lambda \leftrightarrow \nu}{S} \delta_{0\lambda}\delta_{c\nu}\delta^3(x-z)\}\overline{[J_{\mu 0}(z),A(y,P_2)]}$$
$$+ \frac{\kappa}{2} \underset{b \leftrightarrow c}{A} dz_\lambda \overline{R_{0abe}(x,P_1)} \underset{\epsilon \leftrightarrow c}{S} \delta_{\epsilon\lambda}\delta_{c\mu}\delta^3(x-z)[J_{\mu 0}(z),A(y,P_2)].$$

In these equations, z is the coordinate of the element p_2 and dz_λ the length of the element. The commutator between an element of a path and its own operator will have exactly the same form. For instance, corresponding to (3.5b), we may write down the relation,

$$[\dot{\Phi}(x, P_1)]_{\dot{\Phi}, p_1} = -\frac{\kappa}{2} [J_{\mu 0}(z, \vec{x}), dz_\lambda \{ \underset{\lambda \leftrightarrow \mu}{S} \delta_{0\lambda} \delta_{\alpha\mu} \delta^3(x - z) \} \partial_\alpha(x) \Phi(x, P_1)]. \quad (3.7)$$

The angular-momentum operator has been transferred to the beginning of the expression, and has been written $J(z, \vec{x})$ to denote that it only operates on the point x and its associated subscripts. (The z in the brackets denotes, as usual, the point about which we rotate.) We could have written (3.7) simply as

$$-\frac{\kappa}{2} dz_\lambda \delta_{\alpha\lambda}(x - z) \{ \partial_\alpha(x) \Phi(x, P_1) \},$$

but we prefer the longer form, which indicates how it corresponds to (3.5b). Equations for the commutator between an element of a path associated with an operator R, and the operator itself, can similarly be obtained from (3.5b); the angular-momentum operator is simply transferred to the beginning and rewritten $J(z, \vec{x})$, and the $A(y, P_2)$ is omitted.

The commutator between two path elements is finally given by (2.31):

$$[A(x, P_1)B(y, P_2)]_{p_1, p_2}$$

$$= \frac{i\kappa}{2} \delta_{\alpha\lambda} \delta_{\gamma\nu} \underset{\lambda \leftrightarrow \nu}{S} dz_\alpha dz_\lambda' \{ \partial_\mu(z')\delta^3(z - z')\overline{[J_{\gamma 0}(z), A(x, P_1)]}$$

$$\times [J_{\mu\nu}(z'), B(y, P_2)]] - \partial_\beta(z')\delta^3(z - z')\overline{[J_{\beta\gamma}(z), A(x, P_1)][J_{\nu 0}(z'), B(y, P_2)]} \quad (3.8)$$

$$- i\delta^3(z - z')\overline{[J_{\gamma 0}(z), A(x, P_1)]\partial_\nu(z')B(y, P_2)}$$

$$+ i\delta^3(z - z')\overline{\partial_\gamma(z)A(x, P_1)[J_{\nu 0}(z'), B(y, P_2)]} \}.$$

The differential dz_α refers to the path element p_1, dz_λ' to p_2. The commutator $[A(x, P_1)]_{p_{1a}, p_{1b}}$ between two elements of the same path can be obtained from (3.8) in the same way as (3.7) was obtained from (3.5); the operator $J_{\mu\nu}(z')$ is transferred to the beginning of the right-hand side and written $J_{\mu\nu}(z', \vec{x})$ and the operator $B(y, P_2)$ is omitted.

The commutation relations (3.2)–(3.8) involve generalized products containing delta functions and Kronecker deltas. Such functions are considered to have path dependences associated with both their arguments, and an element on the path leading to one argument may be ordered in a different position from an element on the path leading to the other argument, even though the two paths are spatially the same except for their end-portions. The delta functions will have their usual c-number meaning, modified as described in the last section, only if two conditions are satisfied. The elements in the same position associated with the paths leading to the two arguments must be ordered together and the ordering of the elements in the end-portions not common to both paths must be unmixed with other operators or path elements. For other orderings, generalized products involving delta functions are determined by the commutation relations between path elements, like generalized products involving other operators. Thus δ-functions must be treated just like other functions when finding commutation relations between path elements.

One must also be careful about the ordering among themselves of the elements in end-portions of the paths. The delta function will be defined to have its usual meaning if the elements are symmetrically ordered. For other orderings, the delta functions are again defined from the commutation relations between path elements.

The verification of the consistency of the commutation relations with the path-dependence equation and the equations of motion follows the classical case very closely, and the equations need not be rewritten. Path elements will always be ordered in the same way on both sides of the equations, so that ambiguities of ordering give rise to no difficulty. In the classical theory we were only concerned with Poisson brackets between variables, together with their paths. However, as the consistency proofs for equations other than the Einstein equations of motion did not involve cancellations between different parts of the Poisson brackets, the proofs work equally well in the quantized theory, where we are concerned with commutation relations between individual path elements and operators. The special case of the Einstein equations will be treated in the Appendix. The equations of motion (2.19) for the matter field are valid even if Φ forms part of a generalized product, as the left-hand side commutes with all path elements. The Einstein equations, however, are only valid if $\mathfrak{J}_{\lambda\nu}$—the quantity which must be zero—is not ordered between path elements associated with an operator, or between a path element and the operator itself. As we have seen in the last section, the effect of commuting $\mathfrak{J}_{\mu\nu}$ with path elements or operators is to displace them, so that, unless this quantity is commuted with all the elements of a path and with the operator itself, the result cannot be zero.

In the consistency proofs we had to use Eq. (2.38b). As the delta function only has its usual meaning when the elements of the path joining the two arguments are symmetrically ordered, (2.38b) will only be true in the quantum theory if the two differential operators are symmetrically ordered. Thus, for consistency, the two differential operators in the equation of motion (2.14) must be symmetrically ordered.

It is evident that all the consistency proofs can be carried over from the classical to the quantum theory and, since they are guaranteed to work in the classical theory by virtue of the possibility of a Lagrangian formalism, they will work here too. By using ordered path elements and generalized products of operators, ambiguities in ordering are thus avoided. Needless to say, the nonlinearity of the theory will still cause tremendous complication in any applications. However, in the present theory, it does not give rise to ambiguities in the formulation.

Other consistency conditions which should be verified are the Jacobi identities associated with the commutators. Again we can be sure that they work in the classical theory. As long as the verification there does not make use of cancellations between different parts of the Poisson bracket, it can be taken over into the quantum theory.

NON-SPACELIKE PATHS

The treatment so far has been restricted to paths which are spacelike except possibly for infinitesimal portions. It will now be necessary to remove this restriction, since we do not in general know whether or not a particular path is spacelike. For the moment we shall still neglect this complication and shall simply extend the reasoning of the last subsection to apply to general paths.

The concepts of ordering operators and path elements are the same as before, and one must in principle be able to find the commutators between a path element and an operator and between two path elements. For non-spacelike paths one cannot write down these commutation relations explicitly, of course.

If one examines the covariant commutation rules given in the Appendix, one notices that the commutator between an element of a path P_2 and an operator or another element always has the following form:

$$[A(x, P_1)B(y, P_2)]_{X, p_2}$$
$$= \tfrac{1}{2} dz_\lambda (\partial_\mu(\tau) \overline{G_{\lambda\nu, X}\{A(x, P_1), z, P_3\}[J_{\mu\nu}(z)B(y, P_2)]} \qquad (3.9)$$
$$- i\overline{G_{\lambda\nu, X}\{A(x, P_1), z, P_3\}\partial_\nu z)B(y, P_2)}).$$

An equation of such a form is true whether the element p_2 is being commuted with the operator A itself or with an element of its path. This operator or element has been denoted by X. The function G will depend, through A, on P_1, and it will also depend on that portion of P_2 leading to z, which we denote by P_3. It will not depend on the remainder of P_1 or on the operator A. G is symmetrical in λ and ν.

We now postulate that (3.9) holds for general paths. The function G will not be known explicitly except for spacelike paths but, by varying the path P_2 around z and applying the path-dependence equation, we can obtain an equation for it. The manipulations are similar to those leading to (2.35), and the result is

$$\underset{\substack{\kappa \leftrightarrow \lambda \\ \mu \leftrightarrow \nu}}{A} \partial_\kappa(z)\partial_\mu(z) G_{\lambda\nu}\{A(x, P_1), z, P_3\} + \underset{\kappa \leftrightarrow \lambda}{A} R_{\mu\nu\kappa\rho}(z, P_3) G_{\lambda\rho}\{A(x, P_1), z, P_2\}$$
$$\qquad\qquad\qquad\qquad\qquad\qquad\qquad\qquad\qquad\qquad (3.10)$$
$$= 2i[A(x, P_1), R_{\mu\nu\kappa\lambda}(z, P_3)]_{X, R}.$$

The curl of G is thus determined by the path-dependence equation. We also require an equation for its divergence in order to calculate it for arbitrary paths once it is known for spacelike paths.

If we examine the commutation relations—most conveniently in their covariant form given in the Appendix—we observe that, for spacelike paths, the following equation is always satisfied:

$$\partial_\lambda(z) G_{\lambda\nu,A}\{A(x, P_1), z, P_3\} - \tfrac{1}{2}\partial_\nu(z) G_{\lambda\lambda,A}\{A(x, P_1), z, P_3\}$$
$$= \frac{i}{2} \partial_\mu(z)\Delta(x - z)[J_{\mu\nu}(z), A(x, P_1)] + \Delta(x - z)\partial_\nu(z)A(x, P_1). \quad (3.11\text{a})$$

We have written the function as $G_{\lambda\nu,A}$ to indicate that we are interested in the commutation relation between the element dz_λ and the operator A itself. If we were dealing with the commutator between two elements, the equation would be

$$\partial_\lambda(z) G_{\lambda\nu,p_1}\{A(x, P_1), z, P_3\} - \tfrac{1}{2}\partial_\nu(z) G_{\lambda\lambda,p_1}\{A(x, P_1), z, P_3\}$$
$$= -i\, dz_\alpha' \partial_\alpha(z')\{\tfrac{1}{2}\partial_\mu(z)\Delta(x - z)[J_{\mu\nu}(z), A(x, P_1)] \quad (3.11\text{b})$$
$$- i\Delta(x - z)\partial_\nu(z)A(x, P_1)\},$$

where dz' represents the element p_1. It is evident that, if the function Δ in (3.11) were replaced by a δ-function, the effect of the right-hand side would be simply to deform the coordinates in which the operator $A(x, P_1)$ or the element p_1 is being measured. Such a deformation can also be effected by commuting with the energy tensor. In fact, following closely the manipulations leading to (2.44) in the classical theory, one can show from the commutation relations that

$$\overline{[A(x, P_1), \mathfrak{J}_{\mu\nu}(z, P_3)]}_{x,3} = \underset{\mu \leftarrow \nu}{S}\; \partial_\mu(z)\, F_{\nu,x}\{A(x, P_1), z, P_3\}$$
$$\overline{\phantom{= \underset{\mu \leftarrow \nu}{S'}\; \mathfrak{J}_{\mu\sigma}(z, P_3) G_{\sigma\nu,x}\{A(x, P_1), z, P_3\},}} \quad (3.12)$$
$$- \underset{\mu \leftarrow \nu}{S'}\; \mathfrak{J}_{\mu\sigma}(z, P_3) G_{\sigma\nu,x}\{A(x, P_1), z, P_3\},$$

where F is i times the right-hand side of (3.11). S' is the normal symmetrizing operator (2.25b) without the third term. Equation (3.11) can therefore be written in the form

$$\partial_\lambda(z) G_{\lambda\nu,x}\{A(x, P_1), z, P_3\} - \tfrac{1}{2}\partial_\nu(z) G_{\lambda\lambda,x}\{A(x, P_1), z, P_3\}$$
$$= -iF_{\nu,x}\{A(x, P_1), z, P_3\}. \quad (3.13)$$

The analysis so far has applied only to spacelike paths. However, Eq. (3.12) will retain its form when subjected to the path-dependence equation or the equations of motion for A, so that it will be true generally. We can thus define a function F by (3.12), though for non-spacelike paths it will not be known explicitly. If we were dealing with a theory which did not involve path-dependent operators, a divergence condition on a commutator such as (3.13) would be valid in general if it were valid for spacelike separations of x and z. We cannot assert such a statement in the present theory, since the commutation relations involving timelike separated elements do not follow from the spacelike commutation relations, except in so far as they are restricted by (3.10), but must be postulated. The natural postulate would appear to be that the divergence condition (3.13), which is valid for spacelike paths, is valid in general, and we shall adopt this postulate here. Such a postulate is rendered even more necessary by the fact that the spacelike character of the paths is not known *a priori*.

The reasoning leading to (3.13) and the postulate that it is valid for general paths also hold if G is part of a generalized product.

The equations (3.10) and (3.13) combined determine G for general paths from its value for spacelike paths. Thus the equations of motion and the path-dependence equation, when supplemented by the divergence condition (3.13) on the commutator, enable operators and generalized products of operators to be found for all paths once they are known for spacelike paths.

SPACELIKE CHARACTER OF PATHS NOT ASSUMED KNOWN

The difficulty with the commutation relations (3.2)–(3.8) as they have been stated is due to the fact that one can only postulate them to be true when the two operators or elements involved are connected by a spacelike path. However, the determination whether or not a particular path is spacelike is a complicated process, even in the classical theory, and requires a knowledge of the Riemann tensor over a region in space. It can quite easily happen that a path is locally spacelike at all points along its length, but that a signal can nevertheless be propagated between two distant points on it. In the quantum theory, where the Riemann tensor is no longer a c-number, one cannot talk about the "global" spacelike character of a path, and the use of such a concept in the formulation must therefore be avoided.

The commutation relations between two neighboring operators or elements present no difficulty, as they only depend on the local spacelike character of the path. The difficulty concerns the additional assumption implied by Eqs. (3.2)–(3.8) that distant operators or elements commute if they are spacelike separated.

It appears, however, that in a Lorentz-invariant theory one does not have the freedom to state commutation relations for distant points. The local commutation relations, together with the equations of motion and the Lorentz invariance, determine the distant commutation relations.[3] If this is so, there is no difficulty, since we need simply postulate that (3.2)–3.8) are valid for neighboring spacelike separated operators and elements. We cannot of course prove rigorously that the local commutation relations determine the theory, since, in our present

---

[3] We are still referring to an essentially conventional formalism, and are not considering at the moment the more radical lack of freedom suggested by dispersion theory.

state of knowledge, nothing can be maintained with certainty about the existence and uniqueness of solutions of equations governing interacting quantized fields. For noninteracting fields we can easily show that it is sufficient to state the local commutation rules, and the extension to any order of perturbation theory with interacting fields should not prove difficult.

To take a free scalar field as our example, the most general commutation relation which satisfied both the Klein-Gordon equation and Lorentz invariance will be

$$[\phi(x, t), \phi(0, 0)]$$

$$= \int d^3\mathbf{p} \, |\mathbf{p}|^{-1} \{ A(\mathbf{p}, p_0) e^{i(\mathbf{p}\cdot\mathbf{x} - p_0 x_0)} - A(-\mathbf{p}, p_0) e^{i(\mathbf{p}\cdot\mathbf{x} + p_0 t)} \}. \quad (3.14)$$

The symbol p_0 is short for $(\mathbf{p}^2 + \mu^2)^{1/2}$, and A is any operator which depends on the four-vector $\mathbf{p}$, p_0 in a Lorentz-covariant way. As A is certainly defined for all $\mathbf{p}$ once it is known for one $\mathbf{p}$, there is only sufficient freedom in (3.14) to fix one operator, and a statement of the commutation relations between ϕ and $\dot\phi$ for neighboring points will therefore determine it. For, if we are given one set of operators $A_1(\mathbf{p}, p_0)$ and to it add another $A_2(\mathbf{p}, p_0)$, the addition will certainly change the value of the time-derivative of the right-hand side at $x = 0$, $x_0 = 0$; the various contributions to the integral

$$\int d^3\mathbf{p} \, \frac{p_0}{|\mathbf{p}|} A_2(\mathbf{p}, p_0)$$

have different Lorentz properties and cannot cancel.

We shall therefore assume that it is sufficient to state the local commutation relations in our theory. In addition, we shall assume the general form (3.9) for the commutator between a path element and an operator or another element, and we shall assume that it satisfies the divergence condition (3.13) with F defined by (3.12). This equation, together with (3.10), gives us an "equation of motion" for G so that it is reasonable to assume that commutators involving path elements, like commutators between operators, are determined from their local values in a Lorentz-covariant theory.

To sum up, we shall take the path-dependence equation (3.1), the consistency conditions (2.6) and (2.7), the equations of motion (2.19) and (2.20), the commutation relations (2.2)–(2.8) between neighboring spacelike separated operators and elements, and the general equations (3.9) and (3.13) to define our theory of interacting quantized scalar and gravitational fields.

IV. PERTURBATION EXPANSION

Of the treatments given in I, only the perturbation expansion seems capable of being used for gravitation. As in other field theories, one would probably be able in principle to sum subsets of the series by means of integral equations. One would then have a semiperturbation method capable of dealing with bound states. Owing to the nonlinearity of the theory, there does not appear to be any obvious way of obtaining an equivalent of Coulomb gauge from the path-dependent gravitation theory. The analogous classical problem has been solved by Arnowitt et al. (5).

In zeroth order the theory developed here is precisely the same as the linearized theory (1, 6); the results are familiar and need not be restated. As an example of the procedure to be used in higher orders, we shall treat the vertex function in first order. The results will be the same as in the Gupta theory, though now only the coordinate independent functions will be regarded as physically significant.

We have unfortunately not succeeded in finding a definition for Green's functions in the present theory, as the time-ordering operation is not always defined for our path-dependent operators. We shall therefore use instead the vacuum-expectation values of simple products of operators (Wightman functions). One can formulate any field theory in terms of such functions, which contain the same information as Green's functions, and one can obtain perturbation expansions for them. Since there are more Wightman functions than Green's functions owing to the different ways of ordering products, the amount of mathematical manipulation required when formulating perturbation theory in terms of them is slightly greater.

The functions we shall require are

$$W_{\alpha\beta\gamma\delta}(xP_1, yP_2; zP_3) = \overline{\langle \Phi(x, P_1)\Phi(y, P_2)R_{\alpha\beta\gamma\delta}(z, P_3)\rangle}, \quad (4.1a)$$

$$W_{\alpha\beta\gamma\delta,\mu\nu\kappa\lambda}(xP_1, yP_2; zP_3, wP_4)$$
$$= \overline{\langle \Phi(x, P_1)\Phi(y, P_2)R_{\alpha\beta\gamma\delta}(z, P_3)R_{\mu\nu\kappa\lambda}(z, P_4)\rangle}. \quad (4.1b)$$

In addition to specifying the ordering of the operators one must also specify the ordering of the path elements. In the approximation to which we shall work it will be sufficient to give the ordering of the path elements in (4.1a) relative to the operator $R_{\alpha\beta\gamma\delta}(z, P_3)$. For definiteness, we shall suppose that the path elements are ordered after the operator.

In zeroth order the Wightman functions are not path dependent, and are the same as in the linearized theory.

$$W^{(0)}_{\alpha\beta\gamma\delta}(x, y; z) = 0, \quad (4.2a)$$

$$W^{(0)}_{\alpha\beta\gamma\delta,\mu\nu\kappa\lambda}(x, y; z, w) = \frac{\kappa}{2}(2\pi)^{-6} \underset{\substack{\alpha\leftrightarrow\beta\\ \gamma\leftrightarrow\delta\\ \mu\leftrightarrow\nu\\ \kappa\leftrightarrow\lambda}}{A} \underset{\alpha\leftrightarrow\gamma}{S} \int dp\, dq\, e^{ip(y-x)} e^{iq(z-w)} \quad (4.2b)$$

$$\times q_\beta q_\delta q_\nu q_\lambda \delta_{\alpha\mu} \delta_{\gamma\kappa} \delta^-(p^2+m^2)\delta^-(q^2).$$

We use the abbreviation

$$\delta^+(p^2 + m^2) = \theta(p_0)\delta(p^2 + m^2), \quad \delta^-(p^2 + m^2) = -\theta(-p_0)\delta(p^2 + m^2).$$

We shall show how $W_{\alpha\beta\gamma\delta}(xP_1, yP_2; zP_3)$ may be found in first order. Its path dependence can be found from (2.3) and is, in lowest order,

$$\delta_{P_1,w} W^{'(1)}_{\alpha\beta\gamma\delta}(xP_1, yP_2; zP_3)$$
$$= -\frac{i}{4} J_{\mu\nu}(w, \bar{x}) W^{'(0)}_{\alpha\beta\gamma\delta,\mu\nu\kappa\lambda}(xP_1, yP_2; zP_3, wP_4)\sigma_{\kappa\lambda}. \quad (4.3)$$

As before, the symbol $J_{\kappa\lambda}(w, \bar{x})$ indicates that the rotation operator affects the point x (and, of course, the path P_1 leading to it) only. The path P_4 is the path P_1 up to the point w. There will be similar equations giving the changes in $W^{'(1)}_{\alpha\beta\gamma\delta}$ due to variations in P_2 and P_3. We may note that, as path elements are ordered after the operator $R(z, P_3)$, one must order the operator $R_{\mu\nu\kappa\lambda}(w, P_4)$, associated with the variation of the path, after $R_{\alpha\beta\gamma\delta}(z, P_3)$. This has been done in $W^{(0)}_{\alpha\beta\gamma\delta,\mu\nu\kappa\lambda}$ according to the definition (4.1b).

Our method of procedure will be the same as in I. We shall try to find integrals which can be written down explicitly and which satisfy Eq. (4.3), so that the difference between $W^{'(1)}_{\alpha\beta\gamma\delta}$ and these integrals is not path dependent. Accordingly, we write

$$W^{'(1)}_{\alpha\beta\gamma\delta}(xP_1, yP_2; zP_3) = w^{(1)}_{\alpha\beta\gamma\delta}(x, y; z)$$
$$+ \frac{i}{2} \sum_{\substack{x,P_1 \\ y,P_2 \\ z,P_3}} \int_{P_1}^{x} dw_\lambda [J_{\mu\nu}(w, \bar{x}), w^{(1)}_{\alpha\beta\gamma\delta,\mu\nu\lambda}(x, y; z, w)] \quad (4.4)$$
$$- \sum_{\substack{x,P_1 \\ y,P_2 \\ z,P_3}} \int_{P_1}^{x} dw_\lambda (\partial/\partial x_\nu) w^{(1)}_{\alpha\beta\gamma\delta,\nu\lambda}(x, y; z, w).$$

The summation sign is to indicate that to each integral we add corresponding integrals over the paths P_2 and P_3.

If $w^{(1)}_{\alpha\beta\gamma\delta,\mu\nu\lambda}$ satisfies the equation,

$$\frac{\partial}{\partial w_\kappa} w^{(1)}_{\alpha\beta\gamma\delta,\mu\nu\lambda}(x, y; z, w) - \frac{\partial}{\partial w_\lambda} w^{(1)}_{\alpha\beta\gamma\delta,\mu\nu\kappa}(x, y; z, w) = W^{(0)}_{\alpha\beta\gamma\delta,\mu\nu\kappa\lambda}(x, y; z, w) \quad (4.5a)$$

the variation of the integral in the first term of (4.4) on applying Stokes' theorem to w will be given by (4.3), so that $w^{(1)}_{\alpha\beta\gamma\delta}(x, y; z)$ will not be path dependent. There will be an additional variation of the integral in the first term of (4.4) obtained by applying Stokes' theorem to $J_{\mu\nu}(w, \bar{x})$; this will just be canceled by the variation of the second integral of (4.4) provided that

$$\frac{\partial}{\partial w_\nu} w^{(1)}_{\alpha\beta\gamma\delta,\mu\lambda}(x, y; z, w) - \frac{\partial}{\partial w_\mu} w^{(1)}_{\alpha\beta\gamma\delta,\nu\lambda}(x, y; z, w) = w^{(1)}_{\alpha\beta\gamma\delta,\mu\nu\lambda}(x, y; z, w). \quad (4.5b)$$

Our auxiliary functions correspond to the following functions on the Gupta theory:

$$w^{(1)}_{\alpha\beta\gamma\delta}(x, y; z) \sim \langle \phi(x)\phi(y) R_{\alpha\beta\gamma\delta}(z) \rangle,$$

$$w^{(1)}_{\alpha\beta\gamma\delta,\mu\nu\lambda}(x, y; z, w) \sim \left\langle \phi(x)\phi(y) R_{\alpha\beta\gamma\delta}(z) \left\{ \frac{\partial \gamma_{\mu\lambda}}{\partial w_\nu} - \frac{\partial \gamma_{\nu\lambda}}{\partial w_\mu} \right\} \right\rangle,$$

$$w^{(1)}_{\alpha\beta\gamma\delta,\mu\lambda}(x, y; z, w) \sim \langle \phi(x)\phi(y) R_{\alpha\beta\gamma\delta}(z) \gamma_{\mu\lambda}(w) \rangle,$$

and the integral (4.4) for the path-dependent function corresponds to Eq. (2.13). This correspondence does not play a part in our logical deductions, of course.

Functions $w^{(1)}_{\alpha\beta\gamma\delta,\mu\nu\lambda}$ and $w^{(1)}_{\alpha\beta\gamma\delta,\mu\lambda}$ which satisfy (4.5) are

$$w^{(1)}_{\alpha\beta\gamma\delta,\mu\nu\lambda}(x, y; z, w) = \frac{i\kappa}{2}(2\pi)^{-6} \underset{\substack{\alpha\leftrightarrow\beta\\ \gamma\leftrightarrow\delta\\ \mu\leftrightarrow\nu}}{A} \underset{\alpha\leftrightarrow\gamma}{S}$$

$$\cdot \int dp\, dq\, e^{ip(y-x)} e^{iq(z-w)}\, q_\beta q_\delta q_\nu \delta_{\alpha\mu} \delta_{\lambda\gamma}\, \delta^-(p^2 + m^2)\delta^-(q^2) \quad (4.6a)$$

$$w^{(1)}_{\alpha\beta\gamma\delta,\mu\lambda}(x, y; z, w) = -\frac{\kappa}{2}(2\pi)^{-6} \underset{\substack{\alpha\leftrightarrow\beta\\ \gamma\leftrightarrow\delta}}{A} \underset{\alpha\leftrightarrow\gamma}{S}$$

$$\cdot \int dp\, dq\, e^{ip(y-x)} e^{iq(z-w)} q_\beta q_\delta \delta_{\alpha\mu} \delta_{\gamma\lambda}\, \delta^-(p^2 + m^2)\delta^-(q^2). \quad (4.6b)$$

The functions are not defined uniquely by (4.5), and there are other possible choices. The final path-dependent functions will be independent of our choice.

We can now apply the equations of motion to find $w^{(1)}_{\alpha\beta\gamma\delta}$. According to (2.19), $W^{(1)}_{\alpha\beta\gamma\delta}$ will satisfy the equation

$$\left(\frac{\partial^2}{\partial x_\mu^2} - m^2 \right) W^{(1)}_{\alpha\beta\gamma\delta}(xP_1, yP_2; z, w) = 0. \quad (4.7)$$

The differential operator applied to the two integrals in (4.4) gives:

$$\left(\frac{\partial^2}{\partial x_p^2} - m^2 \right) \frac{i}{2} \int_{P_1}^{x} dw_\lambda\, [J_{\mu\nu}(w, \vec{x}), w^{(1)}_{\alpha\beta\gamma\delta,\mu\nu\kappa}(x, y; z, w)]$$

$$= \frac{\partial}{\partial x_\nu} w^{(1)}_{\alpha\beta\gamma\delta,\mu\nu\mu}(x, y; z, w)_{w=x}, \quad (4.8a)$$

$$-\left(\frac{\partial^2}{\partial x_p^2} - m^2 \right) \int_{P_1}^{x} dw_\lambda\, \frac{\partial}{\partial x_\nu} w^{(1)}_{\alpha\beta\gamma\delta,\nu\lambda}(x, y; z, w)$$

$$= -\left\{ 2\frac{\partial}{\partial x_\lambda} + \frac{\partial}{\partial w_\lambda} \right\} \frac{\partial}{\partial x_\nu} w^{(1)}_{\alpha\beta\gamma\delta,\nu\lambda}(x, y; z, w)_{w=x}. \quad (4.8b)$$

From (4.4), (4.7), and (4.8) we can now obtain directly a differential equation for $w^{(1)}_{\alpha\beta\gamma\delta}(x, y; z)$. On inserting into (4.8) the functions $w^{(1)}_{\alpha\beta\gamma\delta,\mu\nu\mu}$ and $w^{(1)}_{\alpha\beta\gamma\delta,\nu\lambda}$ given by (4.6), we obtain

$$\left(\frac{\partial^2}{\partial x_\mu^2} - m^2\right) w^{(1)}_{\alpha\beta\gamma\delta}(x, y; z) = -\kappa(2\pi)^{-6} \underset{\substack{\alpha\leftrightarrow\beta\\\gamma\leftrightarrow\delta}}{A} \underset{\alpha\leftrightarrow\gamma}{S} \int dp\, dq\, e^{i(-p-q)x + ipy + iqz} \quad (4.9)$$

$$\times q_\beta q_\delta \{p_\mu(-p_\nu - q_\nu)\delta_{\alpha\mu}\delta_{\gamma\nu} + p_\mu(-p_\mu - q_\mu)\delta_{\alpha\gamma} + m^2\delta_{\alpha\gamma}\}\delta^-(p^2 + m^2)\delta^-(q^2)$$

This is the same equation we would obtain in the Gupta theory. The solution is

$$w^{(1)}_{\alpha\beta\gamma\delta}(x,y;z) = \kappa(2\pi)^{-6} \underset{\substack{\alpha\leftrightarrow\beta\\\gamma\leftrightarrow\delta}}{A} \underset{\alpha\leftrightarrow\gamma}{S} \int dp\, dq\, e^{i(-p-q)x + ipy + iqz} q_\beta q_\delta$$

$$\cdot \{p_\mu(-p_\nu - q_\nu)\delta_{\alpha\mu}\delta_{\gamma\nu} + p_\mu(-p_\mu - q_\mu)\delta_{\alpha\gamma} + m^2\delta_{\alpha\gamma}\}\{(p+q)^2 + m^2\}^{-1} \quad (4.10)$$

$$\delta^-(p^2 + m^2)\delta^-(q^2) + \int dp\, dq\, e^{i(-p-q)x + ipy + iqz} f(p,q)\delta\{(p+q)^2 + m^2\}.$$

This formula, with the arbitrary function in the second term, which represents the solution of the homogeneous equation corresponding to (4.9), is as far as we can go by looking at the differential equation in x satisfied by $w^{(1)}_{\alpha\beta\gamma\delta}(x, y; z)$. We can complete the solution by taking Wightman functions with the operators in different orders and using the differential equations satisfied by them as functions of x. According to the commutation relations, two three-point functions obtained by interchanging the order of a pair of operators are equal, to first order, if the operators are space-like separated. We thus have enough information to find the functions. We shall not go through the procedure, which is the same as if we were trying to find the Wightman functions from the differential equations in any other field theory. The result is

$$w^{(1)}_{\alpha\beta\gamma\delta}(x,y;z) = \kappa(2\pi)^{-6} \underset{\substack{\alpha\leftrightarrow\beta\\\gamma\leftrightarrow\delta}}{A} \underset{\alpha\leftrightarrow\gamma}{S} \int dp\, dq\, e^{i(-p-q)x + ipy + iqz} q_\beta q_\delta$$

$$\cdot \{p_\mu(-p_\nu - q_\nu)\delta_{\alpha\mu}\delta_{\gamma\nu} + p_\mu(-p_\mu - q_\mu)\delta_{\alpha\gamma} + m^2\delta_{\alpha\gamma}\}(P\{(p^2+q^2)^2 \quad (4.11)$$

$$+ m^2\}^{-1}\delta^-(p^2 + m^2)\delta^-(q^2) + \delta^+\{(-p-q)^2 + m^2\}P(p^2 + m^2)^{-1}$$

$$\cdot \delta^-(q^2) + \delta^+\{(-p-q)^2 + m^2\}\delta^+(p^2 + m^2)\delta^-(q^2)),$$

where the symbol P denotes the principle value.

One can use a similar method to calculate Wightman functions to any order. One may encounter renormalization difficulties. As is well known, a simple counting of the degree of divergence of higher-order perturbation diagrams, according to Dyson's prescription, indicates that the theory is not renormalizable. However, more work is required to find out whether the coordinate invariant functions treated here can be obtained in renormalized form. Feynman has suggested a prescription for calculating the S-matrix elements in any order, and is investigating whether his theory is renormalizable. It is not clear whether Feynman's prescription is equivalent to the prescription which would result from the present theory, or whether his theory can be embedded in one which involves quantities other than the S-matrix. In any case, it may quite easily happen that individual terms of the perturbation series are not renormalizable whereas the series as a whole, or a subseries, is. Such a situation is of course possible in any field theory, but it is much more plausible in gravitational theory. The reason is that repeated applications of the path-dependence equation (2.3) in perturbation theory give higher and higher derivatives associated with the rotation operator, whereas the net effect would be to displace the end-point of the path by a finite distance—a much less singular operation. If such a situation held, it would be rather awkward for calculations; however, it might be possible to sum a subset of the perturbation series, by means of an integral equation or otherwise, in such a way that the sum of the derivatives associated with the various orders became replaced by a finite displacement. One could then hope to obtain finite results.

V. CONCLUDING REMARKS

It is hoped that the present paper has borne out our conviction that Einstein's theory of gravitation can be incorporated into quantum theory, subject to the one important possible limitation of renormalizability. Once one assumes that inertial coordinate systems exist only locally, in other words, that space is curved and is described by Riemann geometry, one has no absolute inertial coordinate system. In classical theory, one has then the alternative of either formulating the laws so that they hold in an arbitrary coordinate system, or of working with quantities which are independent of the coordinate system. The first alternative is usually preferred for reasons of simplicity. In quantum theory, one cannot work with quantities which are dependent on an arbitrary choice of certain variables such as coordinates, and one is forced to the second alternative. Instead of starting with a coordinate system, which does not correspond to anything physical, one must describe how one is going to fix the points in space where measurements are to be performed. By doing so one is led to the path-dependent variables used in this paper.

In the present theory it is never necessary to talk about the $g_{\mu\nu}$'s. Such quantities do not occur in measurements and should therefore be regarded as mathematical aids in the classical theory. It is therefore not surprising that the quantization of such variables present difficulties even on superficial examination. Instead the curvature is introduced directly in our theory. The definition of the curvature is the same as in the conventional theory, and we have seen that such a definition fits naturally into our present formalism without the necessity of introducing intermediate unphysical variables. The quantization of curvature means that we can no longer talk about space–time consisting of a four-dimensional infinity of points. In classical theory one can reach a point by a number of different paths and, nevertheless, the result of a measurement will be independent of which path was taken. This is no longer true in quantum theory. Most physicists will probably not be surprised by such a feature, and would have expected it in a quantized gravitational theory. It appears to be an inescapable result of the general principles of quantum theory, and the absence of an inertial frame.

Thirring (*2*) has shown that a tensor field which interacts with the energy tensor will automatically have the physical significance of a metric tensor in first-order classical theory. Furthermore, the only possible interaction of a massless tensor field is with the energy tensor, since the equation of continuity must be satisfied. These features are a manifestation of the fact that a massless field with nonzero spin can only interact with matter if it is associated with a gauge-invariance or coordinate-invariance principle. Thus a massless tensor field can only interact with matter according to Einstein's ideas. This does not of course imply that it necessarily obeys his actual equations.

The theory developed in the preceding paragraphs may appear complicated. We should like to stress, however, that it does not possess any more *arbitrary* complications than does any other quantized field theory. Once we assume that an inertial frame does not exist and that the curvature of space is caused by matter, we are led uniquely to our theory—except of course in the actual equation of motion, when we followed Einstein and took the simplest possible consistent set. In other words, the situation in quantum theory as regards space–time curvature is not very different from what it is in classical theory.

APPENDIX I. COVARIANT FORMS OF THE POISSON BRACKETS

The uncorrected covariant forms of the Poisson brackets are:

$$[\Phi(x, P_1), \Phi(y, P_2)]_{0,1} = i\Delta(x - y), \tag{A1.1}$$

$$[R_{\alpha\beta\gamma\delta}(x, P_1), \Phi(y, P_2)]_{0,1} = 0, \tag{A1.2a}$$

$$[R_{\alpha\beta\gamma\delta}(x, P_1), \Phi(y, P_2)]_{0,1} = -\frac{i\kappa}{2} \underset{\substack{\alpha \leftrightarrow \beta \\ \gamma \leftrightarrow \delta}}{A} \underset{\beta \leftrightarrow \delta}{S} \partial_\gamma(x)\partial_\alpha(x)\{\delta_{\beta\mu}\delta_{\delta\nu}\Delta(x - y)\}$$

$$\times \partial_\nu(y)\Phi(y, P_2) - \frac{i\kappa}{2} \underset{\gamma \leftrightarrow \delta}{A} R_{\alpha\beta\gamma\epsilon}(x, P_1) \underset{\epsilon \leftrightarrow \delta}{S} \delta_{\epsilon\mu}\delta_{\delta\nu}\Delta(x - y)\partial_\nu(y)\Phi(y, P_2), \tag{A1.2b}$$

$$[R_{\alpha\beta\gamma\delta}(x, P_1), R_{\mu\nu\kappa\lambda}(y, P_2)]_{0,1}$$

$$= \frac{i\kappa}{2} \underset{\substack{\alpha \leftrightarrow \beta \\ \gamma \leftrightarrow \delta \\ \mu \leftrightarrow \nu \\ \kappa \leftrightarrow \lambda}}{A} \underset{\beta \leftrightarrow \delta}{S} \partial_\gamma(x)\partial_\alpha(x)\partial_\kappa(y)\partial_\mu(y)\{\delta_{\beta\nu}\delta_{\delta\lambda}\Delta(x - y)\}$$

$$+ \frac{i\kappa}{2} \underset{\substack{\kappa \leftrightarrow \lambda \\ \alpha \leftrightarrow \beta \\ \gamma \leftrightarrow \delta}}{A} \underset{\beta \leftrightarrow \delta}{S} R_{\mu\nu\kappa\rho}(y, P_2) \partial_\gamma(x) \partial_\alpha(x) \{\delta_{\beta\rho} \delta_{\delta\lambda} \Delta(x - y)\} \qquad (A1.3)$$

$$+ \frac{i\kappa}{2} \underset{\substack{\kappa \leftrightarrow \lambda \\ \mu \leftrightarrow \nu \\ \gamma \leftrightarrow \delta}}{A} \underset{\nu \leftrightarrow \kappa}{S} R_{\alpha\beta\gamma\epsilon}(x, P_1) \times \partial_\kappa(y) \partial_\mu(y) \{\delta_{\epsilon\nu} \delta_{\delta\lambda} \Delta(x - y)\}$$

$$+ \frac{i\kappa}{2} \underset{\substack{\kappa \leftrightarrow \lambda \\ \gamma \leftrightarrow \delta}}{A} R_{\alpha\beta\gamma\epsilon}(x, P_1) R_{\mu\nu\kappa\rho}(y, P_2) \underset{\rho \leftrightarrow \lambda}{S} \delta_{\epsilon\rho} \delta_{\delta\lambda} \Delta(x - y),$$

$$[\Phi(x, P_1), A(y, P_2)]_{0,2\Phi} = 0, \qquad (A1.4a)$$

$$[\partial_\alpha(x)\Phi(x, P_1), A(y, P_2)]_{0,2\partial\Phi}$$

$$= -\frac{\kappa}{2} \int_{P_2} dz_\lambda\, \partial_\beta(x)\Phi(x, P_1)(\partial_\mu(z)\{\underset{\lambda \leftrightarrow \nu}{S} \delta_{\alpha\lambda} \delta_{\beta\nu} \Delta(x - z)\} \qquad (A1.4b)$$

$$\times [J_{\mu\nu}(z), A(y, P_2)] - i\{\underset{\lambda \leftrightarrow \nu}{S} \delta_{\alpha\lambda} \delta_{\beta\nu} \Delta(x - z)\} \partial_\nu(z) A(y, P_2)),$$

$$[R_{\alpha\beta\gamma\delta}(x, P_1), A(y, P_2)]_{0,2R}$$

$$= \frac{\kappa}{2} \underset{\substack{\alpha \leftrightarrow \beta \\ \gamma \leftrightarrow \delta}}{A} \underset{\beta \leftrightarrow \delta}{S} \int_{P_2} dz_\lambda (\partial_\gamma(x) \partial_\alpha(x) \partial_\mu(z) \{\delta_{\beta\lambda} \delta_{\delta\mu} \Delta(x - z)\}$$

$$\times [J_{\mu\nu}(z), A(y, P_2)] - i\partial_\gamma(x) \partial_\alpha(x) \{\delta_{\beta\lambda} \delta_{\delta\nu} \Delta(x - z)\} \partial_\nu(z) A(y, P_2)) \qquad (A1.5)$$

$$+ \frac{\kappa}{2} \underset{\gamma \leftrightarrow \delta}{A} \int_{P_2} dz_\lambda\, R_{\alpha\beta\gamma\epsilon}(x, P_1) \underset{\epsilon \leftrightarrow \delta}{S} (\partial_\mu(z)\{\delta_{\epsilon\lambda} \delta_{\delta\nu} \Delta(x - z)\}[J_{\mu\nu}(z), A(y, P_2)]$$

$$- i\delta_{\epsilon\lambda} \delta_{\delta\nu} \Delta(x - z) \partial_\nu(z) A(y, P_2)),$$

$$[A(x, P_1), B(y, P_2)]_{0,3}$$

$$= -\frac{i\kappa}{2} \int_{P_1}^{x} dz_\alpha \int_{P_2}^{y} dz'_\lambda (\partial_\beta(z) \partial_\mu(z') \{\underset{\alpha \leftrightarrow \gamma}{S} \delta_{\alpha\lambda} \delta_{\gamma\nu} \Delta(z - z')\}$$

$$\times [J_{\beta\gamma}(z), A(x, P_1)][J_{\mu\nu}(z'), B(y, P_2)] + i\partial_\beta(z)\{\underset{\alpha \leftrightarrow \gamma}{S} \delta_{\alpha\lambda} \delta_{\gamma\nu} \Delta(z - z')\} \qquad (A1.6)$$

$$\times [J_{\beta\gamma}(z), A(x, P_1)] \partial_\nu(z') B(y, P_2)$$

$$- i\partial_\mu(z')\{\underset{\alpha \leftrightarrow \gamma}{S} \delta_{\alpha\lambda} \delta_{\gamma\nu} \Delta(z - z')\} \partial_\gamma(z) A(x, P_1)[J_{\mu\nu}(z'), B(y, P_2)]$$

$$+ \{\underset{\alpha \leftrightarrow \gamma}{S} \delta_{\alpha\lambda} \delta_{\gamma\nu} \Delta(x - z)\} \partial_\gamma(z) A(x, P_1) \partial_\nu(z') B(y, P_2)).$$

In writing down these relations we have included several terms involving the function Δ without derivatives acting on it. Such terms are zero for completely spacelike paths and should properly be considered as correction terms. We have included them here for convenience to preserve certain formal properties, e.g., a term $\partial_\mu(z)\Delta(x - z)[J_{\mu\nu}(z), A(y, P_2)]$ is always associated with another term $-i\Delta(x - z)\partial_\nu(z)A(y, P_2)$. Such an association is necessary if the uncorrected terms alone are to be consistent with the path-dependence equations. The relations (A1.2b) and (A1.4b) are obtained by differentiating (A1.5) and (A1.6) with respect to the end-points of the P_2- and P_1-integrations respectively, just as in the derivation of (2.32).

The correction terms to the Poisson brackets are

$$[R_{\alpha\beta\gamma\delta}(x, P_1), \Phi(y, P_2)]_{c,1}$$
$$= -i\kappa \underset{\substack{\alpha \leftrightarrow \beta \\ \gamma \leftrightarrow \delta}}{A} \underset{\beta \leftrightarrow \delta}{S'} \partial_\gamma(x)\partial_\alpha(x)\{\partial_\beta(x)\Gamma(x - y)\partial_\delta(x)\Phi(x, P_1)\} \qquad (A1.7)$$
$$- i\kappa \underset{\gamma \leftrightarrow \delta}{A} R_{\alpha\beta\gamma\epsilon}(x, P_1) \underset{\epsilon \leftrightarrow \delta}{S'} \partial_\epsilon(x)\Gamma(x - y)\partial_\delta(x)\Phi(x, P_1),$$

$$[\Phi(x, P_1), A(y, P_2)]_{c,2\Phi}$$
$$= -\kappa \int_{P_2} dz_\lambda (\partial_\mu(z)\{\underset{\lambda \leftrightarrow \nu}{S'} \partial_\lambda(z)\Gamma(x - z)\partial_\nu(z)\Phi(z, P_3)\} \qquad (A1.8)$$
$$\times [J_{\mu\nu}(z), \Phi(x, P_1)] - i \underset{\lambda \leftrightarrow \nu}{S'} \partial_\lambda(z)\Gamma(x - z)\partial_\nu(z)\Phi(z, P_3)\}\partial_\nu(z)\Phi(x, P_1)).$$

S' is just the normal symmetrizing operator (2.25b) without the last term. The correction terms to all the other relations, except (A1.1) which has no correction term, are obtained by writing the uncorrected relation with the substitution

$$\delta_{\beta\lambda}\delta_{\delta\nu}\Delta(x - y) \to \{-2R_{\epsilon\beta\eta\delta}(x, P_1)\delta_{\eta\lambda}\delta_{\epsilon\nu} - R_{\epsilon\beta\epsilon\eta}(x, P_1)\delta_{\eta\lambda}\delta_{\delta\nu}\}\Gamma(x - y)$$
$$= \{-2R_{\rho\lambda\sigma\nu}(y, P_2)\delta_{\beta\sigma}\delta_{\delta\rho} - R_{\rho\lambda\rho\sigma}(y, P_2)\delta_{\beta\sigma}\delta_{\delta\nu}\}\Gamma(x - y). \qquad (A1.9)$$

APPENDIX II. PROOF OF THE CONSISTENCY OF EINSTEIN'S EQUATIONS WITH THE COMMUTATION RELATIONS

Before we carry out the proof we shall have to establish one or two preliminary results. First, we may write down the covariant form of the commutator between the derivative of an operator and a path element, analogous to equation (A1.4b) of the classical theory

$$\overline{[\partial_\alpha(x)\Phi(x, P_1), \Phi(y, P_2)]}_{\partial\Phi, P_2} = -\frac{\kappa}{2} dz_\lambda \, \partial_\epsilon(x)\Phi(x, P_1)M_{\alpha\epsilon\lambda}\{x, z, \Phi(y, P_2)\}, \qquad (A2.1)$$

where

$$M_{\alpha\epsilon\lambda}\{x, z, \Phi(y, P_2)\} = \partial_\mu(z)\{\underset{\lambda \leftrightarrow \nu}{S} \delta_{\epsilon\lambda}\delta_{\alpha\nu}\Delta(x - z)\}[J_{\mu\nu}(z), \Phi(y, P_2)]$$
$$- i\{\underset{\lambda \leftrightarrow \nu}{S} \delta_{\epsilon\lambda}\delta_{\alpha\nu}\Delta(x - z)\}\partial_\nu(x)\Phi(y, P_2) \qquad (A2.2)$$

A similar result holds for the commutator between the derivative of a delta function and a path element:

$$[\partial_\alpha(x)\delta(x - w)\Phi(y, P_2)]_{\partial\Phi, p_2} = -\frac{\kappa}{2} dz_\lambda\, \partial_\epsilon(x)\delta(x - w) M_{\alpha\epsilon\lambda}\{x, z, \Phi(y, P_2)\}. \quad (\text{A2.3})$$

Equation (A2.3) is most easily derived by showing from (3.8) that the elements of the path joining x and w, when commuted with $\Phi(y, P_2)$, give a contribution equal and opposite to the righthand side of (A.3). Thus Eq. (A2.3) must be postulated in order that $\partial_\alpha(x)\delta(x - w)$, together with all path elements associated with it, commute with $\Phi(y, P_2)$.

As well as the commutator between $\partial_\alpha(x)\Phi(x, P_1)$ and another operator, we shall require the commutator between $\partial_{\nu'}(z')\Phi(x, P_1)$ and another operator, where z' is some point along the path P_1. The operation of differentiation adds an element to P_1, and, by (3.8), there will be a commutator between this element and an element p_2 of P_2 given by

$$[\partial_{\nu'}(z')\Phi(x, P_1), \Phi(y, P_2)]_{\partial, p_2} = \frac{i\kappa}{2} dz_\lambda [\overline{J_{\mu'\rho'}(z'), \Phi(x, P_1)}]$$
$$\times \partial_{\mu'}(z') M_{\nu'\rho'\lambda}\{z', z, \Phi(y, P_2)\} \quad (\text{A2.4})$$
$$- \frac{\kappa}{2} dz_\lambda\, \overline{\partial_{\rho'}(z')\Phi(x, P_1) M_{\nu'\rho'\lambda}\{z', z, \Phi(y, P_2)\}}.$$

We next consider the expression,

$$\partial_{\mu'}(z')\{\delta_{\beta'\nu'}\Delta(x' - z')\}[J_{\mu'\nu'}(z'), \Phi(x, P_1)]$$
$$- i\delta_{\beta'\nu'}\Delta(x' - z')\partial_{\nu'}(z')\Phi(x, P_1), \quad (\text{A2.5})$$

which occurs in the commutation rules. According to (A2.4) we should get a contribution from the differential operator in the last term if we were to commute it with a path element. However, in an expression where differential operators or other space functions (i.e., functions not involving quantum dynamical variables) have repeated subscripts, one must ensure that the commutators associated with them cancel one another out, otherwise some consistency condition will fail. We therefore postulate the following commutation rule between $\delta_{\beta'\nu'}\Delta(x' - z')$, when it occurs in an expression such as (A2.5), and a path element:

$$[\overline{\delta_{\beta\nu'}\,\Delta(x - z')\Phi(y, P_2)}]_{z', p_2} = \frac{\kappa}{2} dz_\lambda\, \delta_{\beta\rho'}\,\Delta(x - z') M_{\nu'\rho'\lambda}\{z', z, \Phi(y, P_2)\}. \quad (\text{A2.6})$$

The subscript z', p_2 indicates that the path element immediately adjacent to z' in the first variable is being commuted with p_2. Now, the expression $\delta_{\beta\nu'}\Delta(x - z')$ as a whole must commute with any operator. We therefore postulate the further commutation relation between the element adjacent to x and p_2:

$$\overline{[\delta_{\beta\nu'}\Delta(x-z')\Phi(y,P_2)]_{x,p_2}} = -\frac{\kappa}{2}\,dz_\lambda\,\delta_{\epsilon\nu'}\,\Delta(x-z')M_{\beta\epsilon\lambda}\{x,z,\Phi(y,P_2)\}. \quad (A2.7)$$

We should emphasize that (A2.6) and (A2.7) are only true when the function $\delta_{\beta\nu'}\Delta(x-z')$ occurs in an expression such as (A2.5). The commutator associated with a Kronecker delta will depend on the expression in which it occurs. The Kronecker delta as a whole must always commute with any operator in the limit that the associated delta function peak becomes infinitely narrow.

In passing, we may remark that the commutation relations such as (A2.6) and (A2.7), together with the other commutation relations, must all be consistent with the Jacobi identity.

We are now ready to show that $\mathfrak{J}_{\alpha\beta}(x,P_1)$, defined by (2.45), commutes with $\Phi(y,P_2)$. For simplicity, we shall suppose that the elements of P_2 are ordered successively according to their spatial positions, with Φ itself at the end. The proof could equally well be carried out with an arbitrary ordering. We commute $\mathfrak{J}$ through the elements of P_2, one by one, and eventually through Φ itself. The elements of P_1 are left in their original positions. Before the process $\mathfrak{J}$ will be zero by Einstein's equations. We then show by induction that, after $\mathfrak{J}$ has been commuted through all the elements up to the point z, the value of the generalized product is

$$\mathfrak{J}_{\alpha\beta}(x,P_1)\Phi(y,P_2) = \underset{\alpha\to\beta}{S}\,\overline{(\partial_\alpha(x)\partial_\mu(z)\{\delta_{\beta\nu}\Delta(x-z)\{[J_{\mu\nu}(z),\Phi(y,P_2)]}$$
$$-i\partial_\alpha(x)\{\delta_{\beta\nu}\Delta(x-z)\}\partial_\nu(z)\Phi(y,P_2)). \quad (A2.8)$$

In (A2.8) the operator $\partial_\alpha(x)$ and the element adjacent to x in $\delta_{\beta\nu}\Delta(x-z)$ are to be ordered at the position we have reached in moving $\mathfrak{J}$, i.e. between the elements adjacent to z. This follows, in fact, from our convention that the ordering of elements on both sides of the equation is always the same.

Now let us commute $\mathfrak{J}$ with the next path element, between the points z and $z+dz$, which we denote by p_2. The commutator has the same form as the corresponding Poisson-bracket in the classical theory, Eqs. (2.41) and (2.44)

$$\overline{[\mathfrak{J}_{\alpha\beta}(x,P_1)\Phi(y,P_2)]_{\mathfrak{J},p_2}} = dz_\lambda\,\underset{\alpha\to\beta}{S}\,\partial_\lambda(z)\,\overline{(\partial_\lambda(x)\partial_\mu(z)\{\delta_{\beta\nu}\Delta(x-z)\}}$$
$$\times [J_{\mu\nu}(z),\Phi(y,P_2)] - i\partial_\alpha(x)\{\delta_{\beta\nu}\Delta(x-z)\}\partial_\nu(z)\Phi(y,P_2) \quad (A2.9)$$
$$- \tfrac{1}{2}\,dz_\lambda\,\underset{\alpha\to\beta}{S}\,\{\mathfrak{J}_{\alpha\epsilon}(x,P_1) - \tfrac{1}{2}\delta_{\alpha\epsilon}\mathfrak{J}_{\delta\delta}(x,P_1)\}M_{\beta\epsilon\gamma}\{x,z,\Phi(y,P_2)\}.$$

The first term in (A2.9) has the effect of replacing z in (A2.8) by $z+dz$, but it leaves the operator $\partial_\alpha(x)$ and the element adjacent to x in $\delta_{\beta\nu}\Delta(x-z)$ ordered in its previous position, i.e., before the path element p_1. According to our induction assumption, $\mathfrak{I}$ in the second term of (A.9) can be found from (A2.8), with the operators $J_{\mu\nu}(z)$ and $\partial_\nu(z)$ acting on the $\Phi(y, P_2)$ within F. It then follows from the commutation relations (A2.3) and (A2.6) that the second term simply represents the effect, on (A.8), of commuting the operator $\partial_\alpha(x)$, and the element adjacent to x in $\delta_{\beta\nu}\Delta(x - z)$, with p_2. The total effect of (A2.9) is thus to replace (A2.8) by an expression with z replaced by $z + dz$, and with elements associated with the point x ordered after p_2, instead of before. In other words, it replaces (A2.8) by the corresponding expression with $\mathfrak{I}$ commuted through all elements up to $z + dz$, and our induction assumption is proved.

After $\mathfrak{I}$ has been commuted with all the path elements, the value of the generalized product (A2.8) will be

$$\overline{\mathfrak{I}_{\alpha\beta}(x, P_1)\Phi(y, P_2)} = -i \underset{\alpha\leftrightarrow\beta}{S} \overline{\partial_\alpha(x)\Delta(x - y)\partial_\beta(x)\Phi(x, P_1)}. \quad (A2.10)$$

The commutator of $\mathfrak{I}$ with Φ itself is just equal and opposite to (A.10) so that, after $\mathfrak{I}$ has been commuted with the operator $\Phi(y, P_2)$ and its path elements, the result is zero.

ACKNOWLEDGMENTS

The author would like to acknowledge critical remarks by Professor R. E. Peierls and Dr. J. R. Oppenheimer which were helpful in writing this paper.

Note: In addition to the papers mentioned in the text, there has also been a paper on the quantization of geometry by DeWitt (The Theory of Gravitation, ed. L. Witten., to be published). In common with the present paper, DeWitt insists on relativistic covariance throughout and does not quantize in "flat space," though he does use the weak field approximation. DeWitt treats a gravitational field interacting with an elastic medium and a system of clocks; such a system is essential to his method in order to fix the points in spacetime. As opposed to him we take the view that it is possible to quantize interacting gravitational and matter fields in curved space, without introducing a further setup for performing measurements which is certainly absent in systems encountered in practice.

RECEIVED February 12, 1962

REFERENCES

1. S. N. GUPTA, *Proc. Phys. Soc.* **A65**, 161, 608 (1952).
2. W. THIRRING, *Fortschr. Physik.* **7**, 79 (1959); *Ann. Phys. (NY)* **16**, 96 (1961).
3. S. MANDELSTAM, *Ann. Phys. (N.Y.)* **19**, 1–24 (1962) (preceding paper).
4. B. S. DEWITT, *Phys. Rev. Letters* **4**, 317 (1960).
5. R. ARNOWITT, S. DESER, AND C. MISNER, *Phys. Rev.* **117**, 1595 (1960).
6. R. ARNOWITT AND S. DESER, *Phys. Rev.* **113**, 745 (1959).

Feynman Rules for the Gravitational Field from the Coordinate-Independent Field-Theoretic Formalism

S. Mandelstam

Phys. Rev. **175** (1968) 1604–1623

Abstract

By using the method developed in the preceding paper, the Feynman-DeWitt perturbation expansion for the gravitational S matrix is shown to follow from the field-theoretic formalism. Again our method is to express the path-dependent Green's functions in terms of auxiliary, path-independent Green's functions, in such a way that the path-dependence equation is automatically satisfied. The formula relating the path-dependent to the path-independent Green's functions will be similar to the classical formula relating the path-dependent Riemann tensor to the metric tensor. The equations for the auxiliary Green's functions are found and solved in a perturbation series. If the result is expressed as a sum of Feymann diagrams, one obtains the expected vertices, together with closed loops of fictitious vector particles.

Dynamics Based on Rising Regge Trajectories*

Stanley Mandelstam

Department of Physics, University of California, Berkeley, California

(Received 5 September 1967)

An outline is given of a dynamical scheme based on rising Regge trajectories. The fundamental approximation is that the scattering amplitude can be approximated by the contribution of a finite number of Regge poles. An additional simplifying assumption is that the Regge trajectories are straight lines or, equivalently, that the scattering amplitude is dominated by narrow resonances. Unitarity is introduced by means of the Cheng-Sharp equations, but, in the narrow-resonance approximation, we adopt a very trivial solution of these equations. Crossing is introduced by means of the generalized superconvergence relations due to Igi and to Horn and Schmid. Levinson's theorem is not used; the bootstrap condition is the absence of Kronecker-δ singularities in the J plane. It is hoped that this scheme avoids some of the disadvantages of conventional schemes. In the narrow-resonance approximation one has to solve numerical equations, not integral equations. The scheme is applied to the pseudoscalar, vector, and axial-vector nonets considered as bound states of the $N\bar{N}$ system. As only one channel is being examined, we have to introduce certain parameters from experiment, but we obtain reasonable values for the other parameters.

I. INTRODUCTION

THE current experimental data provide a number of indications that Regge trajectories rise indefinitely with energy, instead of turning over as they do in potential theory or in single-channel unitarity models. It is usually possible to determine two or three points on the trajectory by measurements of the spin of the appropriate particles or of the asymptotic behavior in the crossed channel. If the trajectory is now projected linearly to higher energies, it often turns out that narrow resonances appear at those energies where the trajectory passes through integers or half-integers. The spin of these high-energy resonances has not yet been measured directly, but, owing to the small width and the large Q value, it is presumably high. It is tempting to assume that the resonances are the higher members of the Regge sequence associated with the trajectory in question.

It may be that the strong interactions are characterized by an energy, large compared with the mass of the nucleon, above which the Regge trajectories do turn over. If all particles are built up of real elementary quarks, the quark mass may correspond to such an energy. However, it appears worthwhile to attempt to construct a theory on the assumption either that such an energy does not exist, or that the limit in which it approaches infinity represents a meaningful approximation.

A dynamical scheme based on rising Regge trajectories possesses several attractive features. The most significant is probably that one may be able to work with it in a narrow-resonance approximation. During the last few years many correlations between masses and coupling constants have been obtained by combining group theory, current commutators, or superconvergence relations with the assumption that scattering amplitudes are dominated by a few narrow

---

* Research supported in part by the Air Force Office of Scientific Research, and Office of Aerospace Research under Grant No. AS-AFOSR 232-66, and in part by the United States Atomic Energy Commission.

resonances at low energy. The success of the Gell-Mann–Okubo mass formula, for instance, indicates that one should attempt to incorporate resonances more directly into a dynamical scheme than has been done hitherto. This point of view has been constantly emphasized by Gell-Mann.

It has been remarked by van Hove[1] and by Durand[2] that one can combine the narrow-resonance approximation with Regge asymptotic behavior if and only if the trajectories rise indefinitely. They showed this to be the case by the use of explicit formulas. Their results may be paraphrased by saying that the narrow-resonance approximation is never valid in the region where the trajectories are falling. Thus, for the narrow-resonance approximation to be universally valid, the trajectories must rise indefinitely.

The assumption that Regge trajectories rise indefinitely casts considerable doubt on the Levinson criterion for determining which particles, if any, are elementary. If this criterion is to be valid, it is essential that the phase shifts begin to fall once we have passed above the resonance region in those channels possessing composite resonances. Intuition would suggest that the energy at which the phase shifts begin to fall is of the same order of magnitude as the energy at which the Regge trajectories begin to fall. In a system with indefinitely rising trajectories, therefore, the phase shifts may never fall. This conclusion may be placed on firmer ground by assuming that an infinite number of trajectories rise indefinitely. At each energy where a trajectory passes through a given integer or half-integer, the corresponding phase shift passes through an odd multiple of $\frac{1}{2}\pi$. It will therefore not approach zero, or a multiple of π, asymptotically. One cannot assert this as a rigorous result, as the imaginary part of the lower trajectories will probably be large. If so, there will be no precise connection between the energies at which the real part of the trajectory passes

---

[1] L. van Hove, Phys. Letters **24B**, 183 (1967).
[2] L. Durand, Phys. Rev. **161**, 1610 (1967).

through an integer and those at which the phase shift passes through an odd multiple of $\frac{1}{2}\pi$. Nevertheless, it is strongly suggested that the phase shifts will not behave in the manner required by Levinson's criterion.

We have argued elsewhere[3] that the experimental decrease of form factors provides further evidence against Levinson's theorem.

The above arguments, together with the absence of any experimental evidence in favor of Levinson's theorem, suggest that one should attempt to construct a scheme where the theorem is not used. If one is attempting to construct a bootstrap theory, one can use the alternative criterion that there are no non-Regge terms in the asymptotic behavior or, in other words, that there are no Kronecker-δ singularities in the angular momentum plane.[4,5] This assumption can be tested experimentally, and all available evidence points to its being valid.

In previous dynamical schemes, such as the N/D scheme, Levinson's theorem has played the essential role of removing the Castillejo-Dalitz-Dyson (CDD) ambiguity. The phase shifts calculated in such dynamical models had the property of rising through the resonance region and then falling. Furthermore, the region where the phase shift was falling was important in the dynamics. Thus, though one might be able to construct narrow resonances within the framework of such calculations, one could not express the dynamical equations in terms of the resonance parameters alone. In certain special cases one could obtain partial correlations between resonance parameters, the Chew-Low effective-range formula being an example. By dispensing with Levinson's theorem, we again raise the possibility of constructing a dynamics which has a narrow-resonance approximation.

We should emphasize that the presence of indefinitely rising Regge trajectories or the nonvalidity of Levinson's theorem does not necessarily imply that a dynamics based on the N/D method is inapplicable. Approximation schemes have been discussed where the trajectory or phase shift turns over at a value of s which increases indefinitely with the order of the approximation.[3,6] Nevertheless, it is certainly desirable to construct a dynamical scheme based on rising trajectories if it is at all possible.

II. OUTLINE OF THE DYNAMICAL SCHEME

The reasoning of the previous section suggests that we attempt to formulate a dynamical scheme based on equations for the Regge parameters themselves. The infinite rise of the trajectories, as well as the absence of Kronecker-δ singularities in the J plane, can easily be inserted directly into the equations. It has been shown by Cheng and Sharp[7] and by Frautschi, Kaus, and Zachariasen[8] that one can treat a Schrödinger-potential problem by using equations for the Regge parameters. The fundamental approximation is that the amplitude is dominated by a finite number of trajectories in the direct channel. In the simplest approximation only one trajectory is taken. One can then write dispersion relations for the Regge parameters. The weight functions in the dispersion relations are determined from unitarity, the subtraction terms from knowledge of the potential.

The scheme proposed in this paper is an application of the Cheng-Sharp scheme to the elementary-particle problem. Naturally, there will be some essential differences between this problem and the potential problem. The infinite rise of the Regge trajectories will be effected by inserting two subtraction terms in the equation for α instead of one. In the narrow-resonance approximation, which we shall use, the imaginary part of α is zero, so that the real part is given simply by the equation $\alpha = as + b$. Experimentally, the Regge trajectories do appear to be roughly linear functions of s. We thus adopt a very trivial solution of the Cheng-Sharp equations to correspond to the narrow-resonance approximation. In fact, the equations are not used explicitly at all. It is well to keep them at the back of one's mind, however, and to regard the linear trajectory as a trivial solution of them; one can then envisage how the scheme will appear when the narrow-resonance approximation is not used and when the imaginary part of α is not neglected. The dynamical scheme which we are suggesting can thus be combined with the narrow-resonance approximation, but it is not tied to this approximation.

Another difference between the potential and relativistic problems lies in the determination of the subtraction terms in the dispersion integrals for the Regge parameters. In the potential problem they are determined from knowledge of the potential while in the relativistic case they will have to be determined from the crossing relation. There is no unique way of applying the crossing conditions, but one attractive possibility is to use the generalized superconvergence relations first proposed by Igi, and discussed more fully by Dolen, Horn, and Schmid, by Logunov, Soloviev, and Tavkhelidze, and by Balázs and Cornwall.[9] These relations

---

[3] S. Mandelstam, in *Proceedings of the 1966 Tokyo Summer Lectures on Theoretical Physics*, edited by G. Takeda (W. A. Benjamin, Inc., New York, 1966).
[4] G. F. Chew and S. C. Frautschi, Phys. Rev. Letters **7**, 394 (1961).
[5] S. Mandelstam, Phys. Rev. **137**, 949 (1965).
[6] P. Carruthers and M. M. Nieto, Phys. Rev. Letters **18**, 297 (1967); P. Carruthers, Phys. Rev. **154**, 1399 (1967).

[7] H. Cheng and D. Sharp, Ann. Phys. (N. Y.) **22**, 481 (1963); Phys. Rev. **132**, 1854 (1963).
[8] S. C. Frautschi, P. Kaus, and F. Zachariasen, Phys. Rev. **133**, B1607 (1964).
[9] K. Igi, Phys. Rev. Letters **9**, 76 (1962); R. Dolen, D. Horn, and C. Schmid, *ibid*. **19**, 402 (1967); A. Logunov, L. D. Soloviev, and A. N. Tavkhelidze, Phys. Letters **24B**, 181 (1967); L. A. P. Balázs and J. M. Cornwall, Phys. Rev. **160**, 1313 (1967).

express the integral

$$\int_{s_1}^{N} ds \, \mathrm{Im} A(s,t)$$

in terms of the Regge parameters in the crossed channel. The sum over all the Regge trajectories is not convergent but is asymptotic in N. When the external particles have spin, we can decrease the high-s contribution by dividing the integrand by appropriate functions of s, just as we do with ordinary superconvergence relations. In the narrow-resonance approximation, the relations give us equations between the Regge parameters in the direct and crossed channels. The ordinary superconvergence relations are particular cases of the generalized superconvergence relations, which are valid when the Regge poles in the crossed channel are sufficiently far to the left. The correlations between resonance parameters that have been obtained from superconvergence relations[10] will thus be incorporated automatically in the dynamics.

The dynamical scheme which we shall outline in the following sections should be regarded simply as suggestive, and further work may well reveal the need for substantial modifications or additions. Examination of one channel alone (together with its crossed channels) will not give us enough equations to determine all the resonance parameters. This feature is to be expected and occurs in all bootstrap schemes, since the same parameters occur in the equations for different channels. We shall not attempt to answer the question whether examination of all channels provides us with a uniquely determined, an underdetermined, or an overdetermined system. The question is closely connected with that of the number of trajectories in each channel. The greater the number of trajectories, the larger will be the number of resonance parameters. On the other hand, each trajectory provides a sequence of external particles which give rise to further channels and hence further equations.

One parameter which cannot be determined in the narrow-resonance approximation is the strength of the coupling. The equations will be linear in the coupling constants, and hence one will at most be able to determine ratios of coupling constants. To complete the scheme one will have to go beyond the narrow-resonance approximation, which we shall not do in this paper.

In Sec. III, we shall discuss the dependence of the Regge parameters α and β on the variable s. We shall concentrate mainly on the narrow-resonance approximation, where we shall be able to obtain the dependence explicitly, but we shall also mention the Cheng-Sharp equation which can be used in more general cases. In Sec. IV we shall give arguments which make it plausible that several trajectories with different quantum numbers, and possibly all trajectories, have the same slope.

Experimentally, all trajectories so far investigated do appear to have approximately the same slope. In Sec. V we shall discuss the use of the Igi-Horn-Schmid generalized superconvergence relations to obtain crossing formulas, and in Sec. VI, we shall outline an approximation scheme based on the foregoing sections. In Sec. VII we shall apply our scheme to a particular problem, the determination of the pseudoscalar, vector, and axial-vector nonets as bound states of the nucleon-antinucleon system. Since we are only investigating one channel, we shall not be able to calculate all the parameters; we shall take the mass of the nucleon and the relative masses of the mesons from experiment. We shall show that there does exist a solution of our equation. The solution predicts the correct sign for the ratios of the coupling constants and gives a reasonable value for the absolute mass of the mesons.

III. DEPENDENCE OF THE REGGE PARAMETERS ON s

In this section we shall obtain equations for the dependence of the Regge parameters α and β on s, the square of the energy. We shall concentrate on the narrow-resonance approximation, which we shall use in the following sections; the parameters will then have a simple, explicit dependence on s. At the end of the section, we shall indicate how we may go beyond the narrow-resonance approximation. We shall then have to use the unitarity condition to give us nonlinear integral equations for α and β.

We begin with the spinless case. The fundamental approximation which we shall make is that the scattering amplitude can be expressed as a contribution from a finite number of Regge poles

$$a(s,t) = \sum_{r=1}^{n} \frac{\beta_r(s)}{l - \alpha_r(s)}. \tag{3.1}$$

Strictly speaking, we require an expression which converges as the number of poles becomes infinite. Equation (3.1) should therefore be replaced by an expression such as the modified Cheng representation.[11] To our knowledge such a representation has only been worked out for the single-channel problem, but it would be surprising if a generalization did not exist. The modified Cheng representation has been applied to the potential problem.[12] The equations are more complicated than those obtained from (3.1), but do not differ from them in any fundamental way. In the narrow-resonance approximation the modified Cheng representation becomes equivalent to (3.1), and we shall not consider it further in this paper.

The reality conditions are

$$\alpha(s), \beta(s)/(4q^2)^{\alpha(s)} \text{ real analytic}, \tag{3.2}$$

---

[10] F. Gilman and H. Harari, Phys. Rev. Letters **18**, 1150 (1967); **19**, 723 (1967); and (to be published).

[11] H. Cheng, Phys. Rev. **144**, 1237 (1966).

[12] W. J. Abbe, P. Kaus, P. Nath, and Y. N. Srivastava, Phys. Rev. **141**, 1513 (1966).

where q is the center-of-mass momentum. For simplicity, we have considered the equal-mass case; the results can easily be generalized. In the single-channel problem where only Regge pole is considered, i.e., where the series on the right of (3.1) is approximated by one term, the unitarity condition is

$$\text{Im}\alpha(s) = k(s)\beta(s), \quad (3.3)$$

where k is the usual kinematic factor in the unitarity condition.[7] Note, in particular, that β is real in this approximation. The reality of β is not preserved in higher approximations where more terms are included on the right of (3.1).

In the narrow-resonance approximation, α is real on the real axis and both sides of (3.3) are small. Furthermore, the function $\beta(s)/(4q^2)^{\alpha(s)}$, which is real analytic by (3.2), is also real above threshold, since α and β are both real. This function therefore has no right-hand cut. The analyticity conditions imply that $\alpha(s)$ and $\beta(s)/(4q^2)^{\alpha(s)}$ have no left-hand cut unless two trajectories intersect. In this paper we shall not consider the intersection of trajectories, though our equations could easily be modified to account for this possibility. The Regge parameters $\alpha(s)$ and $\beta(s)/(4q^2)^{\alpha(s)}$ thus have no left-hand or right-hand cuts and they are real, entire functions of s.

When several terms of (3.1) are taken into account, Eq. (3.3) is valid as long as the separation between trajectories is large compared with their imaginary part. It is thus valid in the narrow-resonance approximation. In the multichannel problem, the right-hand side of (3.3) must be replaced by a sum of terms, but it is not difficult to show that each Regge residue is still real. It thus remains true that $\alpha(s)$ and $\beta(s)/(4q^2)^{\alpha(s)}$ are entire functions of s in the narrow-resonance approximation.

We shall assume that Regge trajectories do not rise more than linearly with s. The parameter α will therefore be given by the simple equation

$$\alpha(s) = as + b. \quad (3.4)$$

The equation for β is not quite so simple. When α passes through a negative half-integer l_1 other than $-\frac{1}{2}$, β must vanish unless there is a compensating trajectory passing through the positive half-integer $-l_1-1$ at the same value of s. When the leading trajectory passes through a negative half-integer there is no trajectory with positive α, and therefore no compensating trajectory. The Regge residue $\beta(s)$ must therefore vanish whenever $\alpha(s)$ passes through a negative half-integer less than $-\frac{1}{2}$. We can ensure this by inserting a factor $1/\Gamma(\alpha(s)+\frac{3}{2})$ into the expression for β. Since $\alpha(s) = as+b$, this factor is an entire function of s. Thus

$$\frac{\beta(s)}{(4q^2)^{\alpha(s)}} = \frac{E_1(s)}{\Gamma(\alpha(s)+\frac{3}{2})},$$

where E_1 is an entire functions of s. Inserting the equation $\alpha(s) = a(s) + b$, we find that

$$\beta(s) = \frac{E_1(s)(4q^2)^{as+b}}{\Gamma(as+b+\frac{3}{2})},$$

or, by redefinition of the entire function,

$$\beta(s) = \frac{E(s)(4aq^2/e)^{as+b}}{\Gamma(as+b+\frac{3}{2})}. \quad (3.5)$$

The factor e^{-as-b} has been inserted in order that the right-hand side of (3.5), without the factor E, should not increase exponentially when s approaches $\pm\infty$. It is easily seen from Stirling's theorem that the expression $(4aq^2/e)^{as+b}/\Gamma(as+b+\frac{3}{2})$ behaves like $1/s$ at infinite s.

There does not appear to be any simple argument for restricting the entire function E to a polynomial of a given degree, or even to a polynomial of arbitrary degree. When s approaches infinity, the narrow-resonance approximation will probably become invalid, and one can therefore not combine Eq. (3.5) with assumptions about the behavior of β at large s. In the approximation scheme which we shall develop by combining (3.5) with the crossing relations, we shall expand E in powers of s; the number of terms kept will depend on the order of the approximation. We shall apparently be able to obtain equations for all the terms, but, as we have pointed out before, questions of this type cannot yet be answered with any degree of certainty.

We now allow the external particles to have spin, and we shall begin with systems where the particles are both bosons or both fermions. As usual, we shall work with states of fixed helicity λ, μ. Equation (3.5) must be modified in several ways. First, the behavior at $q^2=0$ will depend on the orbital rather than on the total angular momentum. We construct states of fixed incoming and outgoing orbital angular momentum at threshold by taking the usual linear combinations of helicity states. These orbital angular momenta will differ from the total angular momentum by integers which we shall denote by τ and $\tau'(|\tau|<|\lambda|,|\tau'|<|\mu|)$. The states with fixed orbital angular momentum will then have a factor

$$(4q^2)^{as+b+\frac{1}{2}(\tau+\tau')} \quad (3.6)$$

instead of $(4q^2)^{as+b}$.

We also require extra factors when $as+b$ takes on integral values where nonsense states are present. For positive integers or zero, the factors will be as follows:

Factors $(as+b-n)^{1/2}$,

$$|\lambda|\leq n<|\mu| \text{ or } |\mu|\leq n<\lambda \quad (3.7a)$$

Factors $(as+b-n)$, $|\lambda|\leq n$, $|\mu|\leq n$ if the trajectory chooses nonsense at $\alpha=n$ (3.7b)

Factors $(as+b-n)$, $|\lambda|>n$, $|\mu|>n$ if the trajectory chooses sense at $\alpha=n$. (3.7c)

There are similar factors when α passes through negative integral values:

Factors $(as+b-n)^{1/2}$,
$$|\lambda| \leq -n-1 < |\mu| \text{ or } |\mu| \leq -n-1 < |\lambda|. \quad (3.8a)$$

Factors $(as+b-n)$ for one of the inequalities
$$|\lambda| \leq -n-1, \quad |\mu| \leq -n-1 \quad (3.8b)$$
or
$$|\lambda| > -n-1, \quad |\mu| > -n-1. \quad (3.8c)$$

If we are dealing with the leading trajectory, the factors $(as+b-n)$ occur in the case (3.8c). This is because the alternative which occurs at negative half-integral values of α also occurs at negative integral values with $|\lambda|$, $|\mu| > -\alpha-1$. Either β must vanish or there must be a compensating trajectory passing through the point $-\alpha-1$ at the same value of s. For the leading trajectory the second alternative is impossible, and β must vanish. We thus have factors $(as+b-n)$, where n is a negative integer satisfying the inequality (3.8c).

In channels where one of the particles is a fermion, the Γ function in (3.5) becomes $\Gamma(as+b+1)$, since we now require the zeros at negative integers. The extra factors due to spin will involve half-integral instead of integral values of n.

We have emphasized that Eq. (3.5) is really the narrow-resonance approximation to a more general set of equations. Since we shall not go beyond the narrow-resonance approximation in the following sections, we shall confine ourselves to the single-channel, spinless, equal-mass problem, and we shall keep only one term of (3.1). We can then write down the equations by making slight modifications to the nonrelativistic equations of Cheng and Sharp.[7] On doing so, we obtain the following:

$$\alpha(s) = as + b + \frac{1}{\pi} \int ds' \frac{\mathrm{Im}\alpha(s')}{s'-s}, \quad (3.9a)$$

$$\beta(s) = \left[\frac{E(s)(4aq^2/e)^{\alpha(s)}}{\Gamma(as+b+\tfrac{3}{2})} \right]$$

$$\times \exp\left(-\frac{1}{\pi} \int ds' \frac{\mathrm{Im}\alpha(s')\ln(4q'^2)}{s'-s}\right)$$

$$\times \prod_{n=1}^{\infty} \frac{a(s-s_n)}{as+b+n+\tfrac{1}{2}}, \quad (3.9b)$$

$$\mathrm{Im}\alpha(s) = k\beta(s). \quad (3.9c)$$

We have assumed that $\mathrm{Im}\alpha \to 0$ as $s \to \infty$. If this is not the case, we must modify the dispersion integrals in (3.9a) and (3.9b) in the usual way. The constants s_n in the infinite products of (3.9b) are those values of s for which $\alpha(s) = -n-\tfrac{1}{2}$. The infinite product, together with the factor $1/\Gamma(as+b+\tfrac{3}{2})$, ensures that β has zeros at those values of s for which α is a negative half-integer less than $-\tfrac{1}{2}$.

We could formally have omitted the factor $1/\Gamma(as+b+\tfrac{3}{2})$ and rewritten infinite product as $\prod[(s_n-s)/s_n]$, but this product does not converge. The infinite product in (3.9b), on the other hand, converges if $\mathrm{Im}\alpha$ approaches zero at infinite s; if $\mathrm{Im}\alpha$ does not approach zero but $\mathrm{Im}\alpha/\alpha$ does, the ratio of infinite products in (3.9b) taken at two values of s converges. Thus the divergence of the infinite product will be cancelled by a factor independent of s in the entire function $E(s)$, and the final result will converge.

Equations (3.9) provide a system of nonlinear integral equations for α and β, and they have been solved by iteration in the potential model.[7] In the narrow-resonance approximation, $\mathrm{Im}\alpha=0$, and Eqs. (3.9) reduce to the explicit form (3.5). One can easily modify (3.9) to the case where the external particles have spin or to the unequal-mass case. The modifications where more terms of (3.1) are included, or where (3.1) is replaced by a better representation such as the modified Cheng representation, are straightforward in principle, but they complicate the numerical work considerably. We refer the reader to the treatment of the potential model.[8,12]

Whether we are using the narrow-resonance approximation or not, the constants a and b, and the entire function E, remain to be determined. In potential theory, these quantities are obtained from a knowledge of the potential; in the present problem we shall have to use the crossing relations. Before we apply the crossing relations to our problem, however, we shall examine the possibility that the constant a is the same for all trajectories.

IV. SLOPE OF THE REGGE TRAJECTORIES

By considering models where the sequence of resonance on a Regge trajectory corresponds to the sequence of external particles on another Regge trajectory, one can obtain plausible results about the slope of Regge trajectories. The work of Carruthers and Nieto is an example of the type of model we have in mind. Carruthers and Nieto examined the ρN channel and obtained the $D_{3/2}$ resonance as a composite system in that channel. They then combined the ρN and $\rho F_{5/2}$ channels, the $F_{5/2}$ being the next member of the nucleon Regge series, and obtained a $G_{7/2}$ resonance in addition to the $D_{3/2}$. As the $G_{7/2}$ is the next member of the Regge series beginning with the $D_{3/2}$, we can now envisage a more complicated model in which the whole Regge sequence corresponding to the nucleon is included in the external particles, and the whole Regge trajectory corresponding to the $D_{3/2}$ resonance appears as a composite system.

Let us now examine a more general case in which an external particle of mass μ is combined with a series of external particles of spin σ and mass m_σ, where

$$am_\sigma^2 + b = \sigma, \qquad (4.1)$$

i.e.,

$$m_\sigma^2 = (\sigma - b)/a. \qquad (4.2)$$

This system is assumed to produce a Regge sequence of particles of spin σ and mass M_σ, where

$$M_\sigma^2 = (\sigma - b')/a'. \qquad (4.3)$$

We may define the binding energy as the quantity $M_\sigma - m_\sigma - \mu$, and we assume that this quantity remains infinite as σ approaches infinity. We can then easily show that

$$a = a'. \qquad (4.4)$$

We obtain the same result from the assumption that the expression $M_{\sigma+n} - m_\sigma - \mu$ remains finite. In other words, we assume that a particle in the M sequence is built up from particles of the m sequence of not too different spin, and that the binding energy remains finite as the spin approaches infinity. We can actually make the weaker assumption that the binding energy increases less rapidly then the mass of the particles.

If we do not use the narrow-resonance approximation, so that the function α is given by (3.9a) rather than (3.4), we can still derive the same result, provided that Imα increases less than linearly with s.

The type of approximation which we are using in our paper is fundamentally different from that used in the Carruthers-Nieto model. Nevertheless, it may still be that the definition of the binding energy which we have just given, and the assumption which we have made about it, are physically reasonable. We would then have to restrict the slopes of our trajectories by (4.4). Such a restriction could well be necessary in the sense that, unless it were applied, our equations would not yield convergent results as we increased the number of resonances considered.

If the above assumptions are correct, the Regge trajectories will occur in groups. Each member of the group can be built from another member by means of a Carruthers-Nieto model, and the trajectories in the same group will have the same slope. The simplest system possessing these features is that in which there is only one group, and all trajectories have the same slope. Now, it is a remarkable empirical fact that all trajectories do have the same slope within the accuracy to which the slope can be defined. One may therefore adopt as a working hypothesis the assumption that the constants a are the same for all trajectories. This universal constant a will then fix a scale of mass and only the constants b and the functions E remain to be determined. The arguments for this hypothesis are obviously far from compelling and, since we do not yet know the extent to which our dynamical equations determine our parameters, we must bear in mind the possibility that it may have to be abandoned.

V. CROSSING RELATIONS

We shall apply crossing with the aid of the generalized superconvergence mentioned in Sec. II.[9] These relations are a consequence of Regge asymptotic behavior and the usual analyticity properties. We assume that a scattering amplitude $A(s,t)$ satisfies dispersion relations with only a right-hand cut and has the asymptotic behavior

$$A(s,t) \sim \sum_r \frac{\gamma_r(s)(-t)^{\alpha_r(s)}}{\sin\pi\alpha_r(s)}, \qquad t \to \infty. \qquad (5.1)$$

The following relation is then asymptotically true as N becomes large:

$$\int^N dt\, \mathrm{Im}A(s,t) \sim \sum_r \frac{\gamma_r(s) N^{\alpha_r(s)+1}}{\alpha_r(s)+1}. \qquad (5.2)$$

By considering the functions $t^n A(s,t)$, we derive the further equations

$$\int^N dt\, t^n\, \mathrm{Im}A(s,t) \sim \sum_r \frac{\gamma_r(s) N^{\alpha_r(s)+n+1}}{\alpha_r(s)+n+1}. \qquad (5.3)$$

We can obtain amplitudes $A(s,t)$ with only a right-hand cut by taking the sum of the positive—and negative—signature amplitudes. The right-hand side of (5.1) will receive contributions from the fixed poles in the J plane at nonsense wrong-signature integers[13,14] as well as from the moving poles. One can evaluate the contribution from the fixed poles in terms of the third double-spectral function. At present, our dynamical scheme is far from the stage where contributions from the third double-spectral function should be included.

If the external particles have spin, one divides the amplitude $A(s,t)$ by the factor $(1+z)^{|\lambda+\mu|/2}(1-z)^{|\lambda-\mu|/2}$ before applying (5.1)–(5.3), since the resulting amplitude will be free of kinematic singularities or zeros at $z = \pm 1$. Thus

$$\frac{A_{\mu\lambda}(s,t)}{(1+z)^{|\lambda+\mu|/2}(1-z)^{|\lambda-\mu|/2}}$$

$$\sim \sum_r \frac{\gamma_r(s)(-t)^{\eta_r(s)}}{\sin\pi\eta_r(s)}, \qquad t \to \infty, \qquad (5.4a)$$

where

$$\eta_r(s) = \alpha_r(s) - \max(|\lambda|, |\mu|) \qquad (5.4b)$$

and

$$z = 1 + t/2q^2. \qquad (5.4c)$$

Then

$$\int^N dt\, \frac{\mathrm{Im}A_{\mu\lambda}(s,t)}{(1+z)^{|\lambda+\mu|/2}(1-z)^{|\lambda-\mu|/2}} \sim \sum_r \frac{\gamma_r(s) N^{\eta_r(s)+1}}{\eta_r(s)+1}, \qquad (5.5)$$

---

[13] C. E. Jones and V. L. Teplitz, Phys. Rev. **159**, 1271 (1967).
[14] S. Mandelstam and L. L. Wang, Phys. Rev. **160**, 1490 (1967).

or, more generally,

$$\int^N dt\, t^n \frac{\mathrm{Im} A_{\mu\lambda}(s,t)}{(1+z)^{|\lambda+\mu|/2}(1+z)^{|\lambda-\mu|/2}}$$
$$\sim \sum_r \frac{\gamma_r(s) N^{\eta_r(s)+n+1}}{\eta_r(s)+n+1}. \quad (5.6)$$

In practice, it is usually more convenient to work with the fixed-parity combinations of helicity states than with the helicity states themselves. The powers of t in the denominator will make the integrals in (5.5) and (5.6) less sensitive to the contributions from high t, where the integrand is less accurately known.

The right-hand side of (5.2) depends on the Regge trajectories in the s channel, the left-hand side on those in the t and u channels. Equation (5.2) therefore provides us with a relation between the parameters in the direct and crossed channels.

We now examine in more detail the form of the two sides of (5.5) in the narrow-resonance approximation. The constants $\gamma_r(s)$ will be proportional to the corresponding Regge residues $\beta_r(s)$; we find that

$$\gamma_r(s) = -\mathrm{sgn}(\mu-\lambda)2^{\max(|\lambda|,|\mu|)+1}(\sqrt{\pi})\Gamma(\alpha(s)+\tfrac{3}{2})$$
$$\times \Gamma(\alpha(s)+1)[\Gamma(\alpha(s)+\lambda)\Gamma(\alpha(s)-\lambda)$$
$$\times \Gamma(\alpha(s)+\mu)\Gamma(\alpha(s)-\mu)]^{-1/2}\beta_r(s)/(q^2)^{\eta_r(s)}. \quad (5.7)$$

The Regge residue $\beta_r(s)$ is expressed in terms of the function E and the constants a and b by using equations (3.5)–(3.8). We note that some of the Γ functions in (3.5) and (5.7) cancel. If we are working to sufficient accuracy, we shall have to include the contributions to the right-hand side of (5.4a) with asymptotic behavior t^{η_r-1}, t^{η_r-2}, etc., from the Regge trajectory α_r. The corresponding functions γ are given by formulas similar to (5.7). Actually we shall use the equations at $s=0$, so that these lower terms should only be included if we also include the daughter and conspirator trajectories.

Turning to the left-hand side of (5.4), we shall confine ourselves to the contributions from the t-channel Regge trajectories; the u-channel trajectories can be handled in a similar way. In the narrow-resonance approximation, a t-channel trajectory with $\alpha(t)=at+b$ will give the following contribution to $\mathrm{Im}B(s,t)$, where B is the amplitude obtained from A by crossing:

$$\mathrm{Im}B_{\mu\lambda}(s,t) = -\frac{1}{a}\sum_J (2J+1)\pi\delta\left(t-\frac{b}{a}-\frac{J}{a}\right)$$
$$\times \beta_{\mu\lambda}(t)d_{\lambda\mu}{}^J\!\left(1+\frac{s}{2qt^2}\right). \quad (5.8)$$

The sum is over integral or half-integral values of J, depending on the spin. After applying the crossing relation, we obtain the equation

$$\frac{\mathrm{Im} A_{\mu\lambda,\alpha}(s,t)}{(1+z)^{|\lambda+\mu|/2}(1-z)^{|\lambda-\mu|/2}} = \sum_J \sum_{\lambda'\mu'\alpha'} C(\lambda,\mu,\alpha,\lambda',\mu',\alpha',s,t)$$
$$\times (2J+1)\pi\delta\!\left(t-\frac{b}{a}-\frac{J}{a}\right)\!\beta_{\mu',\lambda',\alpha'}(t)d_{\lambda'\mu'}{}^J\!\left(1+\frac{s}{2qt^2}\right)$$
$$\times \left(2+\frac{t}{2q^2}\right)^{-|\lambda+\mu|/2}\!\left(-\frac{t}{2q^2}\right)^{-|\lambda-\mu|/2}\!. \quad (5.9)$$

The index α refers to those quantities other than the helicity which characterize the amplitude, e.g., isotopic spin or $SU(3)$ multiplet. The crossing matrix C will be the product of the helicity crossing matrix and the crossing matrix appropriate to the internal symmetry.

The left-hand side of (5.6) can be handled in a very similar way. By substituting (5.7) and (5.9) in (5.5) or (5.6), and using the formulas for β in terms of the function E and the constants b, one can obtain equations for these quantities. As we have already pointed out, we do not know whether the number of equations is sufficient for a complete calculation. In fact, in the narrow-resonance approximation, these equations are linear in the functions E, and we shall at most be able to obtain the ratios of the coupling constants, not their absolute normalization. To determine the normalization factor, one will have to go beyond the narrow-resonance approximation.

VI. APPROXIMATION SCHEME FOR CALCULATING THE REGGE PARAMETERS

There is obviously no unique way of applying the analyticity and crossing formulas in dynamical calculations. The method which we shall propose in this section should be regarded as one possible suggestion. Further work will almost certainly reveal the need for substantial changes.

We assume that each channel has a leading trajectory, together with subsidiary trajectories the number of which will depend upon the stage in the approximation scheme. Since the Q value of a resonance of given J will increase as we go to lower trajectories, most of the resonances on the lower trajectories may be fairly broad and may not appear experimentally as resonances. Nevertheless, it may still be a reasonable approximation to represent the contribution from these resonances by poles near the real axis, i.e., to assume the narrow-resonance approximation. As we explained in Sec. IV, we shall begin by assuming that all the trajectories have the same slope.

The trajectories are now parametrized according to the formulas given in Sec. III, and the crossing relations (5.5) and (5.6) are applied. Several points remain to be specified, including the values of N and s to be taken in applying the crossing relations, the number of terms of the function E which are kept, and the number of relations (5.6) which are used.

To begin with the value of N, we should obviously make this integration limit as high as possible, since Eqs. (5.5) and (5.6) are only assymptotically true. On the other hand, we cannot take N above the value of t for which we know the function $\text{Im}A(s,t)$ in the integrand. In the narrow-resonance approximation the function $\text{Im}A(s,t)$ will consist of a series of δ functions in t at the positions of the resonances. The integrand in (5.5) and (5.6) will therefore be known up to the lowest resonance on the lowest trajectory which is kept; the first unknown contribution will be at the position of the lowest resonance on the highest trajectory which is omitted. The value of N to be taken should therefore be between these resonances. We may estimate the position of the first omitted trajectory on the assumption, based on the Schrödinger or Bethe-Salpeter equation at high energy, that the trajectories are spaced at integral distances in the J plane. If we have kept a number of trajectories, we may alternatively estimate the position of the first omitted trajectory from the spacing between the trajectories which have been kept.

The energy of the lowest resonance on the lowest trajectory included and that of the lowest resonance on the highest trajectory omitted thus represent lower and upper limits for N. If N is large, the result of the calculations will be insensitive to the precise value of N between the two limits. If N is not large, the results would not be expected to be quantitatively accurate, and their sensitivity to the value of N between the two limits represents a lower limit to their uncertainty. Since we have no *a priori* basis for assigning a precise value to N between the two limits, we shall take it to be midway between the two.

A related uncertainty lies in the form of the asymptotic expansion (5.1). We could equally well have used the series

$$\sum \frac{\gamma_i'(s)(t-t_0)^{\alpha_i(s)}}{\sin\pi\alpha_i(s)}. \quad (6.1)$$

The results will be insensitive to t_0 if a sufficiently large number of terms are taken in the asymptotic series, but they will depend on this parameter if only one or two terms are taken. We shall refer again to this uncertainty in the example of the following section.

With regard to the value of s at which Eqs. (5.5) and (5.6) are to be taken, we must bear in mind that the narrow-resonance approximation neglects all partial waves above a certain angular momentum at any particular value of the energy, so that the approximation would not be expected to be accurate at an unphysical value of the scattering angle where the partial-wave series is badly divergent. The nonresonant contributions from the high angular momenta would probably be large at such angles. We should therefore choose a value of s which avoids unphysical angles in the t and u channels as far as possible. For the equal-mass problem the obvious choice is $s=0$, though other sufficiently small values of s should also be adequate. For the unequal-mass problem, it is impossible to avoid unphysical angles completely; we can avoid one unphysical region only at the expense of going farther into another. The choice $s=0$ still probably provides the best mean. For very unequal masses, we may eventually have to subtract off certain contributions in order to reduce the divergence of the partial-wave expansion; these contributions would be calculated by using the double-dispersion relations. At the moment, however, we shall neglect such complications.

We can obtain further crossing relations by differentiating (5.5) and (5.6) with respect to s. The right-hand sides of these equations will then depend on the derivatives of the Regge residues β with respect to s; they in turn will depend on the higher terms of the entire function $E(s)$ of (5.5). In the lowest approximation, this function is taken to be a constant and the s derivatives of Eqs. (5.5) and (5.6) are not used. In higher approximations, we take the first n terms of the function E, expanded as a power series in s, and we use Eq. (5.6) and its first $(n-1)$ derivatives with respect to s.

Thus, in higher approximations, we would use more values of n in (5.6) corresponding to the larger number of trajectories taken into account; we would also use more derivatives of (5.5) and (5.6) corresponding to the larger number of terms kept in the functions $E(s)$. The question of just how many trajectories to take in each channel, and how to assign the choice of sense or nonsense at the low integers, is not one to which we can give a definite answer. For the characteristics of the leading trajectories we can be guided by experiment; for the lower trajectories we may have to use trial and error. We shall have to examine several trajectories simultaneously in order to obtain a complete set of equations, since the Regge residues β for different trajectories are related to one another. For example, the value of β for the $N\bar{N}$ trajectory at $s=m_\pi^2$ and that for the πN trajectory at $s=m_\pi^2$ are both related to the usual pion-nucleon coupling constant g^2.

We also remark that the number of equations (5.6) which we can use, given any choice of N and of the number of trajectories taken into account, is limited by the accuracy of these equations. The larger the value of n, the larger the difference between the two sides of (5.3), and hence the larger the error in (5.6). Estimates based on potential theory and on the asymptotic distribution of Regge poles calculated by Cheng and Wu[16] indicate that the number of equations which we can take is roughly proportional to N, apart from logarithmic factors. This limitation on the number of equations (5.6) means that we must exhaust the content of any given number of equations, applied to all relevant trajectories,

---

[16] H. Cheng and T. T. Wu, Phys. Rev. 144, 1232 (1966).

before adding another equation. In deciding the maximum value of n for each helicity state, we must remember that the factors $(1+z)^{|\lambda+\mu|/2}(1-z)^{|\lambda-\mu|/2}$ decrease the integrand at high values of l, in contrast to the factor l^n which increases it. Thus more values of n can be taken for the states of greater helicity; the maximum value of $n-\max(|\lambda|,|\mu|)$ should be the same for all helicity states.

VII. MESONS AS BOUND STATES OF BARYON-ANTIBARYON SYSTEMS

As a simple application of the foregoing scheme, we shall attempt to obtain the mesons (the pseudoscalar, vector, and axial-vector nonets) as bound states of the baryon-antibaryon system. In other words, we examine the baryon-antibaryon system and include the three trajectories of which the pseudoscalar, vector, and axial-vector nonets are the lowest members. The quantum numbers of the mesons do indicate that they should be regarded as bound states of the baryon-antibaryon system rather than of boson systems. The two S-wave bound states of the baryon-antibaryon system have the spin, parity, and charge-conjugation quantum numbers of the pseudoscalar and vector nonets. This is not the case if we attempt to obtain the mesons as bound states of meson systems. For instance, the ρ would be a P-wave bound state of the pion-pion system, and we encounter the well-known difficulty of the absence of a corresponding S-wave bound state.

In the simple approximation which we shall employ, we cannot expect to obtain quantitatively correct results. Our aim in performing the calculation is to show that we obtain a set of equations which yield consistent, reasonable values for the quantities of interest.

The kinematics of the nucleon-antinucleon system have been worked out by Goldberger, Grisaru, MacDowell, and Wong.[16] There are five independent helicity amplitudes. After dividing by factors of $(1-z)^{1/2}$ and $(1+z)^{1/2}$ to make them analytic at $z=\pm 1$, and taking fixed parity combinations, we may define the following amplitudes:

$$f_1 = \langle ++|\phi|++\rangle - \langle ++|\phi|--\rangle, \quad (7.1a)$$

$$f_2 = \langle ++|\phi|++\rangle + \langle ++|\phi|--\rangle, \quad (7.1b)$$

$$f_3 = (1+z)^{-1}\langle +-|\phi|+-\rangle \\ -(1-z)^{-1}\langle +-|\phi|-+\rangle, \quad (7.1c)$$

$$f_4 = (1+z)^{-1}\langle +-|\phi|+-\rangle \\ +(1-z)^{-1}\langle +-|\phi|-+\rangle, \quad (7.1d)$$

$$f_5 = (4m/\sqrt{s})(1-z^2)^{-1/2}\langle ++|\phi|+-\rangle. \quad (7.1e)$$

We have denoted particles of positive and negative helicity by the symbols $+$ and $-$. The normalization

---
[16] M. L. Goldberger, M. T. Grisaru, S. W. MacDowell, and D. Y. Wong, Phys. Rev. **120**, 2250 (1960).

used will be such that the unitarity condition involves the integral

$$\int \frac{d\Omega}{4\pi}\left(\frac{4q^2}{s}\right)^{1/2}.$$

We may also define fixed partial-wave amplitudes as follows.
Singlet:

$$f_0{}^J = \langle ++|\phi^J|++\rangle - \langle ++|\phi^J|--\rangle, \quad (7.2a)$$

Triplet, $J=l$:

$$f_1{}^J = \langle +-|\phi^J|+-\rangle - \langle +-|\phi^J|-+\rangle, \quad (7.2b)$$

Triplet, $J=l\pm 1$:

$$f_{11}{}^J = \langle ++|\phi^J|++\rangle + \langle ++|\phi^J|--\rangle, \quad (7.2c)$$

$$f_{12}{}^J = 2\langle ++|\phi^J|+-\rangle, \quad (7.2d)$$

$$f_{22}{}^J = \langle +-|\phi^J|+-\rangle + \langle +-|\phi^J|-+\rangle. \quad (7.2e)$$

The pseudoscalar trajectory corresponds to the singlet state, the axial-vector trajectory to triplet state with $J=l$, and the vector trajectory to the triplet state with $J=l\pm 1$. In the complete amplitudes (7.1) the pseudoscalar trajectory will only affect f_1; the vector and axial-vector trajectories will affect the other four amplitudes. At $J=1$, the value corresponding to the lowest member, the vector trajectory will affect only f_2, f_4, and f_5, and the axial-vector trajectory will affect only f_3. As $z\to\infty$, the vector trajectory will again affect only f_2, f_4, and f_5, the axial-vector trajectory only f_3.

In the lowest approximation we shall keep one trajectory in each of the three channels (pseudoscalar, vector, axial-vector), and shall use Eq. (5.5) but not (5.6). Furthermore, we remarked in the previous section that the generalized superconvergence relations for amplitudes with nonzero helicity should be used before those for amplitudes with zero helicity, since the former relations contain a power of z in the denominator. Thus, the superconvergence relations for f_3, f_4, and f_5, but not those for f_1 and f_2, will be used in our approximation.

We have already pointed out that we cannot obtain equations for all relevant parameters by looking at one channel only. We shall therefore take the following quantities from experiment:

(i) The ratio of the mass of the baryon to the constant $1/a$ (a being the slope of the trajectory) which fixes the dimension of mass. Since we are not examining channels with baryon number equal to 1, we would not expect to obtain the baryon mass in this calculation. We shall take the mass of the baryon to be equal to $1/a$.

(ii) The spacings between the three meson trajectories. We shall take the vector trajectory to be one unit above the pseudoscalar trajectory and half a unit above the axial trajectory. We thus do not attempt to calculate ratios of the masses of the pseudoscalar, vector, and axial-vector mesons. In particular, we assume the mass of the pseudoscalar and vector mesons to be equal, as is required by $SU(6)$.

(iii) We shall also make an assumption regarding the ratio of the Regge residues for the three amplitudes $f_{11}{}^J$, $f_{12}{}^J$, and $f_{22}{}^J$ which depend on the vector trajectory. Owing to the factorization theorem, there are only two independent amplitudes in the approximation where one trajectory is kept. The ratio in question is that between the magnetic and electric coupling of the vector meson. We shall discuss the estimation of this ratio below. The electric coupling is, in fact, considerably smaller than the magnetic coupling, and our final results would not be very different if we neglected the electric coupling completely.

In our lowest approximation we shall take the functions E in (3.5) to be constants, so that there is only one known parameter associated with each Regge residue β. Three quantities remain to be determined, the position of one of the trajectories (the constant b) and the two ratios between the Regge residues for the three trajectories. Since we have three generalized superconvergence relations, we should be able to determine these quantities. We remark again that the absolute value of any of the Regge residues cannot be determined in the narrow-resonance approximation.

We next treat the internal symmetry of our problem. We shall assume exact $SU(3)$, and shall further assume that each of our mesons consists of a degenerate octet and singlet. The baryons will be assumed to be octets. The crossing matrix in (5.9) will contain the $SU(3)$ crossing matrix as a factor. Thus, if (5.9) and (5.7) are substituted in (5.4), the equations will have the form

$$x = CMx,$$

where x is a vector in helicity and $SU(3)$ space, C the $SU(3)$ crossing matrix, and M a helicity matrix. Each solution of this equation will be proportional to an eigenfunction of the $SU(3)$ crossing matrix. We emphasize that this is only true in the approximation where we do not allow more than one trajectory for each value of the parity and charge conjugation.

The $SU(3)$ crossing matrix for an octet-octet channel has been given by de Swart.[17] There are two eigenfunctions which involve only singlets (S) and octets (DD, DF, FD, FF):

(i) $DD = 5$, $DF = 0$, $FD = 0$, $FF = 9$,
$S = 16$, eigenvalue 1; (7.3a)

(ii) $DD = 0$, $DF = 1$, $FD = 1$, $FF = 0$,
$S = 0$, eigenvalue -1. (7.3b)

Unfortunately, neither of these eigensolutions is factorizable between the D and F states, as it should be in our approximation where we have only one trajectory.

---

[17] J. J. de Swart, Nuovo Cimento **31**, 420 (1964).

We can obtain a factorizable eigenfunction which resembles (7.3a) by writing

$$DD = 5, \quad DF = FD = 3\sqrt{5}, \quad FF = 9, \quad S = 16. \quad (7.4)$$

This is not far from the result which we would obtain for $N\bar{N}$ scattering through the vector meson by magnetic coupling according to the estimates of Sugawara and von Hippel[18]:

$$DD = 5, \quad DF = FD = 2\sqrt{5}, \quad FF = 4, \quad S = 16. \quad (7.5)$$

We can now resolve (7.4) into components along the directions of the two eigenfunctions (7.3a) and (7.3b) If we weight each amplitude according to the multiplicity of the states involved, the eigenfunctions of the crossing matrix will be orthogonal, so that we can take components in the usual way. We shall demand that the component in the direction (7.3a) satisfy the equations of Sec. V. The component in the direction (7.3b) will then not satisfy our equations. We prefer to demand that our equations be true for the component along the direction (7.3a) than for that along the direction (7.3b), since not every resonance gives a positive-definite contribution to a given partial wave when resolved in the latter direction. We are thus more likely to obtain cancelling contributions from higher resonances when resolving in the direction (7.3b). The D/F ratio for these higher resonances would have to be of the opposite sign to that of the original resonance. Another way of viewing the problem is to regard the single resonance which we have taken as representing the contribution of two or more resonances with D/F ratios of both signs. The resultant resonance could then have the decomposition (7.3a) in $SU(3)$ space.

We shall also use the paper of Sugawara and von Hippel to estimate the ratio of the electric to the magnetic coupling of the vector meson. The magnetic-magnetic transition has been given in (7.5); the electric-electric and electric-magnetic transitions are pure FF and are equal to 1 and 2 on the same scale. If we denote the components in the direction (7.3a) in $SU(3)$ space by $(g_M)^2$, $(g_E)^2$, and $(g_{EM})^2$, we easily find that

$$g_E{}^2 = (3/31)g_M{}^2, \quad (g_{EM})^2 = (6/31)g_M{}^2. \quad (7.6)$$

Throughout our work we shall assume that the only important crossed channel to the nucleon-antinucleon channel is the other nucleon-antinucleon channel. The effect of the nucleon-nucleon channel will be neglected owing to the absence of any low-mass resonances. The existence of only one important crossed channel implies that signature may be neglected in the nucleon-antinucleon channel. In other words, we have exchange degeneracy, and the trajectories of odd and even signature coincide. It has been noted empirically that exchange degeneracy is approximately valid for meson trajectories. For instance, the ρ and A_2 trajectories

---

[18] H. Sugawara and F. von Hippel, Phys. Rev. **145**, 1331 (1966).

appear to coincide within the limits of experimental error.

We can now apply the formulas of Secs. IV and V to set up our equations. We shall work in terms of the coupling constants $g_P{}^2$ for the pseudoscalar meson, $g_A{}^2$ for the axial-vector meson, and $g_M{}^2$, $g_E{}^2$, and $(g_{EM})^2$ for the vector meson. The residues of the functions f have the following values in terms of the coupling constants:

$$\mathrm{Im} f_1(s,t) = \tfrac{1}{2}(\mu_P{}^2/4) g_P{}^2 \delta(s-\mu_P{}^2),$$
$$\mathrm{Im} f_2(s,t) = \tfrac{1}{2} M^2 g_E{}^2 (1+t/2q^2) \delta(s-\mu_V{}^2),$$
$$\mathrm{Im} f_3(s,t) = -\tfrac{1}{2}(M^2-\tfrac{1}{4}\mu_A{}^2) g_A{}^2 \delta(s-\mu_A{}^2),$$
$$\mathrm{Im} f_4(s,t) = \tfrac{1}{2}(\mu_V{}^2/4) g_M{}^2 \delta(s-\mu_V{}^2),$$
$$\mathrm{Im} f_5(s,t) = -\tfrac{1}{2} M^2 (g_{EM})^2 \delta(s-\mu_V{}^2). \quad (7.7)$$

From Eqs. (5.8) and (7.7), we can express the Regge residues $\beta(\mu^2)$, where μ is the mass of the appropriate particle, in terms of the g's. Thus

$$\beta_0(\mu_P{}^2) = -(1/8\pi)\mu_P{}^2 g_P{}^2,$$
$$3\beta_{11}(\mu_V{}^2) = -(1/2\pi) M^2 g_E{}^2,$$
$$3\beta_1(\mu_A{}^2) = (1/\pi)(M^2-\tfrac{1}{4}\mu_A{}^2) g_A{}^2,$$
$$3\beta_{22}(\mu_V{}^2) = -(1/4\pi)\mu_V{}^2 g_M{}^2,$$
$$3\beta_{12}(\mu_V{}^2) = -(1/\sqrt{2\pi}) M^2 (g_{EM})^2.$$

We have defined our unit of mass so that $a=1$. The subscripts on the β's indicate that they are the Regge resonances of the functions $f_0{}^J$, $f_{11}{}^J$, $f_1{}^J$, $f_{22}{}^J$, and $f_{12}{}^J$.

The β's at other values of s can now be found from (3.5)–(3.8). For the pseudoscalar trajectory, corresponding to the function $f_0{}^J$, both helicities are zero and we can use (3.5) directly (with the entire function E set equal to a constant):

$$\beta_0(s) = \frac{\mathrm{const}(4q^2/e)^{s+b_P}}{\Gamma(s+b_P+\tfrac{3}{2})}. \quad (7.8)$$

The constant can be expressed in terms of $g_P{}^2$ by normalizing at $s=-b_P=\mu_P{}^2$ and using (7.8). Thus

$$\beta_0(s) = -\frac{(1/16\sqrt{\pi}) g_P{}^2 \mu_P{}^2 (4q^2/e)^{s+b_P}}{\Gamma(s+b_P+\tfrac{3}{2})}. \quad (7.9a)$$

The formula for the vector trajectory is somewhat more complicated, as three helicity states are involved and the appropriate linear combinations have threshold dependences of $(4q^2)^{s+b_V-1}$, $(4q^2)^{s+b_V}$, and $(4q^2)^{s+b_V+1}$. As the nucleon-nucleon threshold is some distance above the energy region of the mesons, we shall not attempt to obtain the correct threshold behavior but shall assume a uniform threshold dependence of $(4q^2)^{s+b_V-1}$. We also require factors of the form (3.7) and (3.8), since nonsense states are present at $J=0$. We shall assume that the trajectory chooses nonsense at $J=0$, since the contrary assumption would introduce the well-known ghost problem.[19] The helicity matrix element in which we shall be interested is that with $\lambda=1$, $\mu=0$, corresponding to the function $f_{12}{}^J$. According to (3.7) and (3.8), this matrix element will contain a factor $\{(s+b_V)(s+b_V+1)\}^{1/2}$. Hence

$$\beta_{12}(s) = \frac{\mathrm{const}(4q^2/e)^{s+b_V-1}[(s+b_V)(s+b_V+1)]^{1/2}}{\Gamma(s+b_V+\tfrac{3}{2})}.$$

Normalizing at $s=-b_V+1=\mu_V{}^2$ and using (7.7), we find that

$$\beta_{12}(s) = -\frac{(1/8\sqrt{\pi})(g_{EM})^2 M^2 (4q^2/e)^{s+b_V-1}[(s+b_V)(s+b_V+1)]^{1/2}}{\Gamma(s+b_V+\tfrac{3}{2})}. \quad (7.9b)$$

For the axial-vector trajectory, $J=l$, and the threshold factor is $(4q^2)^{s+b_A}$. We assume that the trajectory chooses nonsense at $J=0$, as no appropriate particle has been seen. The factors from (3.7) and (3.8) in β_1 are then $(s+b_V+1)$. Thus, applying (3.5) and normalizing according to (7.8), we find that

$$\beta_1(s) = \frac{(1/8\sqrt{\pi})\tfrac{1}{4} e g_A{}^2 (4q^2/e)^{s+b_A}(s+b_A+1)}{\Gamma(s+b_A+\tfrac{3}{2})}. \quad (7.9c)$$

We must next combine our formulas (7.9) with (5.4) and (5.7) to find the asymptotic behavior of our amplitudes at $s=0$ and large t. Looking first at f_1, we find from (7.9a), (5.4), and (5.7) that

$$f_1(0,t) \sim \frac{\tfrac{1}{8} g_P{}^2 \mu_P{}^2 (4/e)^{b_P}(-t)^{b_P}}{\Gamma(b_P+1)\sin\pi b_P}. \quad (7.10)$$

We are not directly interested in the asymptotic behavior of f_1, as we are only using the superconvergence relations for f_3, f_4, and f_5. However, at $s=0$, there will be a conspiracy associated with the condition

$$f_1 - f_3 - z f_4 = 0. \quad (7.11)$$

We shall assume that the conspiracy is of the type III of Freedman and Wang,[20] since any other type would require the existence of hitherto undiscovered particles. The function f_3 will then not contribute to the leading term of (7.11) as $z\to\infty$, and

$$f_4(0,t) \sim -(2M^2/t) f_1(0,t),$$

---

[19] As we are assuming exchange degeneracy, the ρ and A_2 trajectories are the same.
[20] D. Z. Freedman and J. M. Wang, Phys. Rev. **160**, 1560 (1967).

or, from (7.10),

$$f_4(0,t) \sim -\frac{\tfrac{1}{4}M^2 g_P{}^2 \mu_P{}^2 (4/e)^{b_P}(-t)^{b_P-1}}{\Gamma(b_P+1)\sin\pi(b_P-1)}. \quad (7.12a)$$

Note that the vector trajectory does not contribute to the asymptotic behavior of $f_4(0,t)$, as it contains kinematic factors which vanish in that helicity state at $s=0$. The asymptotic behavior of $f_5(0,t)$ can be found directly from (7.9b), (5.4), and (5.7):

$$f_5(0,t) \sim -\frac{\tfrac{1}{2}(g_{EM})^2 M^2 (4/e)^{b_V-1}(-t)^{b_V-1}}{\Gamma(b_V)\sin\pi(b_V-1)}. \quad (7.12b)$$

Similarly, the asymptotic behavior of $f_3(0,t)$ can be found from (7.9c), (5.4), and (5.7):

$$f_3(0,t) \sim -\frac{\tfrac{1}{2} g_A{}^2 M^2 (4/e)^{b_A-1}(-t)^{b_A-1}}{\Gamma(b_A)\sin\pi(b_A-1)}. \quad (7.12c)$$

Having obtained the asymptotic behavior (7.12), we can easily write down the generalized superconvergence relations for our functions. They are as follows:

$$\int^N dt \, \mathrm{Im} f_4(0,t) \sim -\frac{\tfrac{1}{4}M^2 g_P{}^2 \mu_P{}^2 (4/e)^{b_P} N^{b_P}}{b_P \Gamma(b_P+1)}, \quad (7.13a)$$

$$\int^N dt \, \mathrm{Im} f_5(0,t) \sim -\frac{\tfrac{1}{2}(g_{EM})^2 M^2 (4/e)^{b_V-1} N^{b_V}}{\Gamma(b_V+1)}, \quad (7.13b)$$

$$\int^N dt \, \mathrm{Im} f_3(0,t) \sim -\frac{\tfrac{1}{2} g_A{}^2 M^2 (4/e)^{b_A-1} N^{b_A}}{\Gamma(b_A+1)}. \quad (7.13c)$$

In applying the criteria given in the last section to find a suitable value for N, we observe that our assumptions regarding the Regge trajectories imply that $\mu_P{}^2 = \mu_V{}^2$, while $\mu_A{}^2 = \mu_V{}^2 + \tfrac{1}{2}$ (in units for which $a=1$). If the trajectories in the same channel are spaced at unit distance apart, the next pseudoscalar and vector particles will have the squares of their masses equal to $\mu_P{}^2 + 1$. Hence, if we take the limit of integration to be midway between the position of the first two resonances in the pseudoscalar and vector channels, we obtain $N = \mu_P{}^2 + \tfrac{1}{2}$.

We observe that the limit of integration N is at the position of the lowest resonance in the axial-vector channel. Now, according to our criterion for choosing N, a value of N midway between the nth and the $(n+1)$th resonance corresponds to taking the first n resonances on the left-hand side of the superconvergence relations (5.5). Hence a value of N at the position of the nth resonance should correspond to taking the first $n-1$ resonances together with half the contribution of the nth resonance. In evaluating the left-hand side of (7.13), we shall therefore take only half the contribution of the axial-vector meson. Since the masses of the resonances in this channel are greater than those in the pseudoscalar and vector channels, this would appear to be more reasonable than to take the full contribution of the lowest resonance in all three channels.

The left-hand side of (7.13) can now be evaluated by using (7.7) at $t=0$, together with the helicity crossing matrix. This crossing matrix has been evaluated in Ref. 16[21]; we could also use the general formulas of Trueman and Wick and of Cohen-Tannoudji, Morel, and Navelet.[22] For the particular case $s=0$ and for the elements f_3, f_4, and f_5, the crossing matrix is as follows:

$$\begin{bmatrix} f_4(0,t) \\ f_5(0,t) \\ f_3(0,t) \end{bmatrix} = \begin{bmatrix} -\dfrac{M^2}{t} & \dfrac{M^2}{4M^2-t} & \dfrac{4M^4}{t(4M^2-t)} & \dfrac{4M^4}{t(4M^2-t)} & 0 \\ 0 & \dfrac{2M^2}{4M^2-t} & \dfrac{2M^2}{4M^2-t} & \dfrac{2M^2}{4M^2-t} & \dfrac{4M^2}{4M^2-t} \\ \dfrac{M^2}{t} & \dfrac{M^2}{4M^2-t} & \dfrac{2M^2(2M^2-t)}{t(4M^2-t)} & -\dfrac{2M^2(2M^2-t)}{t(4M^2-t)} & 0 \end{bmatrix} \begin{bmatrix} f_1(t,0) \\ f_2(t,0) \\ f_3(t,0) \\ f_4(t,0) \\ f_5(t,0) \end{bmatrix}. \quad (7.14)$$

If we now insert the expressions (7.7) on the right of (7.14), with a factor $\tfrac{1}{2}$ for the axial-vector function f_3, and integrate, we find that

$$\begin{bmatrix} \int^N dt \, f_4(0,t) \\ \int^N dt \, f_5(0,t) \\ \int^N dt \, f_3(0,t) \end{bmatrix} = \tfrac{1}{2}M^2 \begin{bmatrix} -\dfrac{1}{4} & \dfrac{M^2}{(4M^2-\mu_V{}^2)} & -\dfrac{M^2}{4M^2-\mu_V{}^2} & 0 & -\dfrac{M^2}{2\mu_A{}^2} \\ 0 & \dfrac{\mu_V{}^2}{2(4M^2-\mu_V{}^2)} & \dfrac{2M^2}{4M^2-\mu_V{}^2} & \dfrac{-4M^2}{4M^2-\mu_V{}^2} & \dfrac{1}{4} \\ \dfrac{1}{4} & \dfrac{(2M^2-\mu_V{}^2)}{2(4M^2-\mu_V{}^2)} & -\dfrac{M^2}{4M^2-\mu_V{}^2} & 0 & \dfrac{2M^2-\mu_A{}^2}{4\mu_A{}^2} \end{bmatrix} \begin{bmatrix} g_P{}^2 \\ g_M{}^2 \\ g_E{}^2 \\ (g_{EM})^2 \\ g_A{}^2 \end{bmatrix} \quad (7.15)$$

---

[21] We are interested in the crossing matrix between the s and the t channels, whereas Goldberger, Grisaru, MacDowell, and Wong write down the crossing matrix between the s and the u channels. The elements of our crossing matrix between f_1, f_2, and f_3 on the one hand, and f_4 and f_5 on the other, will therefore have the opposite sign to that of GGMW. We thank Dr. D. Wong for pointing out some misprints in their crossing matrix.

[22] T. L. Trueman and G. C. Wick, Ann. Phys. (N. Y.) **26**, 322 (1964); G. Cohen-Tannoudji, A. Morel, and H. Navelet (to be published). The former paper contains some phase ambiguities.

$$=\tfrac{1}{2}M^2\begin{bmatrix} -\dfrac{1}{4} & \dfrac{M^2}{4M^2-\mu_V^2}\left(1-\dfrac{3}{31}\right) & -\dfrac{M^2}{2\mu_A^2} \\ 0 & \dfrac{M^2}{4M^2-\mu_V^2}\left(\dfrac{\mu_V^2}{2M^2}-\dfrac{18}{31}\right) & -\dfrac{1}{4} \\ \dfrac{1}{4} & \dfrac{M^2}{4M^2-\mu_V^2}\left(-\dfrac{2M^2-\mu_V^2}{2M^2}-\dfrac{3}{31}\right) & \dfrac{2M^2-\mu_A^2}{4\mu_A^2} \end{bmatrix}\begin{bmatrix} g_P^2 \\ g_M^2 \\ g_A^2 \end{bmatrix} \quad (7.16)$$

from (7.6). Equations (7.13) and (7.16) provide us with an eigenvalue equation. From our assumptions about the Regge trajectories, we can put

$$\mu_P^2 = \mu_V^2 = -b_P = -b_V + 1, \quad (7.17a)$$

$$\mu_A^2 = -b_A + 1 = \mu^2 + \tfrac{1}{2}. \quad (7.17b)$$

Also, since we have assumed the nucleon to have unit mass,

$$M^2 = 1, \quad (7.17c)$$

and, finally,

$$N = \mu_V^2 + \tfrac{1}{2}. \quad (7.17d)$$

We thus have an eigenvalue equation for μ_V^2 and, having solved it, we can find the ratio of the coupling constants. There exists a solution with

$$\mu_V^2 = 0.29, \quad g_P^2 = 0.21 g_V^2, \quad g_A^2 = 0.18 g_V^2. \quad (7.18)$$

VIII. CONCLUDING REMARKS

The solution to the equations of the model of the last section does possess the correct qualitative properties. We were able to solve the eigenvalue equation with a value for the mass of the pseudoscalar and vector mesons between 0 and 1, and with the ratios of the coupling constants both positive. The coupling constant g_P^2 appears to be somewhat small, but this is not surprising in view of the very simplified model taken. The error may possibly be due to our assumption that the entire function $E(s)$, which occurs as a factor in β, is constant. Experimentally, the data for backward charge-exchange NN scattering indicate that the residue for the pion trajectory or its conspirator passes through zero near $s=0$. If this is so, the effect of the trajectory at $s=0$, for a given value of g_P^2, would be reduced.

The qualitative nature of our result is not very sensitive to the precise value chosen for N [Eq. (7.17d)] or to the choice $t_0 = 0$ in (6.1). We have seen that these choices represented uncertainties in the lower stages of our approximation scheme.

One important qualitative feature of any dynamical problem is the sign of the force in a particular channel due to the exchange of a particular particle. In this respect the present scheme is similar to more conventional schemes as may be seen by using Eq. (5.9). If the dynamical equations are to possess a solution, the sign of the crossing matrix must be such that the left-hand side is positive. When working with the conventional dynamical schemes, one has to calculate the left-hand discontinuity from an equation similar to (5.9), and, again, the result must be positive in a channel where a resonance is present. When applied to the particular case of the crossing matrix (7.16), the sign requirement is that the elements in the first row must be positive, those in the other two rows must be negative. By examining the matrix we can find the signs of the forces in the three channels due to the exchange of the three mesons, and a sufficiently large number of them are attractive to give a consistent solution. Needless to say, one should not press the analogy between the present scheme and the conventional schemes too far; the quantitative features are completely different.

The calculations performed in the previous section are considerably simpler than conventional calculations of masses and coupling constants. We had to solve a three-by-three eigenvalue problem, whereas even the simplest conventional calculations require the solution of an integral equation. The results which we obtained are much less detailed than results of conventional calculations. We only obtained the position and strength of resonances, whereas conventional calculations give the complete scattering amplitude as a function of energy. However, this extra detail is probably not worth while as long as we do not consider exchange of higher resonances. In the present scheme it may well be feasible to include the exchange of a fairly large number of resonances.

Even in the lowest approximation our calculations were not a complete solution of the bootstrap equations, as we only treated one channel, and we therefore had to take certain quantities from experiment. The next step would be to consider NP scattering, NV scattering, and NA scattering, where P, V, and A represent the pseudoscalar, vector, and axial-vector mesons. The parameters occurring in the lowest approximation would be the same parameters which occurred in our problem, and we would obtain further equations which could be used to determine some, and possibly all, of the parameters which we had to take from experiment. The crossing matrix would not involve the octet and singlet mesons in the same way, and we would no longer have to assume the existence of degenerate nonets. When considering the meson-baryon channels, we would have to include the Δ trajectory and probably the $D_{3/2}$ trajectory in addition to the baryon trajectory. We would in turn have to consider the meson-Δ and meson-$D_{3/2}$ channels, and also channels such as the $\Delta \bar{N}$ and $\Delta \bar{\Delta}$ channels, as these channels would involve the same parameters as the meson-nucleon channels. We might then have a complete set of equations which could be solved without introducing any experimental parameters.

In higher approximations, one would include more than one trajectory with a given set of quantum numbers. One would then have further parameters, but one would also have more equations from which they might be determined. These extra equations would arise, firstly by taking values of n other than zero in (5.6) and secondly by considering channels consisting of the resonances on the new trajectories in combination with the original resonances or with one another. One would also take the higher-spin resonances on the original trajectory as external particles in further possible channels. As long as one had external particles with spin, one would have to decide whether the trajectories chose sense or nonsense at the lower integers or half-integers. One might have to include trajectories with all possibilities. Another question to be decided would be what parity, charge conjugation, and $SU(3)$ states to induce at each stage of our calculation. We need hardly add that the problem of symmetry breaking occurs here as in all other approaches.

We shall not attempt to answer such questions in this paper. One may be able to obtain some guidance from experiment, but it is unlikely that such guidance will be unambiguous, as the resonances associated with the lower trajectories will probably be broad and therefore hard to detect. The quark model may possibly provide further guidance, even if real quarks do not exist, but at the moment, it is not evident whether or how the quark model emerges from our scheme.

In this paper we shall also not attempt to include partially conserved axial-vector current or current commutators within our scheme. It may be that the current-commutation rules can be deduced from the other dynamical equations when the scattering amplitudes are continued off the mass shell; alternatively, it may be necessary to postulate them in order to obtain sufficient equations to complete the scheme. If the latter alternative is true, we would not be able to obtain a complete and rigorous dynamical scheme without going off the mass shell, but we could incorporate the commutation relations in a preliminary approach by making use of the low-energy theorems associated with them.

In view of all the unanswered questions, we must regard the present work as suggesting a method of approach rather than as a complete scheme. It does appear to open the possibility of a treatment free from some of the drawbacks of more conventional methods.

ACKNOWLEDGMENTS

The author would like to acknowledge helpful discussions with G. F. Chew, J. Finkelstein, M. Gell-Mann, D. Horn, and C. Schmid.

Vortices and Quark Confinement in Non-Abelian Gauge Theories

S. Mandelstam

Phys. Lett. B **53** (1975) 476–478

Abstract

It is shown that finite-length vortices in an $SU(n)$ Nielsen-Olesen model require explicit introduction of monopoles, which are confined in multiples of n by the Meissner effect. The model therefore possesses a natural explanation of quark confinement.

Charge-Monopole Duality and the Phases of Non-Abelian Gauge Theories

S. Mandelstam

Phys. Rev. D **19** (1979) 2391–2401

Abstract

It is shown that the O(3) non-Abelian gauge theory possesses charge-monopole duality in the sense that, analogous to the ordinary magnetic vector potentials, one can construct O(3) "electric vector potentials," gauge-invariant combinations of which constitute the physical variables. However, while the Hamiltonian could in principle be expressed in terms of such variables, it would be very complicated; the definition of the potentials is thus not unique. The O(3) magnetic gauge group is separate from the ordinary electric gauge group. One can also formulate the theory in terms of electric (Wilson) or magnetic (Nielsen-Olesen) flux loops; the property of the loops which identifies the electric or magnetic gauge group as O(3) is studied. The possible phases of non-Abelian gauge theories are discussed briefly. The phase with real massless gluons, if it exists at all, is more complicated than the corresponding phase in an Abelian theory in the sense that it is not known how to construct a trial vacuum state without an infrared energy divergence. This phase should not be distinguished from the other phases by the absence of symmetry breaking, again in contradistinction to the Abelian gauge theory. The phase with complete Higgs symmetry breaking and that with confinement are electric-magnetic duals of one another. The relation between the Wilson condition, applied to loops at a fixed time, and the confinement of infinitely massive quarks is studied; it is hoped that the analysis will be helpful in constructing wave functions for hadronic states with confined quarks. It is suggested that Weinberg-Salam-type models may confine due to instanton effects though, for α around $1/137$, this has no practical significance.

General Introduction to Confinement

S. Mandelstam

in the *Proceedings of Les Houches Winter Advanced Study Institute on "Common trends in particle and condensed matter physics"*, February 1980,
Eds: E. Brezin, J.-L. Gervais and G. Toulouse
[*Phys. Rept.* **67**, Issue 1 (1980), 109–121]

1. Preliminaries (There is no abstract in the paper.)
My aim in these lectures will be to give a brief survey of the general features of hadron structure and the confinement problem within the framework of quantum chromodynamics (Q.C.D.). I shall attempt to direct my presentation to the non-specialized audience present at this meeting.

When the quark model was first proposed, and especially when field theories of quarks began to be taken seriously, confinement appeared to be something of a mystery. It is no longer so. We understand the properties which the Q.C.D. vacuum must possess in order to confine, and we know of physical systems which possess analogues of such properties. Indeed, solid-state physicists have long been familiar with systems possessing confinement properties. The question whether the state of lowest energy of the system actually possesses such properties is more difficult. A number of approaches or approximation schemes indicates that the answer is yes but, at present, no explanation has received general support of particle theorists.

I shall divide my lectures into five parts:

1) General nature of hadronic states in confined systems.

2) Characterization of phases of pure Yang–Mills non-Abelian systems. For simplicity we shall treat SU(2); the extension to SU(3) is straightforward.

3) Characterization of such phases for non-Abelian systems with quarks.

4) The 1/N approximation, which simplifies the full Q.C.D. problem in many ways.

5) Survey of schemes for obtaining confinement and calculating hadronic masses.

While I shall mention direct points of contact with the lattice approach, my treatment will be within the framework of the continuum theory. Many of the features are parallel to those of lattice-gauge theory, which will be presented elsewhere at this meeting.

Soliton Operators for the Quantized sine-Gordon Equation

S. Mandelstam

Phys. Rev. D **11** (1975) 3026–3030

Abstract

Operators for the creation and annihilation of quantum sine-Gordon solitons are constructed. The operators satisfy the anticommutation relations and field equations of the massive Thirring model. The results of Coleman are thus reestablished without the use of perturbation theory. It is hoped that the method is more generally applicable to a quantum-mechanical treatment of extended solutions of field theories.

Light-cone Superspace and the Ultraviolet Finiteness of the $N=4$ Model

S. Mandelstam

Nucl. Phys. B **213** (1983) 149–168

Abstract

Superspace in the light-cone frame takes a simple form. No auxiliary fields are necessary, and application to extended supersymmetries is straightforward. It is shown that the $N=4$ model, in a certain form of the light-cone gauge, is completely free of ultraviolet divergences in any order of perturbation theory. It follows that the β-function vanishes in any gauge, to all orders of perturbation theory. Our method differs from the conventional method in that we use only half the number of θ's as there are supersymmetry operators. All fields are unconstrained and independent of the θ's.

Dual-Resonance Models

S. Mandelstam

Phys. Rept. **13** (1974) 259–353

Abstract

Dual-resonance models are treated both as S-matrix theories and as systems of interacting strings. We show how Veneziano was able to construct a dual four-point amplitude with narrow resonances and rising Regge trajectories. The construction is generalized to the N-point amplitude in the manifestly dual manner suggested by Koba and Nielsen. We develop the operator formalism which exhibits the factorization property of the above amplitude. The related questions of ghost elimination and null states are discussed. Models with extra degrees of freedom and, in particular, the Neveu-Schwarz-Ramond model with spin, are treated. The latter model has a quark-line spectrum of mesons, but it possesses massless vector mesons and fermions. It is shown how the operator formalism is related to a quantized string. The theory of such a string is developed, with particular emphasis on the ghost-free "Coulomb-gauge" quantization. By constructing theories of interacting strings, we reproduce the dual-model S-matrix. A brief account is given of the theory of loops, in which one attempts to improve on the narrow-resonance model in a perturbative manner.

The n-loop String Amplitude: Explicit Formulas, Finiteness and Absence of Ambiguities

S. Mandelstam

Phys. Lett. B **277** (1992) 82–88

Abstract

Explicit formulas are obtained for the n-loop scattering amplitudes of N external tachyons in the Bose closed-string theory and in the type-II superstring theory. The superstring amplitudes are shown to be finite and free of ambiguities.